IET MATERIALS, CIRCUITS & DEVICES 33

High Frequency MOSFET Gate Drivers

Other volumes in this series:

Volume 2	**Analogue IC Design: The current-mode approach** C. Toumazou, F.J. Lidgey and D.G. Haigh (Editors)
Volume 3	**Analogue–Digital ASICs: Circuit techniques, design tools and applications** R.S. Soin, F. Maloberti and J. France (Editors)
Volume 4	**Algorithmic and Knowledge-based CAD for VLSI** G.E. Taylor and G. Russell (Editors)
Volume 5	**Switched Currents: An analogue technique for digital technology** C. Toumazou, J.B.C. Hughes and N.C. Battersby (Editors)
Volume 6	**High-frequency Circuit Engineering** F. Nibler *et al.*
Volume 8	**Low-power High-frequency Microelectronics: A unified approach** G. Machado (Editor)
Volume 9	**VLSI Testing: Digital and mixed analogue/digital techniques** S.L. Hurst
Volume 10	**Distributed Feedback Semiconductor Lasers** J.E. Carroll, J.E.A. Whiteaway and R.G.S. Plumb
Volume 11	**Selected Topics in Advanced Solid State and Fibre Optic Sensors** S. M. Vaezi-Nejad (Editor)
Volume 12	**Strained Silicon Heterostructures: Materials and devices** C.K. Maiti, N. B. Chakrabarti and S.K. Ray
Volume 13	**RFIC and MMIC Design and Technology** I.D. Robertson and S. Lucyzyn (Editors)
Volume 14	**Design of High-frequency Integrated Analogue Filters** Y. Sun (Editor)
Volume 15	**Foundations of Digital Signal Processing: Theory, algorithms and hardware design** P. Gaydecki
Volume 16	**Wireless Communications Circuits and Systems** Y. Sun (Editor)
Volume 17	**The Switching Function: Analysis of power electronic circuits** C. Marouchos
Volume 18	**System on Chip: Next generation electronics** B. Al-Hashimi (Editor)
Volume 19	**Test and Diagnosis of Analogue, Mixed-signal and RF Integrated Circuits: The system on chip approach** Y. Sun (Editor)
Volume 20	**Low Power and Low Voltage Circuit Design with the FGMOS Transistor** E. Rodriguez-Villegas
Volume 21	**Technology Computer Aided Design for Si, SiGe and GaAs Integrated Circuits** C.K. Maiti and G.A. Armstrong
Volume 22	**Nanotechnologies** M. Wautelet *et al.*
Volume 23	**Understandable Electric Circuits** M. Wang
Volume 24	**Fundamentals of Electromagnetic Levitation: Engineering sustainability through efficiency** A.J. Sangster
Volume 25	**Optical MEMS for Chemical Analysis and Biomedicine** H. Jiang (Editor)
Volume 26	**High Speed Data Converters** Ahmed M.A. Ali
Volume 27	**Nano-Scaled Semiconductor Devices** E.A. Gutiérrez-D (Editor)
Volume 29	**Nano-CMOS and Post-CMOS Electronics: Devices and modelling** Saraju P. Mohanty and Ashok Srivastava
Volume 30	**Nano-CMOS and Post-CMOS Electronics: Circuits and design** Saraju P. Mohanty and Ashok Srivastava
Volume 32	**Oscillator Circuits: Frontiers in design, analysis and applications** Y. Nishio (Editor)

High Frequency MOSFET Gate Drivers

Technologies and Applications

ZhiLiang Zhang and Yan-Fei Liu

The Institution of Engineering and Technology

Published by The Institution of Engineering and Technology, London, United Kingdom

The Institution of Engineering and Technology is registered as a Charity in England & Wales (no. 211014) and Scotland (no. SC038698).

First published 2017

The Institution of Engineering and Technology
Michael Faraday House
Six Hills Way, Stevenage
Herts, SG1 2AY, United Kingdom

www.theiet.org

British Library Cataloguing in Publication Data
A catalogue record for this product is available from the British Library

ISBN 978-1-78561-365-4 (hardback)
ISBN 978-1-78561-366-1 (PDF)

Typeset in India by MPS Limited
Printed in the UK by CPI Group (UK) Ltd, Croydon

Contents

Authors' biographies **ix**
Preface **xi**
List of symbols **xiii**

1 Introduction **1**
1.1 Introduction 1
1.2 Voltage regulator applications 3
1.3 Voltage source gate driver and high-frequency-dependent loss 4
1.3.1 Switching loss 5
1.3.2 Body diode conduction loss 9
1.3.3 Reverse recovery loss 10
1.3.4 Gate drive loss 10
1.4 Summary 10
References 11

2 Fundamentals of current source driver **13**
2.1 Resonant gate drivers 13
2.2 Concept of current source driver 17
2.3 A practical and accurate switching loss model 19
2.3.1 Introduction 19
2.3.2 Impact of parasitic inductance and load current 20
2.3.3 Proposed switching loss model 24
2.3.4 Turn-on switching loss model 25
2.3.5 Turn-off switching loss model 32
2.3.6 Voltage source drive model verification 37
2.3.7 Experimental validation of the voltage source drive model 38
2.4 Summary 42
References 42

3 Continuous current source driver **45**
3.1 Two-channel low-side continuous current source drivers 45
3.1.1 Operating principle 45
3.1.2 Loss comparison 49
3.1.3 Advantages of the proposed current source driver 52
3.1.4 Applications 54
3.1.5 Experimental results 56

3.2 High-side and low-side continuous current source drivers 59
3.2.1 Operating principle of proposed current source drivers 59
3.2.2 Advantages 63
3.2.3 Loss analysis 65
3.2.4 Experimental results 67
3.3 Accurate switching loss model with current source drivers 69
3.3.1 Proposed MOSFET loss model with current source resonant driver 69
3.3.2 Analytical modeling and simulation results 73
3.3.3 Proposed optimal design with the accurate switching loss model 75
3.3.4 Experimental verification and discussion 79
3.4 High-side and low-side current source drivers 86
3.4.1 Problem of high-side and low-side current source drivers 86
3.4.2 Proposed decoupled high-side and low-side current source drivers 87
3.4.3 New current-source gate driver with integrated magnetics 94
3.4.4 Experimental results and discussion 96
3.5 Summary 101
References 101

4 Discontinuous current source drivers **103**
4.1 Discontinuous current source driver 103
4.1.1 Proposed low-side discontinuous CSD 103
4.1.2 Driver loss analysis 107
4.1.3 CSD design procedure 115
4.1.4 Design example 116
4.1.5 Experimental results 117
4.2 High-side discontinuous CSD 123
4.2.1 Proposed CSD for synchronous buck converter and operation 123
4.2.2 Design example 125
4.2.3 Experimental results 125
4.3 Discontinuous CSD with reduced inductance 132
4.3.1 Proposed CSD and principle of operation 132
4.3.2 Proposed high-side CSD and hybrid gate-drive scheme 137
4.3.3 Experimental results 139
4.4 Current diversion 145
4.4.1 Introduction 145
4.4.2 Proposed switching loss model considering current diversion 147
4.4.3 Experimental results and discussions 154
4.5 Summary 157
References 158

5 Adaptive current source drivers **159**
5.1 Adaptive current source driver for buck converters 159
5.1.1 Introduction 159
5.1.2 Adaptive continuous current source driver 159
5.1.3 Adaptive discontinuous current source drivers 161
5.1.4 Implementation of the adaptive drive voltage for the current source drivers 163
5.1.5 Experimental results and discussion 164
5.2 Adaptive current source driver for power factor correction application 166
5.2.1 Introduction 166
5.2.2 Analysis of current source driver circuits and proposed adaptive current source driver for boost power factor correction converters 166
5.2.3 Loss analysis of proposed adaptive current source driver for boost power factor correction converter 168
5.2.4 Experimental results and discussion 172
5.3 Digital adaptive power factor correction 173
5.3.1 Introduction 173
5.3.2 Principle of the proposed digital adaptive discontinuous current source driver 175
5.3.3 Design procedure and implementation 176
5.3.4 Experimental results and discussion 182
5.4 Summary 186
References 187

6 Resonant gate drivers **189**
6.1 Resonant gate drivers for multi-megahertz isolated resonant converters 189
6.1.1 Introduction 189
6.1.2 Challenges for the SR drive in the multi-MHz isolated resonant converters 190
6.1.3 Proposed self-driven RGD for the SR 192
6.1.4 Experimental verification and discussion 202
6.2 A high-frequency dual-channel isolated resonant gate driver with low gate-drive loss for ZVS full-bridge converters 206
6.2.1 Introduction 206
6.2.2 Review of gate-drive circuits 207
6.2.3 Proposed resonant gate driver for FB converters and principle of operation 208
6.2.4 Loss analysis and optimal design 211
6.2.5 The comparison between the proposed RGD and previous gate-drive circuits 220
6.2.6 Experimental results and discussion 224
6.3 Summary 226
References 227

7 eGaN HEMTs gate drivers **231**
7.1 Three-level gate drivers 231
7.1.1 Introduction 231
7.1.2 Driving requirements for eGaN HEMTs in resonant SEPIC converters 232
7.1.3 Proposed three-level gate drivers for eGaN HEMTs 235
7.1.4 Operation principle of three-level gate driver 237
7.1.5 Rectifier mathematic modeling and design 241
7.1.6 Experimental results and discussion 248
7.2 A digital adaptive driving scheme for eGaN HEMTs in VHF converters 252
7.2.1 Introduction 252
7.2.2 Gate-drive challenges for eGaN VHF converters 254
7.2.3 Proposed digital adaptive driving scheme 256
7.2.4 Relationship analysis between gate-drive timing and input voltage 258
7.2.5 Air-core transformer design 263
7.2.6 Experimental results and discussion 264
7.3 Summary 270
References 270

Index **273**

Authors' biographies

Dr. ZhiLiang Zhang received the B.Sc. and M.Sc. degrees in Electrical Engineering from Nanjing University of Aeronautics and Astronautics (NUAA), Nanjing, P.R. China, in 2002 and 2005, and the Ph.D. degree from the Department of Electrical and Computer Engineering, Queen's University at Kingston, Canada, in 2009. He worked as a Design Engineering Intern from June to September 2007 at Burlington Design Center, VT, Linear Technology Corporation. In June 2009, he joined NUAA and currently, he is a Professor with Aero-Power Sci-Tech Center. His research interests include high-frequency power converters and renewable energy power conversion system.

He has published 29 papers on IEEE Transaction on Power Electronics and published 58 papers on IEEE, ECCE, and APEC conferences. The citation of his papers is over 1,000. He owns one US patent and nine China patents. His work attracts strong interest from the industry and academia, and he successfully obtained the funding from National Natural Science Foundation of China, Aviation Science Fund of China, Delta, and Lite-On, etc. His contribution on the high frequency power converters has been successfully applied to the satellites of China Aerospace Science and Industry Corporation. He was awarded Outstanding Youth Foundation of Jiangsu Province and Fok Ying-Tong Education Foundation of China. He has been selected by "Nanjing Overseas Leading Talent Plan." He was a winner of "United Technologies Corporation Rong Hong Endowment" in 1999. He is an IEEE Senior Member and serves as secretary of PELS Technical Committee on Power and Control Core Technologies since 2013.

Dr. Yan-Fei Liu (IEEE Fellow since 2013) received his Bachelor and Master degrees from the Department of Electrical Engineering from Zhejiang University, China, in 1984 and 1987, and Ph.D. degree from the Department of Electrical and Computer Engineering, Queen's University, Kingston, ON, Canada, in 1994.

He was a Technical Advisor with the Advanced Power System Division, Nortel Networks, in Ottawa, Canada from 1994 to 1999. Since 1999, he has been with Queen's University, where he is currently a Professor with the Department of Electrical and Computer Engineering. His current research interests include digital control technologies for high efficiency, fast dynamic response d.c.–d.c. switching converter and a.c.–d.c. converter with power factor correction, resonant converters, and server power supplies, and LED drivers. He has authored over 200 technical papers in the IEEE Transactions and conferences, and holds 20 US patents. He is also a Principal Contributor for two IEEE standards. He received Premier's

Research Excellence Award in 2000 in Ontario, Canada. He also received the Award of Excellence in Technology in Nortel in 1997.

Currently, he is the Vice President of Technical Operations of IEEE Power Electronics Society (PELS, from 2017 to 2018). He serves as an Editor of *IEEE Journal of Emerging and Selected Topics of Power Electronics* (IEEE JESTPE) since 2013. He will be the General Chair of ECCE 2019 to be held in Baltimore, USA, in 2019. His major services to IEEE are listed below: a Guest Editor-in-Chief for the special issue of Power Supply on Chip of IEEE Transactions on Power Electronics from 2011 to 2013; a Guest Editor for special issues of JESTPE: Miniaturization of Power Electronics Systems in 2014 and Green Power Supplies in 2016; as Co-General Chair of ECCE 2015 held in Montreal, Canada, in September 2015; the chair of PELS Technical Committee (TC1) on Control and Modeling Core Technologies from 2013 to 2016; chair of PELS Technical Committee (TC2) on Power Conversion Systems and Components from 2009 to 2012.

Preface

In recent years, there has been a trend to increase the switching frequency beyond multi-MHz in switching power converters to reduce the passive components and improve the power density significantly. However, this results in high-switching loss and gate-driver loss of the power MOSFETs. This book is about high-frequency power MOSFET gate driver technologies, including the gate drivers for the GaN HEMTs, which have great potential in the next generation switching-power converters. The gate drivers serve a critical role between the control and power devices. Although there are numerous research papers talking about the MOSFET gate drivers, there is actually no such a book focusing on the gate driver topic systematically. The contents mainly cover the state-of-the-art power MOSFET drive technique, the switching-loss model, current source gate drivers (CSDs), resonant gate drivers, adaptive gate drivers, and GaN HEMT gate drivers.

The novel approach in this book is the proposed CSDs including different topologies, control, and applications. The CSD can reduce the switching-transition time and switching loss significantly, and recover high-frequency gate-driver loss compared to the conventional voltage gate drivers. The basic idea can also be extended to other power devices to improve high-frequency switching performance such as SiC MOSFET, IGBT, etc.

It is expected that this book will be very useful to the design engineers in switching power supplies with various backgrounds. It can serve as a fundamental textbook to the undergraduates, graduates, and professionals in the power electronics field.

We would like to thank our colleagues at Queen's Power Group. In particular, we would like to thank Wilson Eberle, Zhihua Yang, and Jizhen Fu for their contribution to the CSDs. We would also like to thank our graduate students in Nanjing University of Aeronautics and Astronautics, and they are Peng-Cheng Xu, Chuan-Gang Xu, Wei Cai, Jing-Ya Lin, Fei-Fei Li, Yuan Zhou, Zhou Dong, Zhi-Wei Xu, and Ke Xu.

Preface

List of symbols

B_{ac_pk}	Magnetic core peak a.c. flux density
C	Capacitor
C_b	Bootstrap capacitance
C_{dsi}	MOSFET parasitic drain-to-source capacitance, where i corresponds to the MOSFET number
C_f	Converter powertrain output filter capacitance
C_g	MOSFET total parasitic gate capacitance
C_{gdi}	MOSFET parasitic gate-to-drain capacitance, where i corresponds to the MOSFET number
C_{gsi}	MOSFET parasitic gate-to-source capacitance, where i corresponds to the MOSFET number
C_{issi}	MOSFET parasitic input capacitance, where i corresponds to the MOSFET number
C_m	Magnetic core loss constant
C_{ossi}	MOSFET parasitic output capacitance, where i corresponds to the MOSFET number
d	MOSFET drain terminal
d_i	MOSFET drain terminal, where i corresponds to the MOSFET number
D	Duty cycle
D_i	Diode, where i corresponds to the diode number
D_f	Bootstrap diode
f_s	Switching frequency
g_i	MOSFET gate terminal, where i corresponds to the MOSFET number
g_{fs}	MOSFET transconductance
i_d	MOSFET drain current
i_g	Gate drive current
i_{Li}	ith inductor current ($i = 1,2,3,\ldots$)
I_o	Load current
I_{off}	Current at turn off
I_{on}	Current at turn on
M_i	ith power MOSFET ($i = 1,2,3,\ldots$)

n	Transformer turns ratio: primary turns over secondary turns
P_{core}	Core loss
P_{cond}	Total conduction loss
P_g	MOSFET gate loss
Q_{rr}	Reverse recovery charge
R_i	*i*th resistor ($i = 1,2,3,\ldots$)
$R_{a.c.}$	a.c. resistance
$R_{d.c.}$	d.c. resistance
$R_{DS(on)}$	MOSFET on-resistance
R_g	MOSFET internal gate resistance
S	MOSFET source terminal
S_i	MOSFET source, where ($i = 1,2,3,\ldots$) is the MOSFET number
t	Time
T_s	Switching period
v_{ds}	Drain-to-source-voltage
v_{gs}	MOSFET gate-to-source voltage
v_o	Output voltage
V_b	Blocking capacitor voltage
V_{cc}	MOSFET driver supply voltage
V_{core}	Volume of the core
V_{ds}	Drain-to-source voltage
V_F	Diode forward voltage drop
V_{in}	Input voltage
V_o	Average output voltage
V_{pl}	MOSFET plateau voltage
V_{th}	MOSFET threshold voltage
x	Core loss switching frequency parameter
y	Core loss peak flux density parameter

Abbreviations

BJT	Bipolar Junction Transistor
VSD	Voltage Source Driver
CSD	Current-Source Driver
FB	Full Bridge
IC	Integrated Circuits
MOSFET	Metal Oxide Silicon Field Effect Transistor

GaN	Gallium Nitride
HEMT	High Electron Mobility Transistor
PCB	Printed Circuit Board
PWM	Pulse Width Modulation
RMS	Root Mean Square
SR	Synchronous Rectifier
VR	Voltage Regulator
VRM	Voltage Regulator Modulator
ZCS	Zero Current Switching
ZVS	Zero Voltage Switching

Prefixes for SI units

p	Pico (10^{-12})
n	Nano (10^{-9})
μ	Micro (10^{-6})
m	Milli (10^{-3})
k	Kilo (10^{3})
M	Mega (10^{6})

SI units

A	Ampere
F	Farad
H	Henry
Hz	Hertz
s	second
V	Volt
W	Watt
Ω	Ohm

Chapter 1
Introduction

1.1 Introduction

Modern power electronics applications expect low power converters to achieve extremely high power density and high efficiency. In such applications, the switching frequency can increase to a few megahertz (MHz). In recent years, there has been a trend to increase the switching frequency beyond multi-MHz in power converters. The objective of this trend is to increase the power density by decreasing the size of passive components and to improve the dynamic performance, since the passive components normally dominate the converter volume. Increasing the switching frequency is an effective way to reduce the energy storage requirement, thus permitting use of smaller passive components and paving the way toward fully integrated power converters. However, it is known that as the switching frequency increases, both switching loss and gate loss increase seriously. Figure 1.1 shows the commercial non-isolated power integrated circuits (ICs) with the different output voltage and current for different applications. Typically, these power converters are realized as power system on a chip (PowerSoC), including the power metal oxide semiconductor field-effect transistors (MOSFETs), gate drivers and control circuits in a package with a few MHz switching frequency to achieve high power density. Particularly, EN23F0QI from Altera's Enpirion, LTM8042 and LTM8026 from Linear Technology also integrate the inductors into the package.

Recently, the topic of the power converters operating at very high frequencies (VHFs, 30–300 MHz) has attracted many research efforts [1–3]. As the switching frequency is over tens of megahertz, the air-core inductors or the printed circuit board (PCB) embedded inductors or transformers can be adopted and integrated efficiently, so that the power density can be increased significantly [4,5]. Moreover, VHF operation lowers the requirement of the energy buffer in the converter, which leads to very fast transient response (compared to the conventional converter, the transient response of the resonant converter is decreased from 1 ms to 50 us for the load steps from 10% to 90% of full load) [6,7].

In the past two decades, much work has been done on reducing or eliminating turn-on switching loss through soft switching techniques. However, in general, these techniques require additional components and require snubber capacitors to reduce turn-off loss. Unfortunately, adding snubber capacitors to reduce turn-off loss makes achieving zero voltage switching (ZVS) at turn-on transitions more

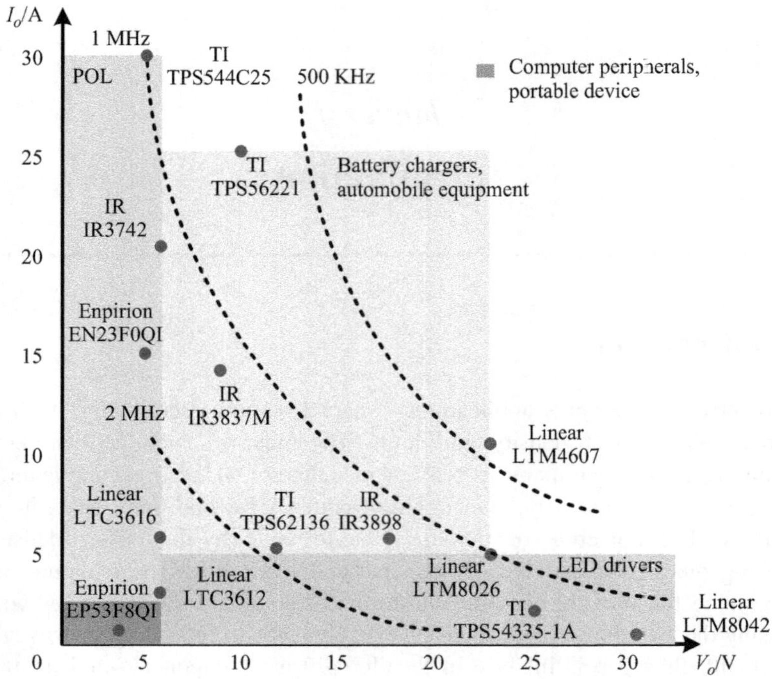

Figure 1.1 Output current versus output voltage for state-of-the-art power ICs

challenging for all line and load conditions. Furthermore, these techniques do not reduce gate loss. Therefore, with some improvements achieved through reduction in turn-on switching loss, it is now essential to focus some effort on reducing turn-off switching loss and gate loss in order to continue to achieve high switching frequency and high power density.

With the excellent figure of merit (FOM), the enhancement mode gallium nitride (eGaN) high electron mobility transistors (HEMTs) are expected to improve the power converters to even higher efficiency, higher frequency, and higher power density [8–10]. The lateral structure of the eGaN HEMTs makes them very low charge devices. They can switch hundreds of volts in nanoseconds, giving them multi-MHz capability [11,12]. However, the eGaN HEMTs could hardly replace the Si MOSFETs in the power circuits directly. When the eGaN field-effect transistors (FETs) are used, the considerations are as follows: (1) the gate-to-source voltage cannot exceed the maximum rating of 6 V; (2) it is important to prevent the false turn-on transitions of the eGaN HEMTs due to the low threshold voltage V_{th}; (3) there is no body-diode, so the reverse current has to flow via the channel by the reverse conduction mechanism, resulting in high reverse conduction loss. Therefore, it is a serious challenge to drive the eGaN HEMTs with precise signal timing to minimize high frequency loss in multi-MHz converters.

In summary, the power MOSFETs are so popular in numerous high frequency power converters applications extensively. With high frequency trend of the power converters, the high frequency switching loss and gate drive loss need to be minimized, which causes serious challenge on advanced high-speed gate driver technology since the gate drivers serve as a critical role between the control and power devices. This book focuses on the gate driver topic systematically providing the fundamental knowledge of high frequency gate driver technologies.

1.2 Voltage regulator applications

The fast development in VLSI (very large scale integration) technologies imposes an increasing challenge to supply state-of-the-art microprocessors with high quality power. As Moore's law predicts that the number of transistors per chip will double every 18 months, the number of transistors has kept increasing steadily in the past decades. Meantime, the clock frequencies of microprocessors have also increased significantly.

The increase in the microprocessor's speed and transistor number has led to high current demands and fast transient load change requirement, since the supply voltage of the microprocessor is decreased to reduce the power consumption of the chip. In the point-of-load regulation system, a dedicated d.c./d.c. converter, a voltage regulator (VR), is used to deliver a highly accurate supply voltage to the microprocessor. VRs are placed in close proximity to the microprocessor in order to minimize the impact of the parasitic impedance between the VRs and microprocessors.

According to Intel microprocessor's road map in [13], the microprocessors operate at extremely low voltage (<1 V) and high current (>100 A) with the continuous increase of the speed and transistors within the chips. Meanwhile, the strict transient requirement of the microprocessors is another serious challenge to VRs.

In order to meet the strict transient requirement, large amount of expensive output capacitors are required to reduce the output-voltage deviations during a transient event. As a result, the cost of a VR will increase while the power density is reduced. However, the real estate on the motherboard for VRs is so limited that the future VRs should have much higher power density than today's version. To increase power density, high switching frequency operation is an effective approach. More importantly, high switching frequency leads to high bandwidth and improves the dynamic response. Therefore, high switching frequency (>1 MHz) operation of VRs is strongly desired and becomes the trend of new generation VRs. Unfortunately, high frequency operation normally increases the frequency-dependent loss in the power supply system. Therefore, high efficiency is of critical importance to achieve good thermal management and thus high power density. Overall, the challenges for future voltage regulator modules (VRMs) are to provide a precisely regulated output with high efficiency, high power density, and fast dynamic response.

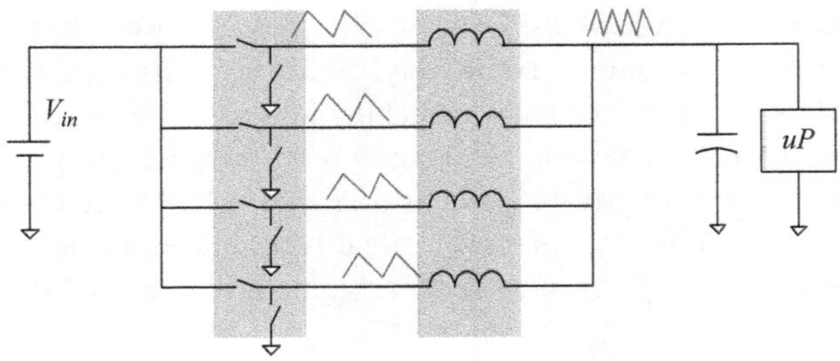

Figure 1.2 Multiphase interleaving buck converters

For 12 V input voltage VRs, most popular topologies are multiphase interleaving synchronous buck converters owing to the simple structure and control as shown in Figure 1.2. The basic interleaving idea was proposed by Virginia Power Electronics Center (VPEC) (Center for Power Electronics Systems (CPES)) in 1997 and now has been adopted widely in industrial applications. The advantages of this topology are: (1) simple and low cost; (2) reduced current ripples to the output capacitors due to the interleaving; (3) small output inductances to improve transient response; (4) better thermal management and package flexibility are also achieved due to the distributed structure of multiphase. But the problem of a phase buck converter is high output filter inductance that limits the transient response. In order to reduce the VR output capacitance, a fast inductor current slew rate is preferred. Because a smaller inductance gives a larger current slew rate, the inductance should be minimized so that the transient requirement can be satisfied. Nevertheless, small inductances and extremely low duty cycle result in large current ripple, which causes high turn-off current and then high switching loss.

High switching frequency shrinks the passive components and leaves more space for power semiconductor devices, thus yields higher power density. Moreover, the reduction of the filter components also makes it possible to increase the bandwidth of the converter and output current slew rate, which improves the dynamic response of VRs.

In recent years, MHz VRs show great advantages over the conventional 300 kHz VRs in terms of cost, power density, dynamic response, etc. It is known that the higher the bandwidth of VRs, the smaller the output capacitance. Unfortunately, an increase in switching frequency will lead to lower efficiency and thermal problems due to excessive frequency-related loss, and result in degradation in overall system performance. Therefore, it is of great importance to understand the loss mechanism in a high-frequency VR and then reduce the loss.

1.3 Voltage source gate driver and high-frequency-dependent loss

The loss of a synchronous buck converter includes the frequency-dependent loss and non-frequency-dependent loss. The conversion efficiency could be degraded greatly due to excessive frequency-dependent loss causing serious thermal

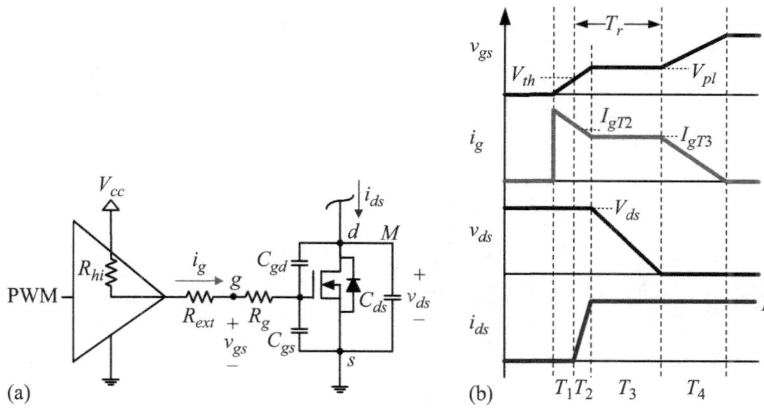

Figure 1.3 MOSFET turn-on transition: (a) simplified turn-on transition equivalent circuit and (b) waveforms

problems. Actually, the switching loss, the body diode loss, the reverse recovery loss, and the gate charge loss are the major frequency-dependent loss associated with the power MOSFETs in a VR. Reducing these losses is critical to achieve high frequency and high power density for new generation VRs.

1.3.1 Switching loss

Switching loss is a most important frequency-dependent loss and the dominant loss in a synchronous buck converter. Therefore, the way of calculating switching loss is critical to design a VR. Normally, the piecewise linear approximation of the MOSFET turn-on and turn-off waveforms is usually made for simplifying the switching loss model [14,15]. The turn-on transition equivalent circuit and its associated waveforms are illustrated in Figure 1.3.

During T_1 interval, the MOSFET driver turns on and begins to supply current to the MOSFET gate to charge its input capacitance C_{gs} and C_{gd}. There is no switching loss until the gate voltage v_{gs} reaches the threshold voltage V_{th}. Entering T_2 interval, v_{gs} reaches V_{th}, the input capacitance is being charged and the MOSFET's drain current i_{ds} is rising linearly until it reaches the load current. During T_2, the MOSFET is sustaining the entire input voltage across it. During T_3 interval, the load current is flowing through MOSFET, and the drain voltage v_{ds} begins to fall. All of the gate current will be going to charge the capacitance C_{gd} and v_{gs} comes into Miller plateau level. During this interval, i_{ds} is constant as the load current and the drain voltage is falling fairly linear from the input voltage to zero. During T_4 interval, the MOSFET is just fully enhancing the channel to obtain its rated on-resistance $R_{DS(ON)}$ at a rated V_{gs}.

Therefore, the turn-on switching loss happens during T_2 and T_3 intervals and can be calculated based on the linear current and voltage waveforms by (1.1)

$$P_{turn\_on} = \frac{V_{ds} \cdot I_{ds}}{2} \cdot (T_2 + T_3) \cdot f_s \tag{1.1}$$

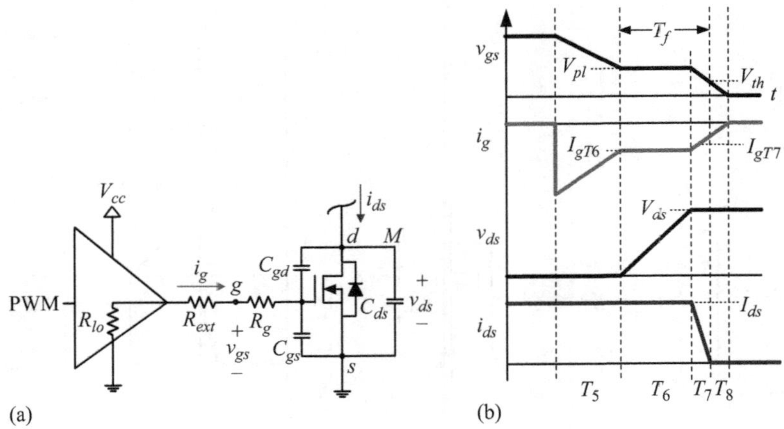

Figure 1.4 MOSFET turn-off transition: (a) simplified turn-off equivalent circuit and (b) waveforms

where V_{ds} is the block voltage and I_{ds} is the load current and f_s is the switching frequency.

The turn-off transition equivalent circuit and its associated waveforms are illustrated in Figure 1.4. During T_5 interval, the input capacitances C_{gs} and C_{gd} are discharged from the initial value to the Miller plateau level. During this interval, the gate current is supplied by the input capacitance. During T_6 interval, the drain voltage v_{ds} begins to rise. All of the gate current will be going to discharge the capacitance C_{gd}, which corresponds to the Miller plateau in the waveform. During T_7 interval, the gate voltage resumes falling from V_{pl} to V_{th}. The MOSFET is in linear operation and the drain current i_{ds} decreases to zero. During T_8, the input capacitor is fully discharged and the MOSFET is turned off completely.

Therefore, the turn-off switching loss happens during T_6 and T_7 intervals and can be calculated based on the linear current and voltage waveforms by (1.2):

$$P_{turn\_off} = \frac{V_{ds} \cdot I_{ds}}{2} \cdot (T_6 + T_7) \cdot f_s \tag{1.2}$$

Unfortunately, this simplifying switching loss model does not consider the common source inductance and the nonlinear characteristics of the parasitic capacitors of the MOSFET. The common source inductance L_{s1} of the control MOSFET, as illustrated in Figure 1.5, is the inductance shared by the main current path and the gate driver loop, which has negative impact on switching transition and increases switching loss at high switching frequency (>1 MHz).

Figure 1.6 illustrates the equivalent circuit of turn-on transition with the common source inductance. During the turn-on transition, the drain current i_{ds} increases, which will induce a positive voltage across L_s. The turn-on gate current is

$$i_{g\_on} = \frac{v_{gs} - v_{Cgs} - L_s \frac{di_{ds}}{dt}}{R_g + R_{hi}} \tag{1.3}$$

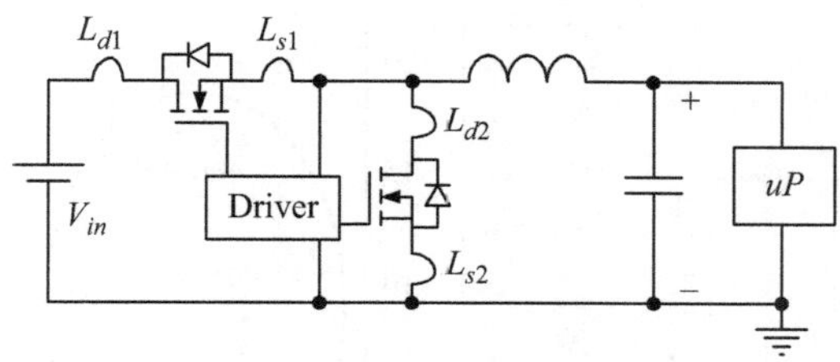

Figure 1.5 The buck converter with parasitic inductance

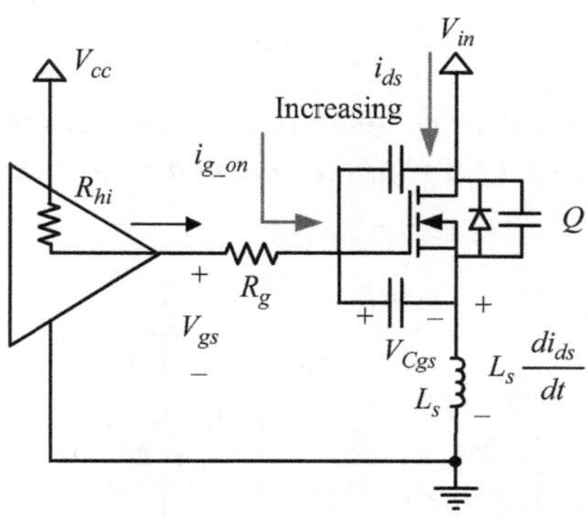

Figure 1.6 Equivalent circuit of turn-on transition with the common source inductance

From (1.3), because of $L_s \frac{di_{ds}}{dt}$, the effective turn-on gate current i_{g_on} is reduced, which increases the turn-on transition time and turn-on loss.

Figure 1.7 illustrates that with the conventional voltage source driver, after the common source inductance of 2 nH is considered, the turn-on transition time dramatically increases from 7 ns to 17 ns, which increases the turn-on loss. The circuit parameters are as follows: $V_{in} = 12$ V, $V_o = 1.2$ V, $I_o = 20$ A, control MOSFET: HAT2168, SR: HAT2165 [16,17].

Figure 1.8 illustrates the equivalent circuit of turn-off transition with the common source inductance. During the turn-off transition, the drain current i_{ds} decreases, which will induce a negative voltage across L_s. Therefore, the turn-off gate current is

$$i_{g\_off} = \frac{v_{Cgs} - L_s \frac{di_{ds}}{dt}}{R_g + R_{lo}} \tag{1.4}$$

From (1.4), because of $L_s \frac{di_{ds}}{dt}$, the effective turn-off gate current i_{g_off} is reduced, which increases the turn-off transition time and turn-off loss.

Figure 1.9 illustrates that after the common source inductor of 2 nH is considered, the turn-off transition time dramatically increases from 6 ns to 23 ns, which increases the turn-off loss greatly.

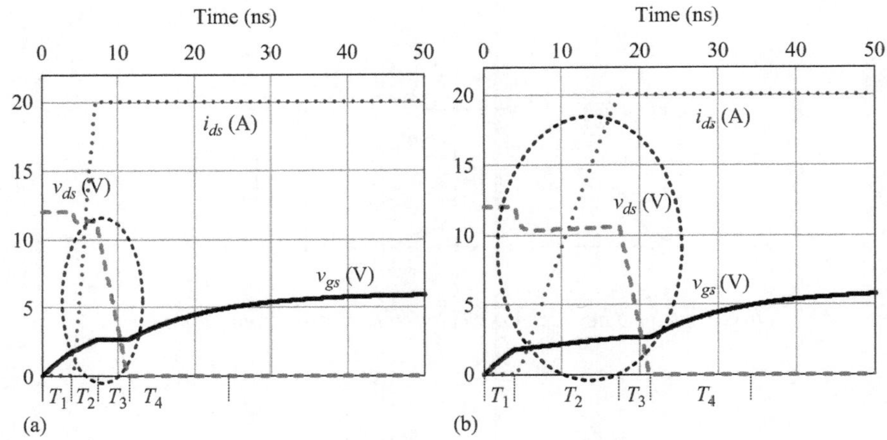

Figure 1.7 Turn-on transition comparison: (a) without the common source inductance and (b) with the common source inductance ($L_{s1} = 2$ nH)

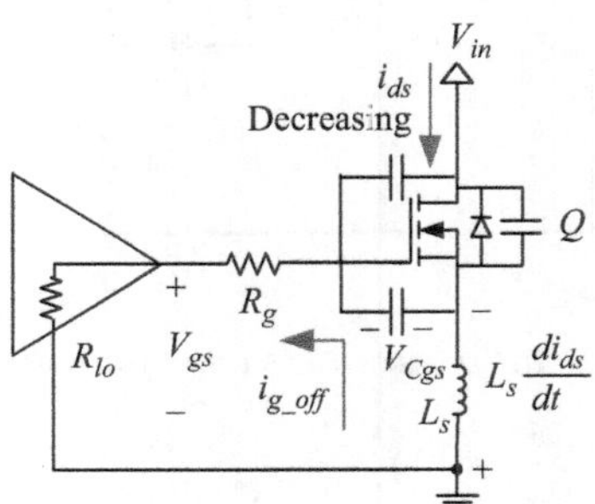

Figure 1.8 Equivalent circuit of turn-off transition with the common source inductance

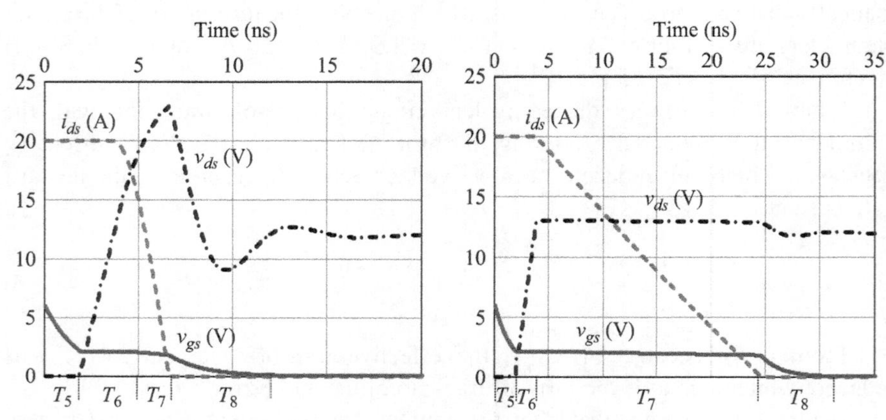

Figure 1.9 Turn-off transition comparison: (a) without the common source inductance and (b) with the common source inductance ($L_{s1} = 2$ nH)

In summary, fast change of the drain-source current induces a voltage across the common source inductance L_{s1}, which always reduces the effective gate drive currents of the control MOSFET and prolongs the switching time, and consequently increases the switching loss significantly at high switching frequency. Therefore, the loss calculation based on the piecewise linear model is much lower than the experimental results obtained from synchronous buck converters with MHz switching frequency.

In order to predict the switching loss at high frequency more accurately, the analytical switching loss model was proposed in [16] considering the common source inductance, which is suitable for massive data processing of some applications that need good accuracy and short simulation time. The bus voltage for 12 V two-stage VR was optimized based on the above accurate analytical loss model in [17] and the loss calculation matches the experimental results at the switching frequency of 2 MHz well. In [18], the common source inductance and switching loop inductance were modeled simultaneously and the MOSFET switching characteristics are explored based on closed-form analytical equations, which provides design criteria for optimizing switching performance of packaged power electronics.

The impact of the common source inductance L_{s2} on synchronous rectifier (SR) Q_2 was investigated in [19]. The common source inductance could help to reduce power loss because of the reverse current direction in the SR compared to the control MOSFET. In addition, with common source inductance, $C_{dv/dt}$ immunity and zero dead time switching could be achieved, which could reduce power loss of the SRs.

New packaging technology, such as PowerPAK SO-8 from Vishay and DirectFET from International Rectifier, reduces the parasitic influence to improve the efficiency and provide better thermal characteristics for a high-frequency VR [20,21].

1.3.2 Body diode conduction loss

In high current VR applications, SR technique is widely used to reduce the conduction losses of the diode rectifiers. In a synchronous buck converter, the dead time is the amount of time that both the control MOSFET and SR are off to prevent "shoot-through." The body diode conduction loss is

$$P_{body} = V_{body} \cdot I_o \cdot f_s \cdot t_{body} \tag{1.5}$$

where V_{body} is the body diode forward voltage drop and t_{body} is the body diode conduction time.

During the dead time, the SR body diode carries the load current, resulting in high conduction loss due to high forward voltage drop of the body diode. At high switching frequency around 1 MHz, the body diode conduction loss becomes high due to the relative ratio of the dead time in one switching cycle. In order to achieve high efficiency of a buck converter, it is so important to minimize the conduction time of the body diode.

The driver chips with adaptive control and predictive control to minimize the dead time are available from different companies such as TPS2832 [22] and UCC27222 [23] from Texas Instrument. In [24], a new resonant gate driver with a transformer and Zener diode was proposed to reduce the body diode conduction loss. As long as the crossover level of the gate driver signals of the two MOSFETs is equal to or less than the threshold voltage of the devices, the dead time can be eliminated and the shoot-through can also be avoided. However, the Zener diode will induce additional gate charge loss and the leakage inductance of the gate drive transformer may deteriorate the overall performance at high frequency.

1.3.3 Reverse recovery loss

The reverse recovery loss for the SR body diode is

$$P_{QRR} = Q_{RR} \cdot V_R \cdot f_s \tag{1.6}$$

where Q_{RR} is the body diode reverse recovery charge and V_R is the reverse block voltage.

The reverse recovery current introduces voltage oscillation, decreasing efficiency and causing electromagnetic interference (EMI) problems. Soft-switching technique may be used to eliminate the reverse recovery when the forward current through the diode decreases to zero before the reverse voltage blocks across the diode.

1.3.4 Gate drive loss

Energy is required to charge and discharge the MOSFET gate capacitance each switching cycle. For the conventional voltage source driver, this energy is dissipated through resistances in series with the gate within the gate drive circuit. The power to charge the gate of MOSFET is

$$P_{GATE} = Q_g \cdot V_c \cdot f_s \tag{1.7}$$

where Q_g is the total gate charge and V_c is the amplitude of the gate drive voltage.

The gate drive loss becomes a concern at switching frequency beyond 1 MHz, which decreases the overall efficiency and causes thermal problems on the driver chips.

1.4 Summary

Recently, there has been a trend to increase the switching frequency beyond multi-MHz in switching power supplies. The eGaN HEMTs with much improved FOM provide a significant switching performance, which even allows the switching frequency above tens of MHz, such as VHF (30–300 MHz) converters.

Unfortunately, the conventional gate drive scheme is a voltage source drive approach, and all the drive energy is dissipated on the resistance in the charge and discharge path. As the switching frequency rises, the gate drive loss of the power MOSFETs rises as well, which is proportional to the switching frequency. The high current MOSFETs with low on-resistance typically own large gate-source capacitance. The gate drive loss becomes more serious, especially in low voltage and high current applications. More importantly, high gate capacitance also increases the switching time and thus the switching loss. Therefore, the performance of the gate drive circuits for the high frequency power MOSFETs on the converter becomes so critical.

References

[1] D. J. Perreault, J. Hu, J. M. Rivas, *et al.*, "Opportunities and challenges in very high frequency power conversion," in *Proceedings on IEEE Applied Power Electronics Conference (APEC)*, pp. 1–14, 2009.

[2] T. M. Andersen, S. K. Christensen, A. Knott, and M. Andersen, "A VHF class E DC–DC converter with self-oscillating gate driver," in *Proceedings on IEEE Applied Power Electronics Conference (APEC)*, pp. 885–891, 2011.

[3] J. S. Glaser and J. M. Rivas, "A 500 W push–pull DC–DC power converter with a 30 MHz switching frequency," in *Proceedings on IEEE Applied Power Electronics Conference (APEC)*, pp. 654–661, 2010.

[4] C. D. Meyer, S. S. Bedair, B. C. Morgan, and D. P. Arnold, "High-inductance-density, air-core, power inductors, and transformers designed for operation at 100–500 MHz," *IEEE Transactions on Magnetics*, vol. 46, no. 6, pp. 2236–2239, 2010.

[5] P. Hazucha, G. Schrom, J. Hahn, *et al.*, "A 233-MHz 80–87% efficient four-phase DC–DC converter utilizing air-core inductors on package," *IEEE Journal of Solid-State Circuits*, vol. 40, no. 4, pp. 838–845, 2005.

[6] W. Liang, J. Glaser, and J. Rivas, "13.56 MHz high density DC–DC converter with PCB inductors," in *Proceedings on IEEE Applied Power Electronics Conference (APEC)*, pp. 633–640, 2013.

[7] J. M. Rivas, R. S. Wahby, J. S. Shafran, and D. J. Perreault, "New architectures for radio-frequency DC–DC power conversion," *IEEE Transactions on Power Electron.*, vol. 21, no. 2, pp. 380–393, 2006.

[8] A. Lidow, J. Strydom, M. D. Rooij, and D. Reusch, "GaN transistors for efficient power conversion," 2nd ed., John Wiley & Sons, 2014.

[9] A. Lidow, D. Reusch, and J. Strydom, "GaN integration for higher DC–DC efficiency and power density," Application Note, EPC Co., 2015.

[10] D. Reusch and J. Strydom, "Evaluation of gallium nitride transistors in high frequency resonant and soft-switching DC–DC converters," *IEEE Transactions on Power Electron.*, vol. 30, no. 9, pp. 5151–5158, 2015.

[11] J. Strydom and D. Reusch, "Design and evaluation of a 10 MHz gallium nitride based 42 V DC–DC converter," in *Proceedings on IEEE Applied Power Electronics Conference (APEC)*, pp. 1510–1516, 2014.

[12] Y. Wang, W. Kim, Z. Zhang, J. Calata, and K. D. T. Ngo, "Experience with 1 to 3 megahertz power conversion using eGaN FETs," in *Proceedings on IEEE Applied Power Electronics Conference (APEC)*, pp. 532–539, 2013.

[13] Voltage Regulator Module (VRM) and Enterprise Voltage Regulator-Down (EVRD) 11.0 Design Guidelines, Apr. 2008.

[14] J. Klein, "Synchronous buck MOSFET loss calculation with Excel model," Application note-6005, 2004, www.fairchildsemi.com.

[15] Laszlo Balogh, "Design and application guide for high speed MOSFET gate drive circuits," Seminar 1400 Topic 2, 2002, www.ti.com.

[16] Y. Ren, M. Xu, J. Zhou, and F. C. Lee, "Analytical loss model of power MOSFET," *IEEE Transactions on Power Electron.*, vol. 21, no. 2, pp. 310–319, 2006.

[17] Y. Ren, M. Xu, F. C. Lee, and P. Xu, "The optimal bus voltage study for 12 V two-stage VR based on an accurate analytical loss model," in *Proceedings on IEEE Power Electronics Specialists Conference (PESC)*, pp. 4319–4324, 2004.

[18] Y. Xiao, H. Shah, T. P. Chow, and R. J. Gutmann, "Analytical modeling and experimental evaluation of interconnect parasitic inductance on MOSFET switching characteristics," in *Proceedings on IEEE Applied Power Electronics Conference (APEC)*, pp. 516–521, 2004.

[19] B. Yang and J. Zhang, "Effect and utilization of common source inductance in synchronous rectification," in *Proceedings on IEEE Applied Power Electronics Conference (APEC)*, pp. 1407–1411, 2005.

[20] J. Lee, "Package parasitics influence efficiency," *Power Electronic Technology*, pp. 14–21, 2005.

[21] R. Monteiro, C. Blake, A. Sawle, and A. Woodworth, "High current voltage regulator module (VRM) uses DirectFET^{TM} MOSFETs to achieve current densities of 25A/in2 at 1 MHz to power 32-bit servers," International Rectifier, Technical Papers, 2003.

[22] Datasheet of TPS2832, Texas Instrument, http://focus.ti.com/docs/prod/folders/tps2832.html

[23] Datasheet of UCC27222, Texas Instrument, http://focus.ti.com/lit/ds/symlink/ucc27222.pdf

[24] Y. Ren, M. Xu, Y. Meng, and F. C. Lee, "12V VR efficiency improvement based on two-stage approach and a novel gate driver," in *Proceedings on IEEE Power Electronics Specialists Conference (PESC)*, pp. 2635–2641, 2005.

Chapter 2
Fundamentals of current source driver

2.1 Resonant gate drivers

Resonant gate-driver technique was originally developed to recover part of metal oxide semiconductor field-effect transistor (MOSFET) gate-drive loss when operating at high switching frequency (above 1 MHz) [1–3]. The self-oscillating resonant gate driver with a resonant network was used in the application of radio-frequency (RF) power amplifier featuring sinusoidal waveforms generally [4,5]. The self-oscillating resonant gate driver (soft-gating driver) is also applied to a high-frequency (>30 MHz) d.c./d.c. converter to reduce the gate-drive loss [6].

In order to recover a portion of the gate energy, otherwise lost in conventional drivers, several papers have been published since the early 1990s proposing resonant gate drive techniques. Several of the proposed drivers are critically reviewed in this subsection. Following the review, a brief summary is presented highlighting the drawbacks of the existing circuits.

One of the first resonant gate drive circuits was proposed in [7]. The circuit and its associated waveforms are given in Figure 2.1. In Figure 2.1, S_1 and S_2 represent the gating waveforms for MOSFETs S_1 and S_2; v_{gs} represents the gate-to-source voltage of power MOSFET, M; i_L represents the inductor current; and i_g represents the gate current. This circuit uses an auxiliary inductor, L, which operates in continuous conduction mode and a resonant capacitor, C_{res}, along with two control switches, S_1 and S_2, to recover a portion of the gate energy. There are two significant advantages of this technique. First, the switch, M, is driven by an essentially constant current source during its transitions, so its associated turn-on and turn-off times can be quick. In addition, switches S_1 and S_2 clamp the gate high and low during the on and off states, respectively, thereby eliminating false triggering of the MOSFET gate. The disadvantage of this technique is the existence of significant circulating current during the on and off states of M, so the control switches must be selected with relatively low R_{ds} ratings and therefore high gate charge. In addition, the inductor required is relatively large since the current is continuous. The circuit in [7] achieved a 3% efficiency improvement for a 20 W converter operating at 2 MHz in comparison to a conventional driver.

Following the resonant transition technique, the resonant pulse technique was proposed in [8]. The circuit and its associated waveforms are illustrated

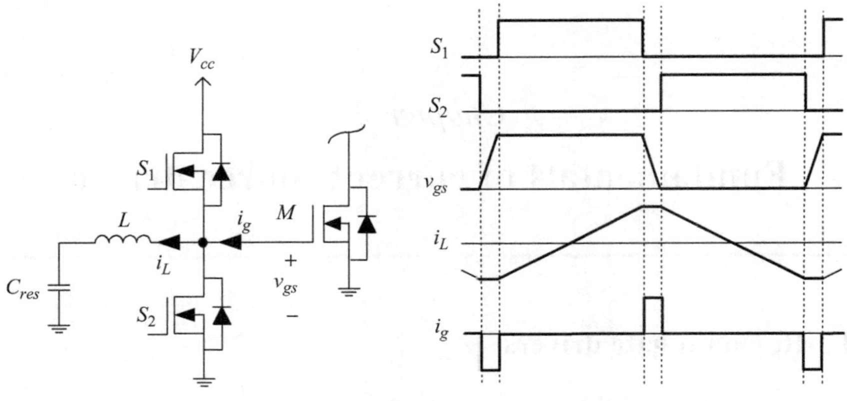

(a) (b)

Figure 2.1 Resonant transition gate driver [7]; (a) circuit and (b) waveforms

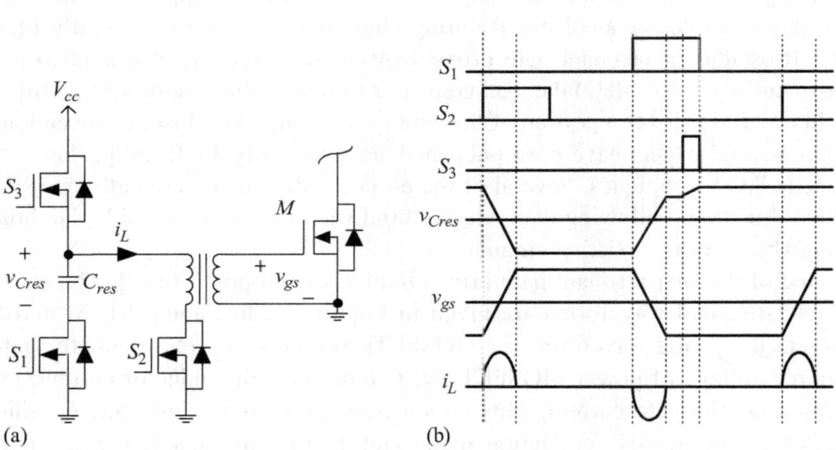

(a) (b)

Figure 2.2 Resonant pulse gate driver [8]; (a) circuit and (b) waveforms

in Figure 2.2. In Figure 2.2, S_1, S_2, and S_3 represent the gating waveforms for MOSFETs S_1, S_2, and S_3; v_{gs} represents the gate-to-source voltage of power MOSFET, M; i_L represents the transformer and gate current; and v_{Cres} represents the voltage across the resonant capacitor, C_{res}. The leakage inductance of the transformer is used to resonate with C_{res} to form the energy recovery circuit. Switch S_3 is gated simultaneously with S_1 to reset the resonant capacitor. The advantages of the resonant transition technique are low conduction losses and gate drive isolation. The disadvantage is slow turn-on and turn-off since the power MOSFET gate must be charged and discharged through the leakage inductance with initially zero current. Furthermore, the power MOSFET gate is driven without a d.c. component, so there is excess negative charge energy.

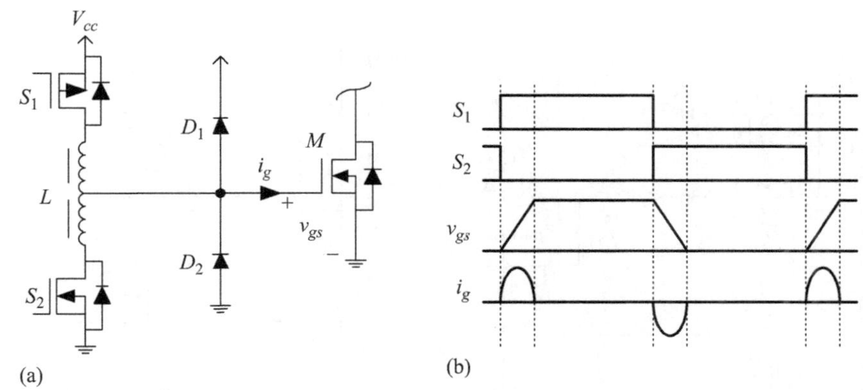

Figure 2.3 Coupled inductor resonant gate driver [9]; (a) circuit and (b) waveforms

A resonant gate driver was proposed utilizing a coupled inductor in series with two control switches to eliminate cross conduction and reduce linear operation losses in the control switches in addition to providing partial gate energy recovery [9]. The circuit and waveforms are illustrated in Figure 2.3. In Figure 2.3, S_1 and S_2 represent the gating waveforms for MOSFETs S_1 and S_2; v_{gs} represents the gate-to-source voltage of power MOSFET, M; and i_g represents gate current. The diodes provide clamping for the gate voltage to prevent over voltage and negative voltage conditions at the gate. However, the diodes do not provide a low impedance path to eliminate spontaneous turn-on due to noise spikes. The most significant drawback of this circuit is the inherent slow turn-on and turn-off since the gate must be charged and discharged through the inductance with zero initial current.

A dual high-side (HS)/low-side resonant gate driver was proposed [10]. The circuit and its associated waveforms are illustrated in Figure 2.4. In Figure 2.4, S_1–S_5 represent the gating signals for MOSFETs S_1–S_5; i_{L1} and i_{L2} represent the coupled inductor currents; and v_{gs1} and v_{gs2} represent the gate-to-source voltages of M_1 and M_2, respectively. The circuit promises low conduction losses since the inductor current is discontinuous. Only one core is used to form the coupled inductor pair, which helps to reduce the size of the circuit. In addition, control switch S_5 actively clamps the gate of the bottom switch, M_2, low during its off time, thereby eliminating spontaneous turn-on of M_2 caused by $C_{dv/dt}$ after it is normally turned off. The disadvantages of the proposed method are slow turn-off of the top and bottom switches in addition to the complexity of the logic required to generate the control switch gating signals. The circuit in [10] achieved a 3% efficiency improvement at full load for a 20 W synchronous buck converter operating at 1 MHz in comparison to a conventional driver.

Another resonant gate drive circuit was proposed in [11]. The circuit and its waveforms are given in Figure 2.5. In Figure 2.5, S_1–S_3 represent the gating

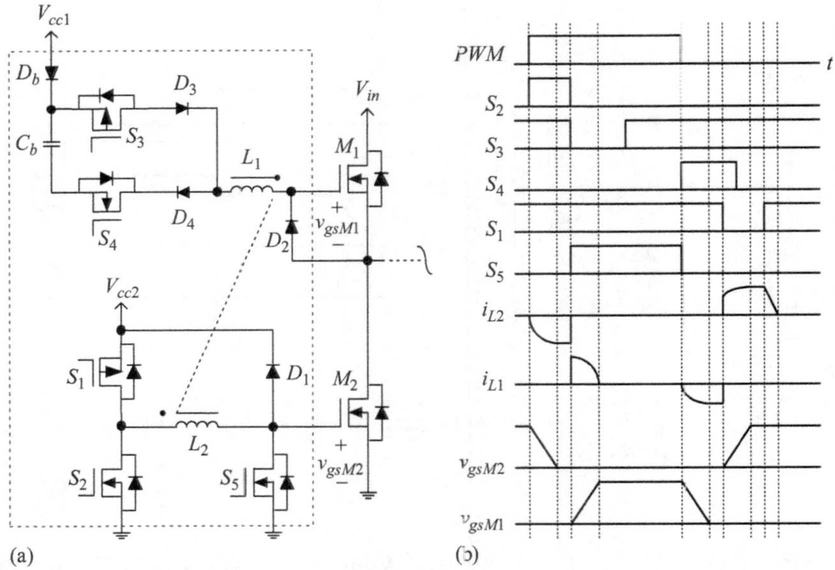

Figure 2.4 High/low coupled inductor resonant gate driver [10]; (a) circuit and (b) waveforms

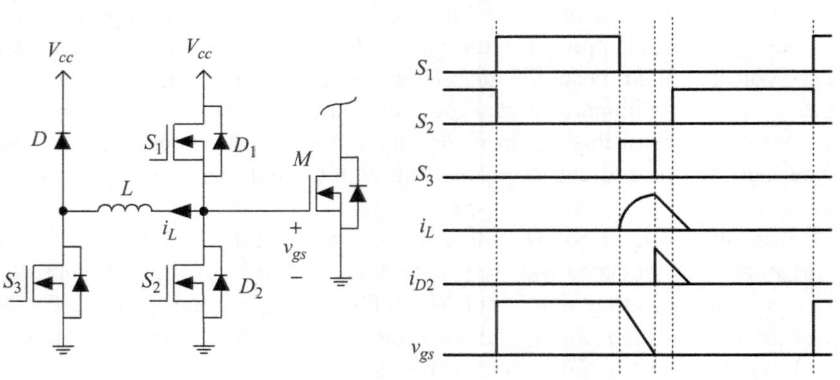

Figure 2.5 Resonant gate driver proposed in [11]; (a) circuit and (b) waveforms

waveforms for MOSFETs S_1–S_3; i_L represents the inductor current; i_{D2} represents the diode current in D_2; and v_{gs} represents the gate-to-source voltage of power MOSFET, M. Control switches S_1 and S_2 actively clamp the gate of M, high and low during its respective on and off times. The circuit only recovers the gate energy during turning off and it does so with a slow transition. The turn-on gate loss is the same as a conventional gate driver.

As a conclusion, among the resonant gate drive circuits previously proposed, all of them suffer from at least one of the following driver-specific problems:

1. Only low side, ground referenced drive [7,8,12–14].
2. The leakage inductance of bulky transformer, or coupled inductance [10,15,16].
3. Slow turn-on and/or turn-off transition times, which increase(s) both conduction and switching losses in the power MOSFET due to charging the power MOSFET gate beginning at zero current [8,10,12,13,15,16].
4. The inability to actively clamp the power MOSFET gate to the supply voltage during the on time and/or to ground during the off time, which can lead to undesired false triggering of the power MOSFET gate, i.e., lack of $C_{dv/dt}$ immunity [8,10,12,13,15,16].

A comparison of fundamental gate-driver topologies for high-frequency applications was discussed in terms of suitable operating conditions, construction and losses in [17]. An assessment of resonant drive techniques for use in low power d.c./d.c. converters was presented in [18] and the mathematical model was built to estimate the power loss of the drive circuit in [19]. The effects of internal parasitic inductance and the parasitic output capacitance of the driver switches were analyzed focusing on their impact on the driver loss and on the synchronous rectifier (SR) gate voltage in different non-isolated resonant driver topologies [20].

Unfortunately, all the above investigations are generally emphasizing gate energy savings by the resonant driver and concentrating on the drive topologies, but ignore the potential switching loss savings that are much more dominant in MHz switching power converter.

2.2 Concept of current source driver

To solve the above problems of existing resonant drivers and reduce switching loss, current source drivers (CSDs) are proposed to achieve high switching speed and reduce switching loss. The basic concept of CSDs is shown in Figure 2.6. The comparison of the gate current waveform shapes between the CSD and conventional voltage driver is illustrated in Figure 2.7. The operation of CSDs is in such a way that the peak value, or close to the peak value of i_g is used to charge and discharge the gate capacitors C_{gs}. S_1 and S_2 form a totem structure to provide a low impedance path to clamp the gate terminal at either zero or V_c during the turn-on or turn-off conditions, respectively. Two major advantages of this circuit are: (1) the gate-drive loss of the power switching device can be significantly reduced and (2) C_{gs} can be charged and discharged quickly to achieve fast switching speed. Based on the above mechanism, different CSD topologies can be developed for different types of power converters to reduce the high-frequency gate-drive loss and switching loss of the MOSFETs.

In hard switching converters, neglecting any body diode reverse recovery and drain and common source inductance, the turn-on switching loss can be

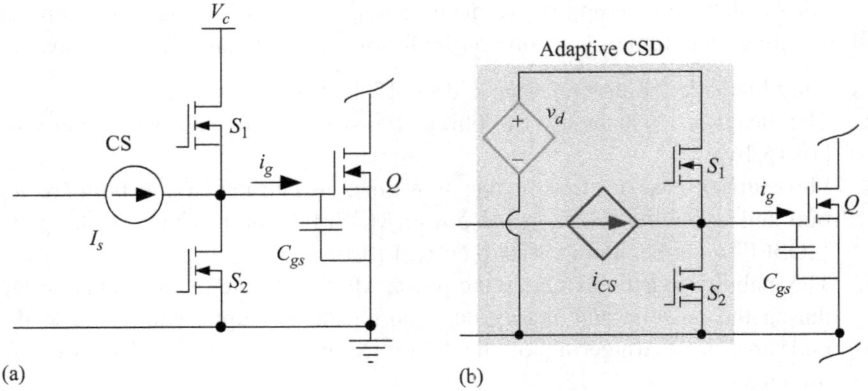

Figure 2.6 Concept of (a) current source driver and (b) adaptive CSD

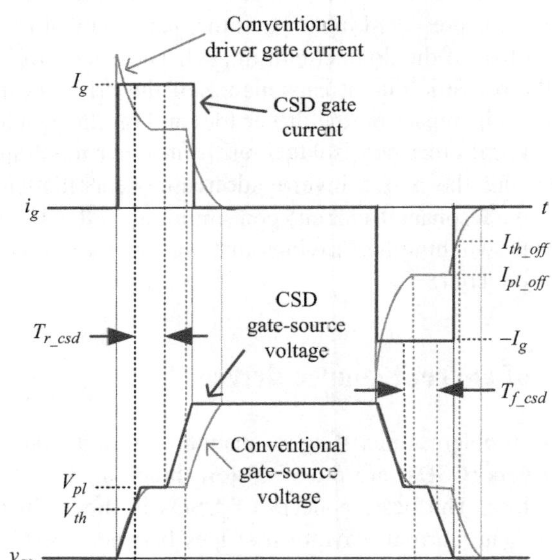

Figure 2.7 Comparison between the gate current waveform shapes for the conventional driver and an ideal CSD

approximated by (2.1), and the turn-off switching loss by (2.2), where V_{ds} represents the voltage across the switch. In (2.1), I_{on} represents the current through the switch at turn on and T_r represents the rise time. In (2.2), I_{off} represents the current through the switch at turn-off and T_f represents the fall time. Since both T_r and T_f are inversely proportional to the average gate current (during the switching interval), I_g, it is clear that switching losses can be reduced by increasing gate current. However, in conventional gate drivers, this is generally not possible since the peak current is already limited by the current handling capability of the driver switches. In hard switched converters, the actual potential total switching loss reduction can be much more significant than the gate loss savings. This statement can hold true for most soft switching converters, since zero voltage switching (ZVS) converters typically eliminate turn-on loss only, or with the addition of snubber capacitors, they are not able to maintain ZVS across varying line and load conditions:

$$P_{on} = \frac{1}{2} f_s V_{ds} I_{on} T_r \tag{2.1}$$

$$P_{off} = \frac{1}{2} f_s V_{ds} I_{off} T_f \tag{2.2}$$

In the conventional driver, the gate current decays significantly from its peak value due to the RC type charging and discharging. On the other hand, the proposed CSD behaves like a nearly constant current source and the gate current actually increases slightly during the rise and fall times as illustrated in Figure 2.7 where it is clear that the magnitude of i_g is greater than I_{pl_on} and I_{pl_off}, respectively. The result is that the turn-on and turn-off times decrease significantly and more importantly, the rise and fall times decrease significantly, which reduces switching loss.

In Figure 2.6(b), a general structure for the adaptive CSDs is given. The adaptive parameters are defined as follows: (1) V_d is a controlled voltage source as the CSD adaptive drive voltage from drive supply voltage V_c. It can be controlled with the variables in the power circuits, such as load current, drain current of the switching devices, input voltage, etc. Higher drive voltage results in lower $R_{DS(on)}$ and thus lower conduction losses, but increases the drive losses. (2) i_{CS} is a controlled current source as the drive current for Q. It can also be controlled with the loads, drain current, and voltage in the power circuit. Higher drive current reduces the switching loss but increases the gate-drive loss. Compared to other CSD circuits, the controllable drive current and drive voltage are beneficial to the optimal design between the switching loss and drive loss during the wide range operation of the main power MOSFETs. This concept can be extended to most of the CSD circuit topologies.

2.3 A practical and accurate switching loss model

2.3.1 Introduction

In order to optimally design a high-frequency switching converter, engineers and researchers begin their design by estimating the losses in a design file that is typically created using a spreadsheet, or other mathematical software. Device data sheet values and analytical models are used to calculate the losses. Using the loss models, many design parameters and components are compared to achieve a design with the optimal combination of efficiency and cost.

With analytical models, most often piecewise linear turn-on and turn-off waveforms are used, or simplified equivalent circuits are used to derive switching loss equations. These methods yield closed form mathematical expressions that can be easily used to produce optimization curves within a design file; however, the challenge is to improve accuracy while minimizing complexity. Most often, piecewise linear turn-on and turn-off waveforms are used, or simplified equivalent circuits are used to derive switching loss equations.

One of the most popular analytical models is the piecewise linear model presented in [21]. This model is referred to as the conventional model and is used as a benchmark later for comparison purposes with the proposed model. This model enables simple and rapid estimation of switching loss; however, the main drawback is that it neglects the switching loss dependences due to common source inductance and other parasitic inductances. Typically, this model predicts that turn-on and turn-off losses are nearly similar in magnitude; however, in a real converter operating at a high switching frequency, the model is highly inaccurate since turn-off loss is much larger due to common source inductance and other parasitic inductances.

A comprehensive analytical model is presented in [22]. This model is an extension of the model presented in [21], with the advantage that it provides accurate characterization of switching loss when common source inductance is included. The main drawback of the models in [22] is their complexity.

The synchronous buck remains the topology of choice for voltage regulators (VRs) in today's computers. However, in order to properly model switching loss in a buck VR, a detailed understanding of the impact of MOSFET gate capacitance, common source inductance, other parasitic inductance, and load current on switching loss is necessary. This is most easily accomplished through careful examination of waveforms through simulation and experiments.

2.3.2 Impact of parasitic inductance and load current

A synchronous buck converter is illustrated in Figure 2.8. In a synchronous buck VR, it is well known that the input voltage, load current, and HS MOSFET gate-to-drain charge influence switching loss in the HS MOSFET. However, it is not well known that the inductances associated with the device packaging and printed circuit board (PCB) traces also contribute significantly to HS MOSFET switching loss. It is worth noting, but generally well known that with proper dead time, the SR switches with near zero switching loss.

The synchronous buck in Figure 2.8 includes parasitic drain and source inductances for the HS MOSFET, M_1 and SR MOSFET, M_2. It can be assumed that the source inductances, L_{s1} and L_{s2}, are common to their respective drive signals. Any other inductance in the source that is not common to the source is assumed to be lumped with the drain inductances, L_{d1} and L_{d2}. These inductances have a significant impact on the switching loss behavior in high-frequency synchronous buck VRs.

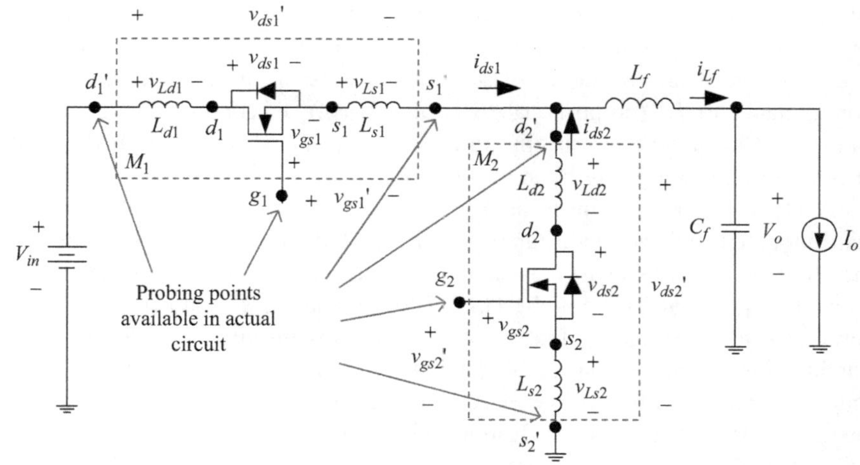

Figure 2.8 Synchronous buck voltage regulator with parasitic inductances

During the switching transitions, the HS MOSFET operates in the saturation (linear) mode as a dependent current source simultaneously supporting the current through the device and voltage across it. At turn on and turn off, the gate-source voltage, v_{gs1}, is held at the plateau voltage, V_{pl}, by the feedback mechanism provided by the voltage across the common source inductance, v_{Ls1}.

As can be observed from the circuit in Figure 2.8, at turn on, as the HS MOSFET current increases, v_{Ls1} is positive in the direction noted, so this voltage subtracts from the V_{cc} voltage applied to the gate, enabling $v_{gs1} = V_{pl}$ while the MOSFET operates in the saturation mode. At the same time, the four parasitic inductances provide a current snubbing effect, which virtually eliminates turn-on switching loss enabling a near zero current switching (ZCS) turn on. During this transition, the rise time, T_r, is dictated by the gate driver's ability to charge the MOSFET gate capacitances (C_{iss} from V_{th} to V_{pl} and C_{gd} to V_{in}), which is the time for v_{ds1} to fall to zero. Then, it is assumed that this time is independent of the time it takes i_{ds1} to rise to its final value equal to the buck inductor current, i.e., after T_r, i_{ds1} can be less than the buck inductor current.

At turn off, as the HS MOSFET current decreases, v_{Ls1} is negative in the direction noted in Figure 2.8, so this voltage subtracts from the low impedance source voltage (ideally zero volts) applied to the gate enabling $v_{gs1} = V_{pl}$ while the MOSFET operates in the saturation mode. During this transition, the fall time, T_f, is the time for the HS MOSFET current to fall from the buck inductor current to zero. This time is dictated by both the gate driver's ability to discharge the MOSFET gate capacitances (C_{gd} from V_{in}, and C_{iss} from V_{pl} to V_{th}) and by the four parasitic inductances, which prolong the time for i_{ds1} to fall to zero by limiting the di_{ds}/dt.

As alluded to in the two paragraphs above, the MOSFET and trace parasitic inductances have vastly different effects at turn on and turn off. At turn on, the inductances provide a current snubbing effect, which decreases turn-on switching loss. At turn off, the inductances increase the turn-off loss by prolonging T_f. In addition, as load current increases, T_f increases, so turn-off losses increase proportionally to I_o^2 (proportional to I_o and $T_f(I_o)$). In contrast, at turn on, the load current magnitude has ideally no effect on the T_r. Therefore, in real circuits, turn-off loss is much greater than turn-on loss.

Another important point to note from Figure 2.8 is that in a real circuit, the board-mounted packaged inductances are distributed within the MOSFET devices. Therefore, when probing in the laboratory, only one has access to the external terminals, g_1, s_1', and d_1' for the HS MOSFET and g_2, s_2', and d_2' for the SR. However, the actual nodes that provide waveform information relevant to the switching loss are at the unavailable internal nodes s_1 and d_1 for the HS MOSFET. Using the plateau portion of the measured gate-source voltage, v_{gs1}' to determine the switching loss times is misleading since the induced voltage across L_{s1} is included. Probing v_{gs1}' in the laboratory, one would observe a negligible T_r at turn on, and a turn-off T_f less than one-half of the actual T_f. The actual v_{gs1} waveform, which cannot be measured in a real circuit, more clearly illustrates the plateau portions in the rise and fall times.

To demonstrate the effects of load current and common source inductance, experimental testing was done at a reduced frequency of 200 kHz, with the source connection cut and a wire inserted in the common source path to measure the MOSFET current. Measurement waveforms are illustrated in Figures 2.9 and 2.10, where the load current has been increased from 0 to 5 A. With this method, the inductance of the wire (approximately 20 nH) is much greater than the approximate total package inductance of 1 nH, so the package inductance can be neglected allowing for measurement of v_{gs1} and v_{ds1}. As stated previously, it is noted that as load current increases from 0 to 5 A, T_r remains nearly unchanged from 20 to 22 ns, but T_f increases significantly from 48 to 96 ns. In addition, at a constant load current of 5 A, as illustrated in Figure 2.11, as L_{s1} increases to 30 nH (using a longer 3-inch wire), T_r remains relatively unchanged from 22 to 24 ns, while T_f further increases from 96 to 160 ns.

From knowledge of the circuit operation and observation of the experimental results presented, three important observations and conclusions can be made:

1. In a practical synchronous buck VR, turn-off loss is much greater than turn-on loss since the circuit inductances provide a current snubbing effect, which decreases and virtually eliminates turn-on switching loss, but increases the turn-off loss by prolonging T_f.
2. T_r is dictated by the time for the voltage to fall to zero and is independent of the final value of the current. In addition, load current has negligible impact on T_r, while common source inductance has only a small impact, since as L_{s1} increases, the current di_{ds}/dt decreases.
3. T_f is dictated by the time for the current to fall to zero. Load current, commons source inductance, and other circuit parasitic inductances (i.e., L_{d1}, L_{s2}, and L_{d2}) increase T_f.

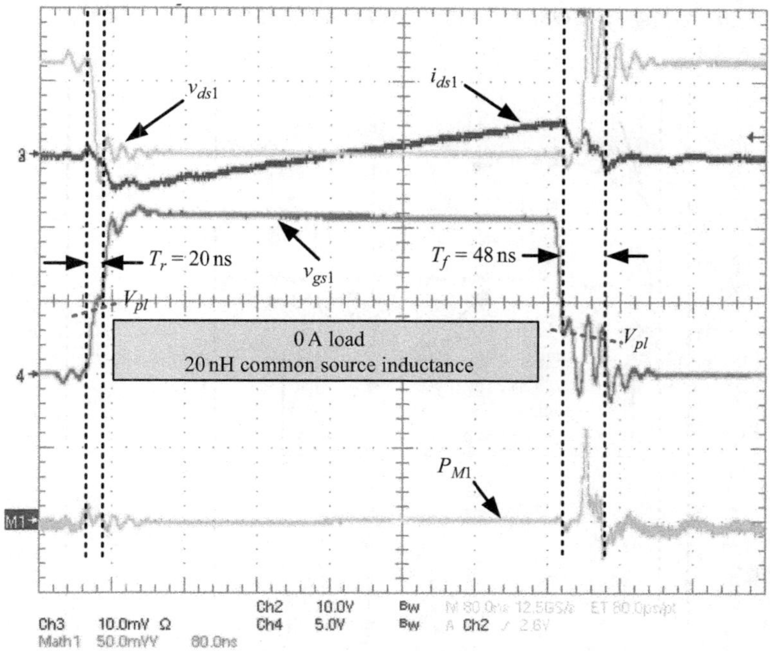

Figure 2.9 Switching waveforms at 0 A load and 20 nH common source inductance (80 ns/div; v_{ds1}: 10 V/div; i_{ds1}: 5 A/div; v_{gs1}: 5 V/div; P_{M1}: 50 W/div)

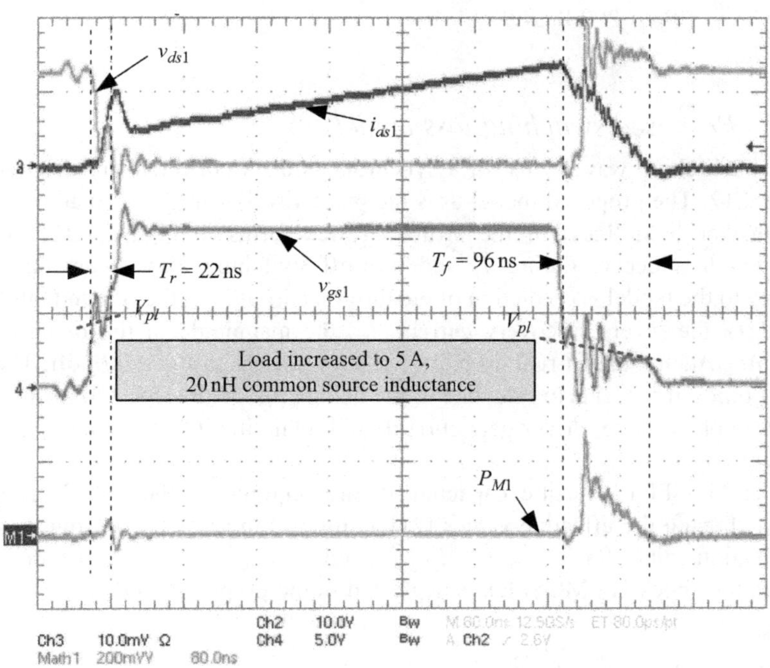

Figure 2.10 Switching waveforms at 5 A load and 20 nH common source inductance (80 ns/div; v_{ds1}: 10 V/div; i_{ds1}: 5 A/div; v_{gs1}: 5 V/div; P_{M1}: 200 W/div)

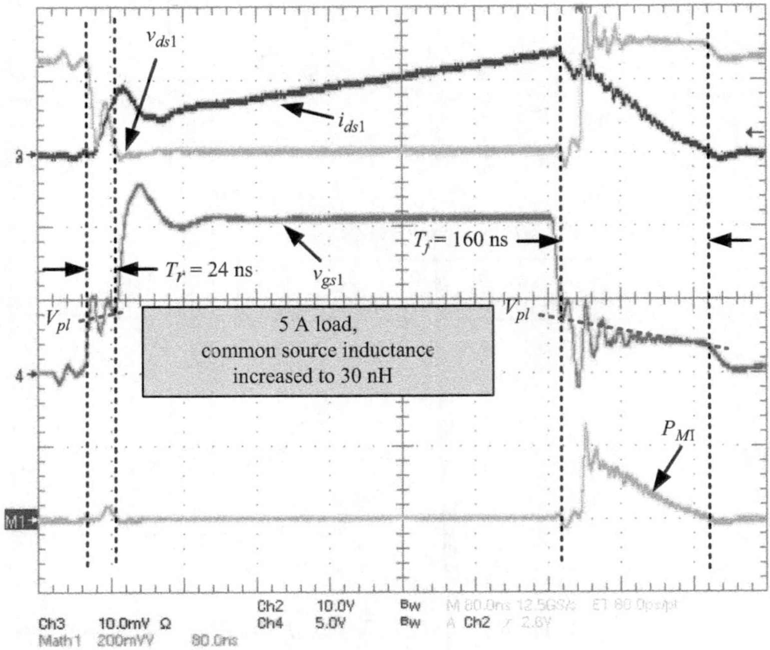

Figure 2.11 Switching waveforms at 5 A load with 30 nH common source inductance (80 ns/div; v_{ds1}: 10 V/div; i_{ds1}: 5 A/div; v_{gs1}: 5 V/div; P_{M1}: 200 W/div)

2.3.3 Proposed switching loss model

Typical switching waveforms for a synchronous buck converter are illustrated in Figure 2.12. The proposed model uses the piecewise linear approximations (noted with thicker bold lines) of the switching waveforms in Figure 2.12. Turn-on switching loss occurs during T_r and turn-off switching loss occurs during T_f. The key to the model is prediction of the turn-on current, I_{on}, the rise and fall times, T_r and T_f, the reverse recovery current, I_{rr}, the magnitude of the rising current slope, $\Delta i_{ds}/\Delta t$, and the current drop, Δi_{1f}, when v_{ds1} rises to V_{in} at turn off. The goal of the proposed model is to calculate the switching loss with respect to load current, driver supply voltage, driver gate current, and total circuit inductance in a simple manner.

The MOSFET parasitic capacitances are required in the model. They are estimated using the effective values [21] as follows in (2.3)–(2.5) using datasheet specification values for V_{ds1_spec}, C_{rss1_spec}, and C_{iss1_spec}. The C_{ds1} capacitor of the synchronous buck HS MOSFET is neglected in the proposed model:

$$C_{gd1} = 2C_{rss1\_spec}\sqrt{\frac{V_{ds1\_spec}}{V_{in}}} \tag{2.3}$$

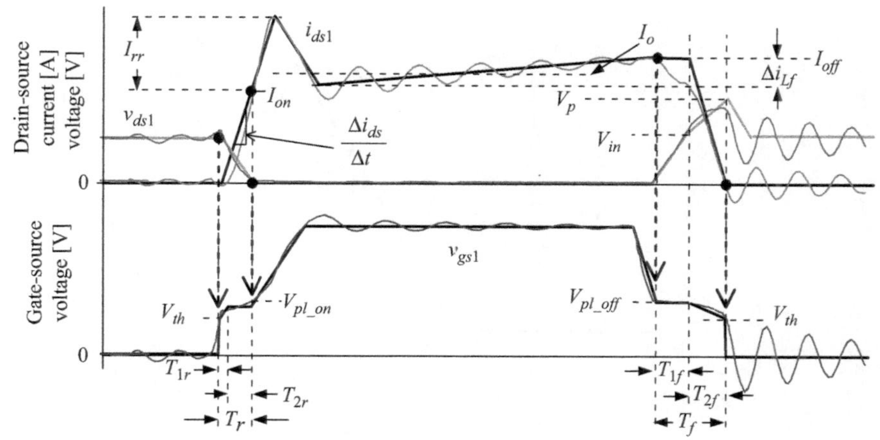

Figure 2.12 Synchronous buck HS MOSFET waveforms with piecewise linear approximations of these waveforms in bold

$$C_{iss1} = C_{iss1\_spec} \tag{2.4}$$

$$C_{gs1} = C_{iss1} - C_{gd1} \tag{2.5}$$

In the following three subsections, derivations of the model for the turn-on, turn-off, and the total switching loss are presented.

2.3.4 Turn-on switching loss model

Piecewise linear turn-on waveforms of i_{ds1}, v_{ds1}, v_{gs1} and the power loss in M_1, P_{M1} are provided in Figure 2.13. These waveforms and knowledge of the circuit operation are used extensively in this subsection in order to derive the turn-on loss, P_{on}.

By definition, P_{on} is derived using the simple integral in (2.6), representing the average power over one switching period:

$$P_{on} = f_s \int_0^{T_r} v_{ds1} i_{ds} dt \tag{2.6}$$

Using the piecewise linear geometry for the voltage and current waveforms at turn on in Figure 2.13, v_{ds1} is given by (2.7) and i_{ds} can be expressed by (2.8), allowing P_{on} to be given by (2.9), which evaluates to the expression given in (2.10):

$$v_{ds1} = V_{in} - \frac{V_{in}}{T_r} t \tag{2.7}$$

$$i_{ds} = \frac{I_{on}}{T_r} t \tag{2.8}$$

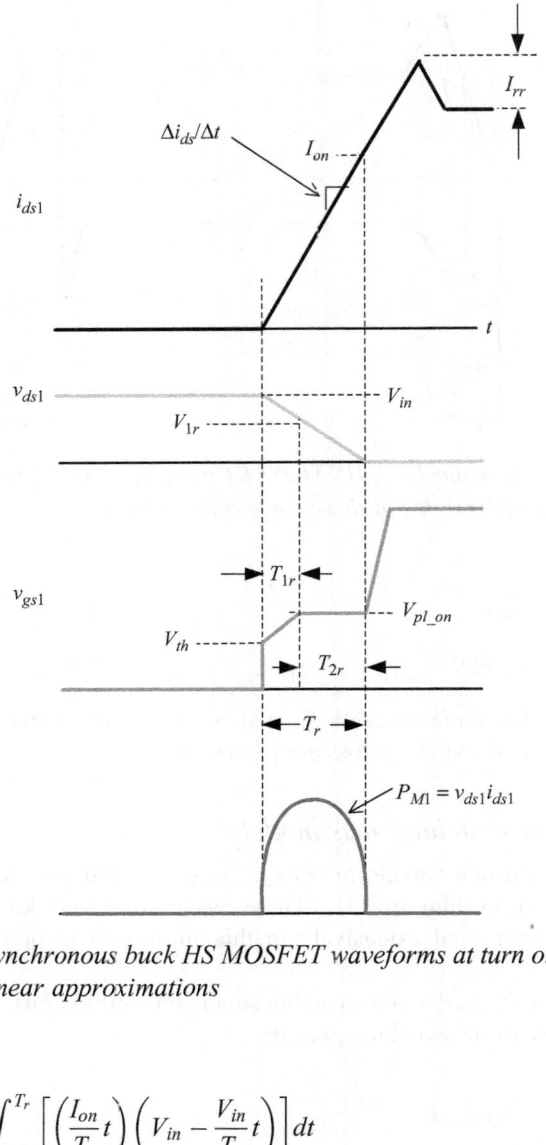

Figure 2.13 Synchronous buck HS MOSFET waveforms at turn on with piecewise linear approximations

$$P_{on} = f_s \int_0^{T_r} \left[\left(\frac{I_{on}}{T_r} t \right) \left(V_{in} - \frac{V_{in}}{T_r} t \right) \right] dt \tag{2.9}$$

$$P_{on} = \frac{1}{6} V_{in} I_{on} T_r f_s \tag{2.10}$$

The power loss in (2.10) is the product of V_{in}, I_{on}, f_s, and T_r. The two parameters that are keys to accurate prediction of P_{on} are the current at turn on, I_{on} and T_r. The turn-on current, I_{on}, is the HS MOSFET drain current when $v_{ds1} = 0$. The remainder of this subsection provides a simple procedure to calculate I_{on} and T_r, to enable calculation of P_{on}.

As discussed in Section 2.3.2, T_r is dictated by the gate driver's ability to charge the MOSFET gate capacitances, which is the time for v_{ds1} to fall to zero. This time is assumed independent of the time it takes i_{ds1} to rise to its final value. Under this assumption, T_r consists of two intervals, T_{1r} and T_{2r} which are discussed in the following subsections.

2.3.4.1 Rise time interval T_{1r}: charging HS MOSFET C_{gs1} and C_{gd1} gate capacitances

The HS MOSFET equivalent circuit during T_{1r} is given in Figure 2.14. The gate resistance, R_r, represents the total series resistance in the gate drive path, i.e., $R_r = R_{hi} + R_{ext} + R_g$, where R_{hi} is the resistance of the driver switch, R_{ext} is any external resistance, and R_g represents the internal gate resistance of the MOSFET.

During T_{1r}, the C_{gs1} capacitance is charged from V_{th} to V_{pl_on}, while the gate side of C_{gd1} charges from V_{th} to V_{pl_on} and the drain side of the C_{gd1} capacitance discharges from V_{in} to V_{1r}. Therefore, the change in voltage across C_{gd1} during T_{1r} is $[(V_{in} - V_{1r}) + (V_{pl\_on} - V_{th})]$. Then, T_{1r} is given by (2.11), assuming an average gate charging current I_{g1r}. V_{pl_on} represents the plateau voltage at turn on and is given by (2.12), where Δi_{Lf} represents the buck output inductor ripple current. Since the peak MOSFET current at turn on is lower than at turn off, the plateau voltage at turn on differs slightly than at turn off. In (2.11), $\Delta V_{gsr} = V_{pl\_on} - V_{th}$:

$$T_{1r} = \frac{C_{gs1}\Delta V_{gsr} + C_{gd1}\left[\Delta V_{gsr} + (V_{in} - V_{1r})\right]}{I_{g1r}} \tag{2.11}$$

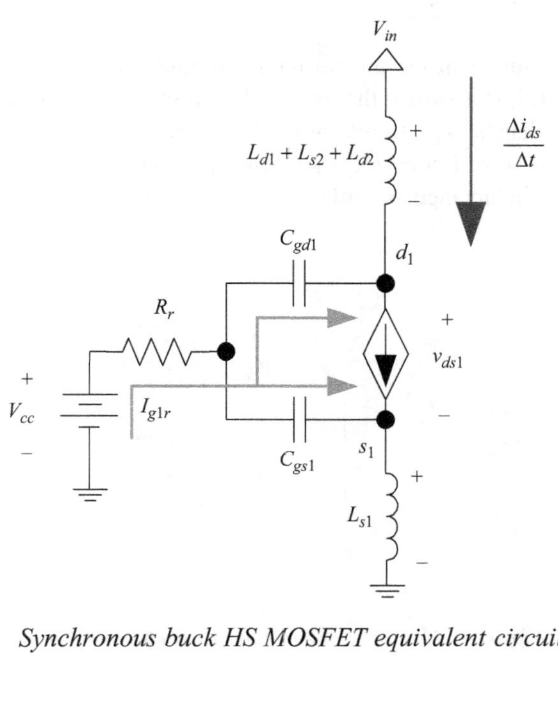

Figure 2.14 Synchronous buck HS MOSFET equivalent circuit during T_{1r}

$$V_{pl\_on} = V_{th} + \frac{I_o - 0.5\Delta i_{Lf}}{g_{fs}} \tag{2.12}$$

The drain-source voltage during T_{1r} is given by (2.13), where $L_{loop} = L_{s1} + L_{d1} + L_{s2} + L_{d2}$:

$$v_{ds1} = V_{in} - L_{loop}\frac{di_{ds1}}{dt} \tag{2.13}$$

Neglecting the gate current through the inductances, the rate of change of drain current in (2.13) is given by (2.14), which is approximated by (2.16) using (2.15) and the piecewise linear approximation of the gate-source voltage waveform during T_{1r}:

$$\frac{di_{ds1}}{dt} = \frac{dg_{fs}(v_{gs1} - V_{th})}{dt} = \frac{g_{fs}dv_{gs1}}{dt} \tag{2.14}$$

$$\frac{di_{ds}}{dt} \approx \frac{\Delta i_{ds}}{\Delta t} \tag{2.15}$$

$$\frac{\Delta i_{ds}}{\Delta t} = \frac{g_{fs}\Delta v_{gs1}}{\Delta t} = \frac{g_{fs}\Delta V_{gsr}}{T_{1r}} \tag{2.16}$$

Using (2.14)–(2.16), the intermediate voltage, V_{1r} is given by (2.17).

$$V_{1r} = V_{in} - L_{loop}\frac{g_{fs}\,\Delta V_{gsr}}{T_{1r}} \tag{2.17}$$

The driver equivalent circuit during T_{1r} is illustrated in Figure 2.15. During this time interval, it is assumed that v_{gs1} is the average value of the plateau, V_{pl_on}, and threshold voltages, V_{th}. In addition, in the proposed model, the slope of the drain current is assumed constant; therefore, the voltage $v_{Ls1} = L_{s1}\Delta i_{ds}/\Delta t$ is constant, so the L_{s1} inductance is replaced by an ideal voltage source in the drive

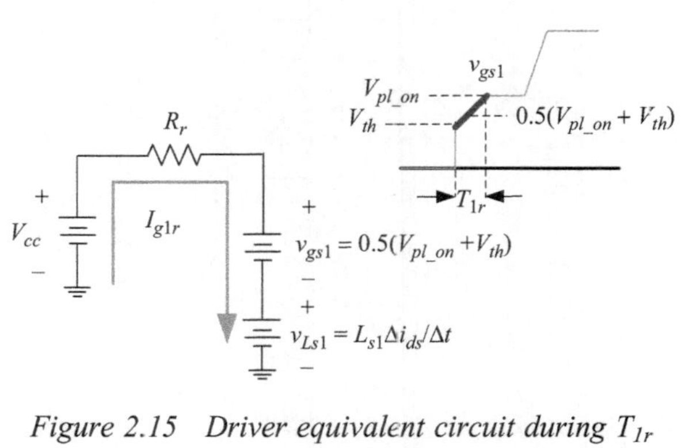

Figure 2.15 Driver equivalent circuit during T_{1r}

circuit. The average gate current, during T_{1r}, using the linearized v_{gs1} waveform is given by (2.18).

$$I_{g1r} = \frac{V_{cc} - 0.5(V_{pl\_on} + V_{th}) - L_{s1}\frac{\Delta i_{ds}}{\Delta t}}{R_r} \tag{2.18}$$

Using (2.11) and (2.16)–(2.18), solving for T_{1r} yields (2.19), where $V_{gs1r} = 0.5(V_{pl\_on} + V_{th})$.

$$T_{1r} = \frac{\Delta V_{gsr}(L_{s1}g_{fs} + R_r C_{iss1}) + \sqrt{\left[\Delta V_{gsr}(L_{s1}g_{fs} + R_r C_{iss1})\right]^2 + 4\Delta V_{gsr}(V_{cc} - V_{gs1r})R_r C_{gd1} L_{loop} g_{fs}}}{2V_{gs1r}} \tag{2.19}$$

2.3.4.2 Rise time interval T_{2r}: charging the HS MOSFET C_{gd1} gate capacitance

The HS MOSFET equivalent circuit during T_{2r} is given in Figure 2.16.

During T_{2r}, the gate voltage of the C_{gd1} capacitance remains constant at V_{pl_on}, while the drain node of C_{gd1} is discharged by current I_{g2r}, allowing T_{2r} to be given by (2.20).

$$T_{2r} = \frac{C_{gd1} V_{1r}}{I_{g2r}} \tag{2.20}$$

The driver equivalent circuit during T_{2r} is illustrated in Figure 2.17. Due to the assumed constant $\Delta i_{ds}/\Delta t$, the L_{s1} inductance is replaced by an ideal voltage source.

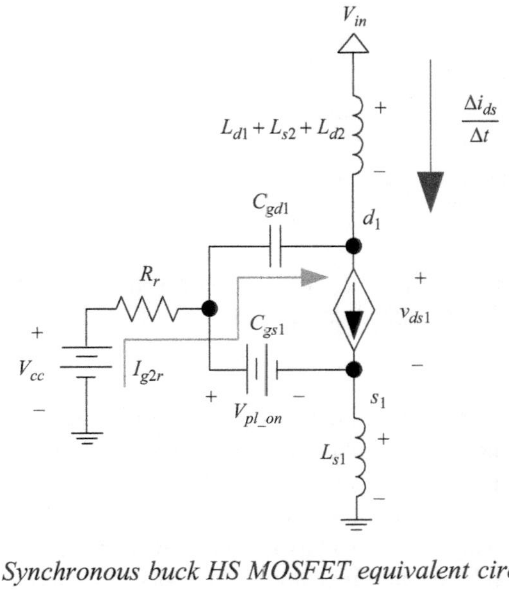

Figure 2.16 Synchronous buck HS MOSFET equivalent circuit during T_{2r}

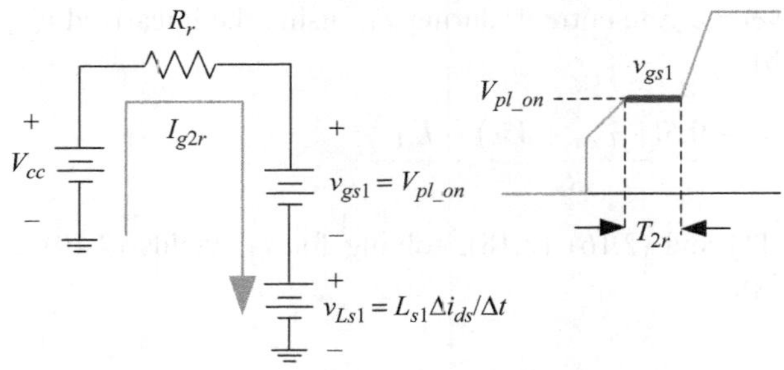

Figure 2.17 Driver equivalent circuit during T_{2r}

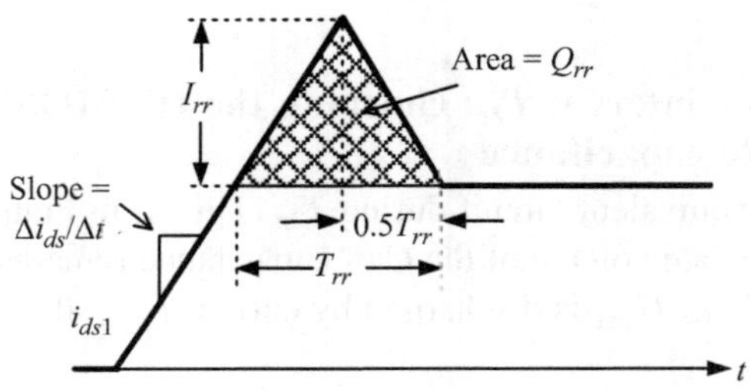

Figure 2.18 Synchronous buck HS MOSFET current waveform approximation during reverse recovery at turn on

Under these assumptions, the gate current is given by (2.21).

$$I_{g2r} = \frac{V_{cc} - V_{pl\_on} - L_{s1}\frac{\Delta i_{ds}}{\Delta t}}{R_r} \tag{2.21}$$

Using (2.16), (2.17), (2.20), and (2.21), solving for T_{2r} yields (2.22).

$$T_{2r} = \frac{R_r C_{gd1}\left(V_{in} - L_{loop}\, g_{fs}\frac{\Delta V_{gsr}}{T_{1r}}\right)}{V_{cc} - V_{pl\_on} - L_{s1} g_{fs}\frac{\Delta V_{gsr}}{T_{1r}}} \tag{2.22}$$

The total T_r is the sum of T_{1r} and T_{2r} as given by (2.23). The total turn-on switching loss can be calculated using (2.28) and (2.23).

$$T_r = T_{1r} + T_{2r} \tag{2.23}$$

The final step to determine the turn-on loss is to estimate the current I_{on} at the end of T_r. Depending on the load current and parasitic inductances, calculating I_{on} can require estimation of the reverse recovery current, I_{rr} of the SR MOSFET. The waveform in Figure 2.18 is used to estimate I_{rr}. When the HS MOSFET turns on, the SR body diode cannot reverse block, so the SR current goes negative and the HS current spikes by the same magnitude. The total reverse recovery time is T_{rr}.

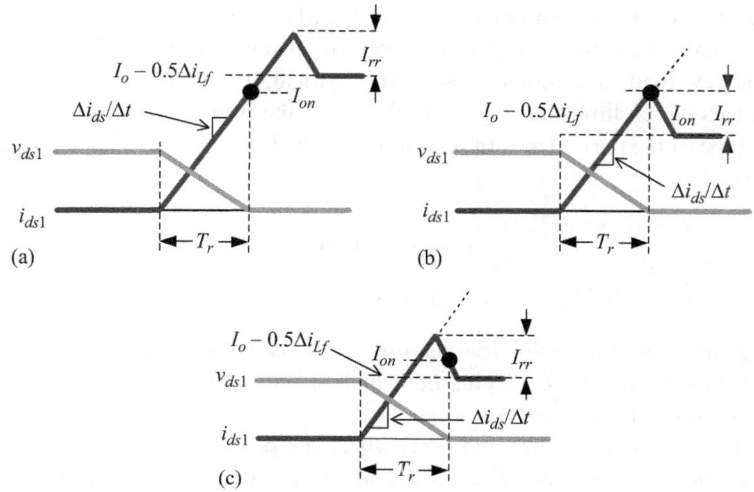

Figure 2.19 Three possible cases of turn-on current when $V_{ds1} = 0$: (a) I_{on} less than I_{ds1} peak value, (b) I_{on} is equal to the I_{ds1} peak value, and (c) I_{on} occurs after the I_{ds1} peak value

The rising slope magnitude is $\Delta i_{ds}/\Delta t$ and the reverse recovery charge is Q_{rr}, which represents the shaded area as given by (2.24). Using the geometry, the reverse recovery current as a function of T_{rr} is given by (2.25). Then, eliminating T_{rr} from (2.24) and (2.25), (2.27) is derived, which represents I_{rr} as a function of Q_{rr} and the known slope. In addition, since reverse recovery charge increases with load current, Q_{rr} is approximated using (2.26), where Q_{rr_spec} and I_{rr_spec} are the datasheet specification values:

$$Q_{rr} = \frac{1}{2} I_{rr} T_{rr} \tag{2.24}$$

$$I_{rr} = \frac{\Delta I_{ds}}{\Delta t} \frac{1}{2} T_{rr} \tag{2.25}$$

$$Q_{rr} = \frac{Q_{rr\_spec}}{I_{rr\_spec}} I_o \tag{2.26}$$

$$I_{rr} = \sqrt{\frac{\Delta I_{ds}}{\Delta t} Q_{rr}} \tag{2.27}$$

Since the rise time is dictated by the time for the HS MOSFET voltage, v_{ds1}, to fall to zero, the current at the end of T_r can be at any value equal to, or less than the inductor current plus the reverse recovery current (i.e., I_{on} is not necessarily equal to the inductor current, as in the conventional model [21], or the inductor current plus the reverse recovery current). There are three cases for I_{on} as illustrated in Figure 2.19.

The first and most common case is illustrated in Figure 2.19(a) where I_{on} is less than the peak of the turn-on current waveform at the end of T_r. This case occurs under heavy load conditions and/or with typical, or large values of parasitic inductances which limit the $\Delta i_{ds}/\Delta t$. In this case, the turn-on current is determined by the slope of the current at turn on multiplied by T_r as given by the first condition in (2.28).

$$I_{on} = \begin{cases} \frac{\Delta i_{ds}}{\Delta t}T_r & \text{if} \quad \frac{\Delta i_{ds}}{\Delta t}T_r < I_o - 0.5\Delta i_{Lf} + I_{rr} \\ I_o - 0.5\Delta i_{Lf} + I_{rr} & \text{otherwise} \end{cases} \tag{2.28}$$

The first condition holds true as long as the calculated value is less than the inductor current ($I_o - 0.5\Delta i_{Lf}$) plus I_{rr}, which leads to the second and third conditions in Figure 2.19(b) and (c).

Under light load and/or conditions where the parasitic inductances are small, using the current slope times T_r would yield a turn-on current greater than the peak current and somewhere on the dotted line extensions in Figure 2.19(b) and (c). In this case, the current is capped at maximum value of the inductor current plus reverse recovery current as given by the second condition in (2.28). The third case, illustrated in Figure 2.19(c), occurs under very light load conditions, and/or when the parasitic inductances are very small. To simplify the model, this case is neglected and if it occurs, the second case in Figure 2.19(b) is used as given by the second condition in (2.28).

2.3.5 Turn-off switching loss model

Piecewise linear turn-off waveforms of i_{ds1}, v_{ds1}, v_{gs1} and the power loss in M_1, P_{M1} are provided in Figure 2.20. These waveforms and knowledge of the circuit operation are used extensively in this subsection in order to derive the turn-off loss, P_{off}. The turn-off transition consists of two intervals, T_{1f} and T_{2f}.

During T_{1f}, the Miller capacitor, C_{gd1} is discharged while v_{gs1} remains at V_{pl_off} and i_{ds1} is assumed to remain constant. In a real circuit, it is noted that i_{ds1} begins to fall during T_{1f}; however, the current slope is limited due to the discharging of the C_{gd2} and C_{ds2} capacitors of the SR. During this interval, v_{ds1} increases from zero to V_{in}. Therefore, from the geometry, the turn-off power loss, P_{1off}, during T_{1f} is given by (2.29).

$$P_{1off} = \frac{1}{2}V_{in}I_{off}T_{1f}f_s \tag{2.29}$$

During T_{2f}, C_{gs1} is discharged from V_{pl_off} to V_{th}, while the gate node of C_{gd1} is also discharged from V_{pl_off} to V_{th} and the drain node of C_{gd1} is charged from V_{in} to the peak voltage at turn off, V_p. During this interval, i_{ds1} falls from I_{off} to zero, while v_{ds1} rises from V_{in} to V_p. Using a simple integral following the procedure presented in Section 2.3.4, the turn-off loss during T_{2f} is approximated as P_{2off}, given by (2.30).

$$P_{2off} = \left(\frac{1}{6}(V_p - V_{in}) + I_{off} + \frac{1}{2}V_{in}I_{off}\right)T_{2f}f_s \tag{2.30}$$

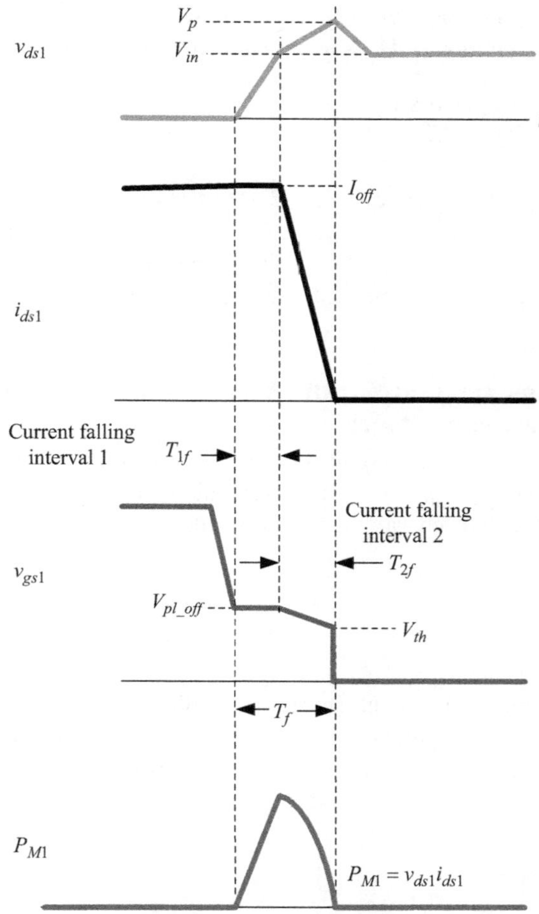

Figure 2.20 Synchronous buck HS MOSFET waveforms at turn off with piecewise linear approximations

The total turn-off loss, P_{off}, is the sum of P_{1off} and P_{2off} as given by (2.31).

$$P_{off} = P_{1off} + P_{2off} \tag{2.31}$$

The parameters that are keys to accurate prediction of the turn-off loss are the HS MOSFET current at turn off, I_{off}, the intervals T_{1f} and T_{2f}, and the peak overshoot voltage of v_{ds1}, V_p. The turn-off current is the load current, I_o, plus half of the filter inductor peak-to-peak ripple current, Δi_{Lf}, as given by (2.32).

$$I_{off} = I_o + \frac{1}{2}\Delta i_{Lf} \tag{2.32}$$

The plateau voltage at turn off, V_{pl_off}, is given by (2.34). It differs slightly from V_{pl_on} at turn on due to the larger switch current during the transition:

$$V_{pl\_off} = V_{th} + \frac{I_o + 0.5\Delta i_{Lf}}{g_{fs}} \tag{2.33}$$

The turn-off loss estimated using (2.31) is a function of T_f. During T_f, the current falls from I_{off} to zero and v_{ds1} rises from zero to V_p. It is a function of the driver's capability to discharge C_{gd1} and C_{iss1}, but in addition, it is a function of circuit parasitic inductances, which limit the current falling slope, and therefore, the falling time.

2.3.5.1 Fall time interval T_{1f}: discharging the HS MOSFET C_{gd1} gate capacitance

The HS MOSFET equivalent circuit during T_{1f} is given in Figure 2.21. Since i_{ds1} remains constant at I_{off}, $\Delta i_{ds}/\Delta t = 0$, so the parasitic inductors can be neglected. T_{1f} is the time required to discharge the C_{gd1} capacitance by gate current I_{g1f} as given by (2.34).

$$T_{1f} = \frac{C_{gd1} V_{in}}{I_{g1f}} \tag{2.34}$$

The driver equivalent circuit during T_{1f} is illustrated in Figure 2.22. During this time interval, it is assumed that v_{gs1} remains constant at the plateau, V_{pl_off}. With this assumption, the gate current is easily derived as given by (2.35), where $R_f = R_{lo} + R_{ext} + R_g$, and V_{pl_off} is given by (2.33):

$$I_{g1f} = \frac{V_{pl\_off}}{R_f} \tag{2.35}$$

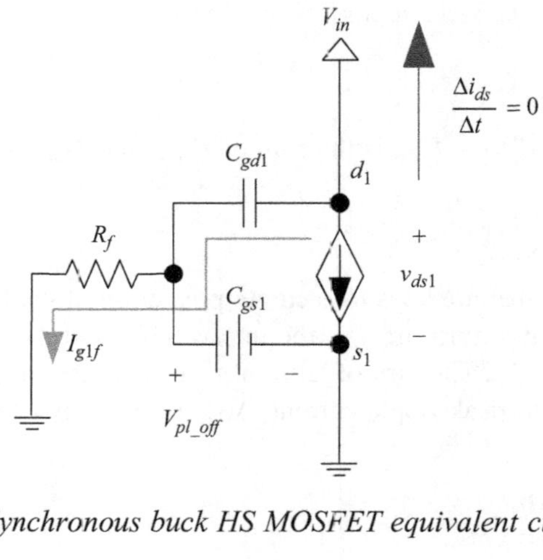

Figure 2.21 Synchronous buck HS MOSFET equivalent circuit during T_{1f}

Using (2.34) and (2.35), T_{1f} is given by (2.36).

$$T_{1f} = \frac{C_{gd1} V_{in} R_f}{V_{pl\_off}} \tag{2.36}$$

2.3.5.2 Fall time interval T_{2f}: current falling and discharging the HS MOSFET C_{gs1} and C_{gd1} gate capacitances

The HS MOSFET equivalent circuit during T_{2f} is given in Figure 2.23. During T_{2f}, the C_{gs1} capacitance is discharged from V_{pl_off} to V_{th}, while the voltage at the drain side of the C_{gd1} capacitance charges from V_{in} to V_p and the voltage at the gate side of C_{gd1} discharges from V_{pl_off} to V_{th}. Therefore, the change in voltage

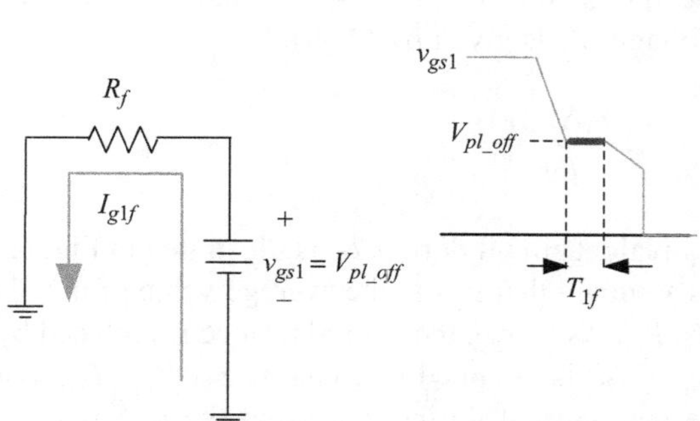

Figure 2.22 Driver equivalent circuit during T_{1f}

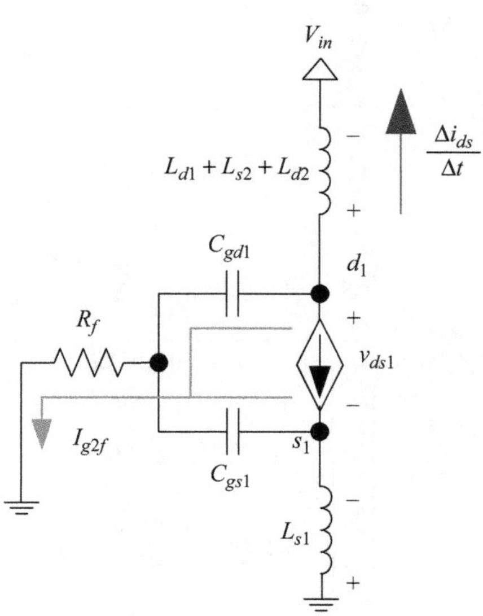

Figure 2.23 Synchronous buck HS MOSFET equivalent circuit during T_{2f}

across C_{gd1} during T_{2f} is $[(V_p - V_{in}) + (V_{pl\_off} - V_{th})]$. Then, T_{2f} is given by (2.37), where $\Delta V_{gsf} = V_{pl\_off} - V_{th}$.

$$T_{2f} = \frac{C_{gs1}\Delta V_{gsf} + C_{gd1}\left[(V_p - V_{in}) + \Delta V_{gsf}\right]}{I_{g2f}} \tag{2.37}$$

The drain-source voltage during T_{2f} is given by (2.38), where $L_{loop} = L_{s1} + L_{d1} + L_{s2} + L_{d2}$.

$$v_{ds} = V_{in} + L_{loop}\frac{di_{ds1}}{dt} \tag{2.38}$$

Following the approach of the approximations made in (2.14) and (2.15), the peak overshoot voltage, V_p is given by (2.39).

$$V_p = V_{in} + L_{loop}\frac{g_{fs}\Delta V_{gsf}}{T_{2f}} \tag{2.39}$$

The driver equivalent circuit during T_{2f} is illustrated in Figure 2.24. During this time interval, it is assumed that v_{gs1} is the average value of the plateau, V_{pl_off}, and threshold voltages, V_{th}. As above, the L_{s1} inductance is replaced by an ideal voltage source, where the di_{ds}/dt is assumed constant at $g_{fs}\Delta V_{gsf}/T_{2f}$. With these assumptions, the average gate current during T_{2f}, using the linearized v_{gs1} waveform, is given by (2.40).

$$I_{g2f} = \frac{\frac{1}{2}(V_{pl} + V_{th}) - L_{s1}\frac{g_{fs}\Delta V_{gsf}}{T_{2f}}}{R_f} \tag{2.40}$$

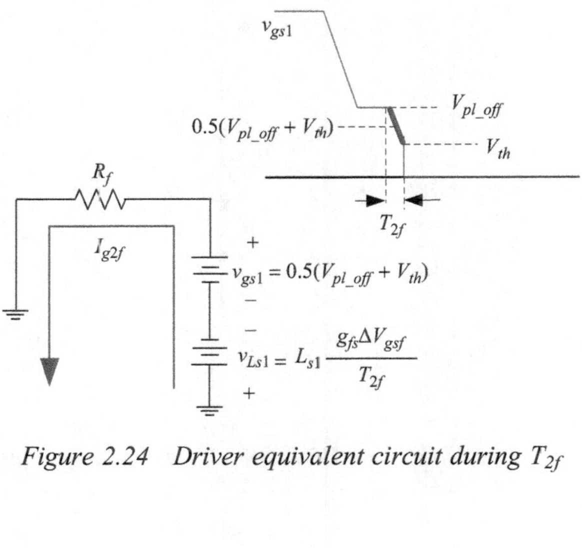

Figure 2.24 Driver equivalent circuit during T_{2f}

Using (2.37), (2.39), and (2.40), solving for T_{2f} yields (2.41), where $V_{gs2f}= 0.5(V_{pl\_off}+ V_{th})$.

$$T_{2f} = \frac{\Delta V_{gs2f}(L_{s1}g_{fs}+R_f C_{iss1}) + \sqrt{\Delta V_{gs2f}{}^2\left[L_{s1}g_{fs}+R_f C_{iss1}\right]^2 + 4\Delta V_{gs2f}V_{gs2f}R_f C_{gd1}L_{loop}g_{fs}}}{2V_{gs2f}} \tag{2.41}$$

T_f is the sum of T_{1f} and T_{2f} as given by (2.42). The total turn-off switching loss can be calculated using (2.29)–(2.33), (2.36), (2.39), (2.41), and (2.42).

$$T_f = T_{1f} + T_{2f} \tag{2.42}$$

The total switching loss, given by (2.43), is the sum of the turn-on loss, P_{on}, given by (2.10) and turn-off loss, P_{off}, given by (2.31):

$$P_{tot\_sw} = P_{on} + P_{off} \tag{2.43}$$

2.3.6 Voltage source drive model verification

The analytical switching loss model with a voltage source drive was compared to SIMetrix Spice simulation and the conventional model in [21].

Simulation results were conducted at 12 V input, 1 MHz switching frequency, and 10 A peak-to-peak buck output inductor ripple (100 nH), $R_{hi} = 2\ \Omega$, $R_{lo} = 2\ \Omega$, $R_g = 1\ \Omega$, $R_{ext} = 0\ \Omega$. Results are included in the following subsections for both models. MOSFET parameters: M_1: Si7860DP, $g_{fs} = 60$ S, $V_{th} = 2$ V, $C_{iss1\_spec} = 1{,}800$ pF (@$V_{ds1\_spec} = 15$ V), $C_{oss1\_spec} = 600$ pF (@$V_{ds1\_spec} = 15$ V), $C_{rss1\_spec} = 200$ pF (@$V_{ds1\_spec} = 15$ V) and M_2: Si7336ADP SR, $Q_{rr\_spec} = 30$ nC, $I_{rr\_spec} = 25$ A.

Curves of total switching loss as a function of: (a) load current, (b) driver supply voltage and (c) common source inductance (assuming matched inductances, i.e., $L_{s1} = L_{d1} = L_{s2} = L_{d2}$) for the proposed model, Spice simulation and the conventional model are given in Figure 2.25(a)–(c). The proposed model follows the trends of the Spice simulation results very well. The accuracy of the proposed model total switching loss is within 0.5 W for all conditions. In Figure 2.25, it is noted that the conventional model does a very poor job predicting the total switching loss in all three cases, but in particular as total circuit inductance increases. Specifically, at 1,000 pH in Figure 2.25(c), the conventional model predicts 2.0 W loss, while the Spice results indicate total switching loss of 6.3 W – a difference of 4.3 W. The results also show that the total switching loss can be reduced by increasing V_{cc}. However, for $V_{cc} > 8$ V, the reduction is not significant.

Curves of total switching loss as a function of: (a) load current, (b) driver supply voltage, and (c) common source inductance for the proposed model, Spice simulation and the conventional model are given in Figures 2.26(a)–(c) and 2.27(a)–(c), respectively. The proposed model follows the trends of the Spice simulation results. In particular, the proposed model correctly predicts that the

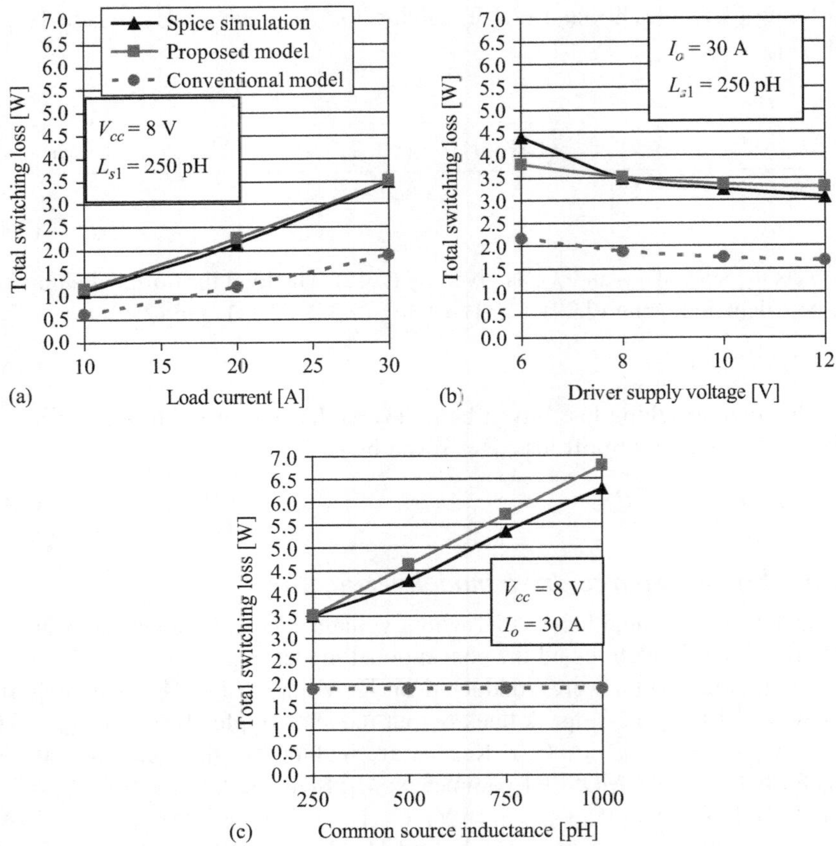

Figure 2.25 Total switching loss at 1 MHz, 12 V input as a function of: (a) load current ($V_{cc} = 8$ V, $L_{s1} = 250$ pH), (b) driver supply voltage ($I_o = 30$ A, $L_{s1} = 250$ pH), and (c) common source inductance ($V_{cc} = 8$ V, $I_o = 30$ A)

turn-off loss increases with common source inductance since T_f increases significantly with L_{s1}. In the conventional model, turn-off switching loss remains constant with common source inductance leading to an error in predicted loss of over 4.6 W at 1,000 pH in Figure 2.27(c). The results also show that turn-on loss decreases with V_{cc} as expected, since increasing V_{cc} provides increasing driver source current. In contrast, the turn-off loss remains constant with V_{cc}, since the driver sink current is determined by V_{pl_off}.

2.3.7 Experimental validation of the voltage source drive model

Experimental results were presented to aid in demonstrating the switching loss characteristics as load and common source inductance change. These results were presented at low frequency with a large inductance wire introduced in series

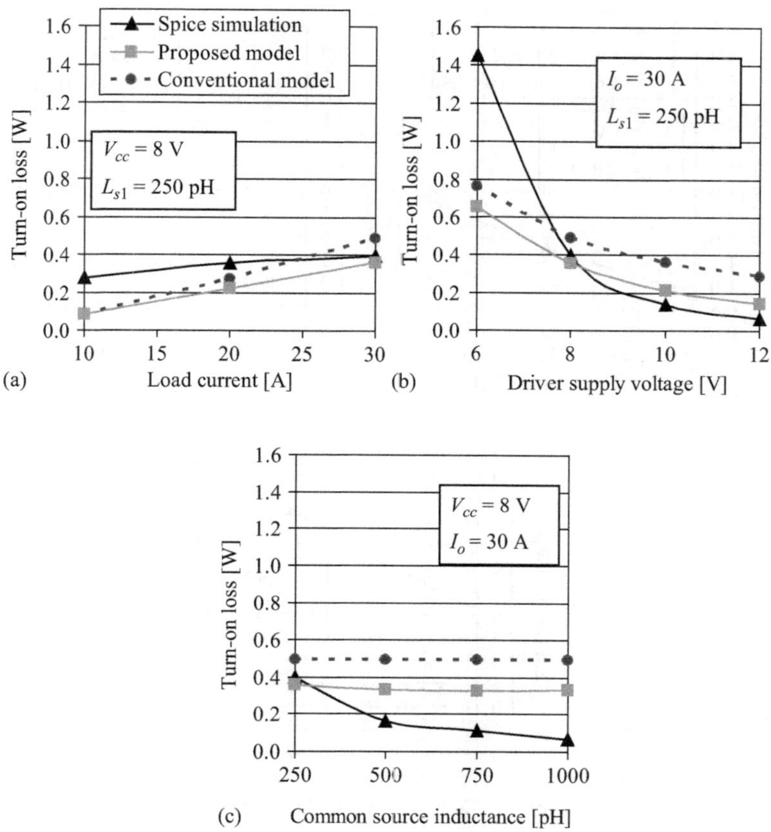

Figure 2.26 Turn-on switching loss at 1 MHz, 12 V input as a function of: (a) load current ($V_{cc}=8$ V, $L_{s1}=250$ pH), (b) driver supply voltage ($I_o=30$ A, $L_{s1}=250$ pH), and (c) common source inductance ($V_{cc}=8$ V, $I_o=30$ A)

between the source and common point on the driver in order to measure the HS MOSFET current and demonstrate the trends. However, since probing the MOSFET current is impractical in a real VR with good layout, therefore, the actual switching loss in the prototype cannot be measured, so a direct comparison between the modeled switching loss and actual switching loss cannot be made.

Given the constraints on measuring actual switching loss, another method to gauge the accuracy of the proposed model is to use it in a loss analysis file that estimates the switching loss, other loss and total loss for a synchronous buck VR and compare it to the total loss in the real circuit. This analysis has been completed for the voltage source drive and the total loss in the design file has been compared to the total measured loss of the circuit by subtracting the load power from the input power. These comparisons are included in the following subsections.

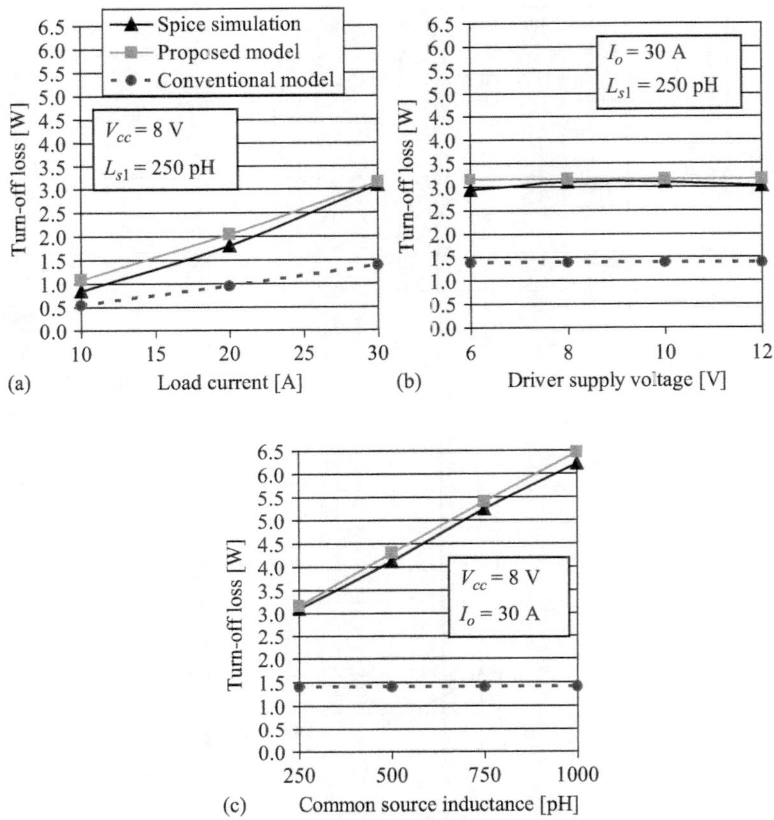

Figure 2.27 Turn-off switching loss at 1 MHz, 12 V input as a function of: (a) load current (V_{cc} = 8 V, L_{s1} = 250 pH), (b) driver supply voltage (I_o = 30 A, L_{s1} = 250 pH), and (c) common source inductance (V_{cc} = 8 V, I_o = 30 A)

Circuit parameters: 1 MHz switching frequency; 12 V input; 1.3 V output; 330 nH buck inductor; $V_{cc} = 10$ V; IRF6617 HS MOSFET: $g_{fs} = 39$ S, $V_{th} = 1.85$ V, $C_{rss1\_spec} = 160$ pF (@$V_{ds1\_spec} = 15$ V), $C_{oss\_spec1} = 450$ pF (@$V_{ds1\_spec} = 15$ V), $C_{iss1\_spec} = 1{,}300$ pF (@$V_{ds1\_spec} = 15$ V); and IRF6691 SR MOSFET; $L_{s1} = L_{d1} = L_{s2} = L_{d2} = 500$ pH (model). Driver parameters: UCC27222 driver (experimental); $R_{hi} = 1.8\ \Omega$, $R_{lo} = 1.8\ \Omega$, $R_g = 1\ \Omega$, $R_{ext} = 0\ \Omega$ (model).

The estimated synchronous buck VR losses and model predicted switching loss as a function of load current is compared to the experimentally measured loss in Figure 2.28 for the voltage source driver. Good agreement is achieved between the loss predicted by the model and the actual loss of the VR, with the accuracy within 0.7 W over the entire load range.

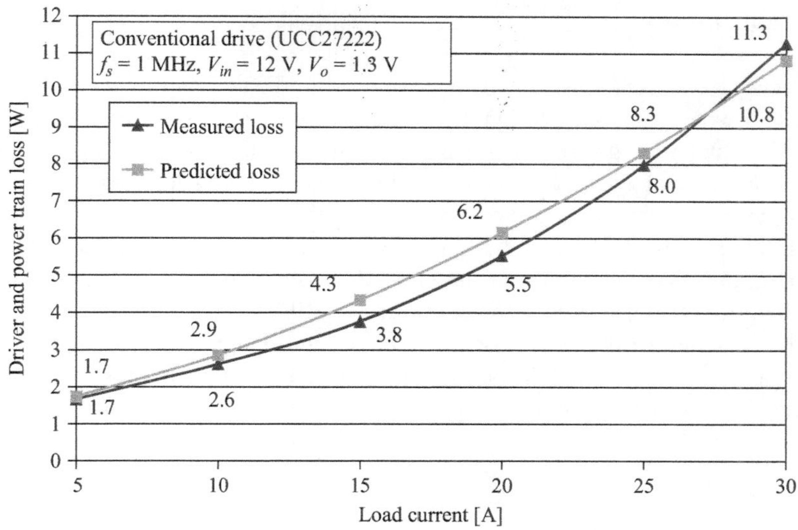

Figure 2.28 Comparison of total loss predicted and measured for voltage source drive (UCC27222, $f_s = 1$ MHz, $V_{in} = 12$ V, and $V_o = 1.3$ V)

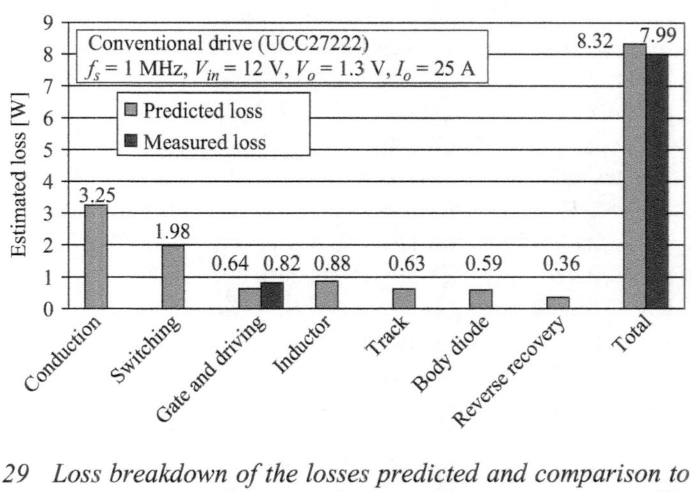

Figure 2.29 Loss breakdown of the losses predicted and comparison to the measured gate and total losses for voltage source drive (UCC27222, $f_s = 1$ MHz, $V_{in} = 12$ V, and $V_o = 1.3$ V)

A loss breakdown of the estimated losses used to generate the model predicted loss in Figure 2.28 is given in Figure 2.29 for 25 A load current. The only losses that can be experimentally measured are the gate and driving power and the input power, which includes the remaining losses (i.e., all except gate and driving) in Figure 2.29.

2.4 Summary

Resonant gate drivers were originally developed to recover part of MOSFET gate-drive loss when operating at high switching frequency. Most of investigations are generally emphasizing gate energy savings by the resonant driver and concentrating on the drive topologies, but ignore the potential switching loss savings that are much more dominant in MHz switching power converter.

The idea of the CSDs is to build strong gate-drive current during the switching transition to realize quick turn-on and turn-off transition and reduce high-frequency switching loss and gate-drive loss in power MOSFETs.

The switching loss characteristics and behavior in a high-frequency synchronous buck VR is investigated and the parasitic inductances act as a current snubber at turn on to reduce turn-on loss, but prolong the turn-off time and increase turn-off loss.

References

[1] W. A. Tabisz and F. C. Lee, "Zero-voltage-switching multiresonant technique – a novel approach to improve performance of high-frequency quasi-resonant converters," *IEEE Transactions on Power Electron*, vol. 4, no. 4, pp. 450–458, 1989.

[2] D. Maksimovic and S. Cuk, "Constant-frequency control of quasi-resonant converters," *IEEE Transactions on Power Electron*, vol. 6, no. 1, pp. 141–150, 1991.

[3] W. A. Tabisz, P. M. Gradzki, and F. C. Lee, "Zero-voltage-switched quasi-resonant buck and flyback converters – experimental results at 10 MHz," *IEEE Transactions on Power Electron*, vol. 4, no. 2, pp. 194–204, 1989.

[4] J. M. Rivas, R. S. Wahby, J. S. Shafran, and D. J. Perreault, "New architectures for radio-frequency DC–DC power conversion," *IEEE Transactions on Power Electron*, vol. 21, no. 2, pp. 380–393, 2006.

[5] Y. Han, O. Leitermann, D. A. Jackson, J. M. Rivas, and D. J. Perreault, "Resistance compression networks for radio-frequency power conversion," *IEEE Transactions on Power Electron*, vol. 22, no. 1, pp. 41–53, 2007.

[6] J. R. Warren, K. A. Rosowski, and D. J. Perreault, "Transistor selection and design of a VHF DC-DC power converter," *IEEE Transactions on Power Electron*, vol. 23, no. 1, pp. 27–37, 2008.

[7] D. Maksimovic, "A MOS gate drive with resonant transitions," in *Proceedings on IEEE Power Electronics Specialists Conference (PESC)*, pp. 527–532, 1991.

[8] H. L. N. Wiegman, "A resonant pulse gate drive for high frequency applications," in *Proceedings on IEEE Applied Power Electronics Conference (APEC)*, pp. 738–743, 1992.

[9] B. S. Jacobson, "High frequency resonant gate drive for a power MOSFET," in *Proceedings on High Frequency Power Conversion Conference*, pp. 133–141, 1993.

[10] K. Yao and F. C. Lee, "A novel resonant gate driver for high frequency synchronous buck converters," *IEEE Transactions on Power Electronics*, vol. 17, no. 2, pp. 180–186, 2002.

[11] L. Faye and Q. Jinrong, "Gate driver apparatus having an energy recovering circuit," United States Patent No. US 6,650,169 B2, 2003.

[12] S. H. Weinberg, "A novel lossless resonant MOSFET driver," in *Proceedings on IEEE Power Electronics Specialists Conference (PESC)*, pp. 1003–1010, 1992.

[13] Y. Chen, F. C. Lee, L. Amoros, and H. Wu, "A resonant MOSFET gate driver with efficient energy recovery," *IEEE Transactions on Power Electron*, vol. 19, no. 2, pp. 470–477, 2004.

[14] Q. Li and P. Wolfs, "The power loss optimization of a current fed ZVS two-inductor boost converter with a resonant transition gate drive," *IEEE Transactions on Power Electron*, vol. 21, no. 5, pp. 1253–1263, 2006.

[15] J. Diaz, M. A. Perez, F. M. Linera, and F. Aldana, "A new lossless power MOSFET driver based on simple DC/DC converters," in *Proceedings on IEEE Power Electronics Specialists Conference (PESC)*, pp. 37–43, 1995.

[16] B. Arntzen and D. Maksimovic, "Switched-capacitor DC/DC converters with resonant gate drive," *IEEE Transactions on Power Electron*, vol. 13, no. 5, pp. 892–902, 1998.

[17] J. T. Strydom, M. A. de Rooij, and J. D. van Wyk, "A comparison of fundamental gate-driver topologies for high frequency applications," in *Proceedings on IEEE Applied Power Electronics Conference (APEC)*, pp. 1045–1052, 2004.

[18] P. Dwane, D. O' Sullivan, and M. G. Egan, "An assessment of resonant gate drive techniques for use in modern low power dc–dc converters," in *Proceedings on IEEE Applied Power Electronics Conference (APEC)*, pp. 1572–1580, 2005.

[19] T. Lopez, G. Sauerlaender, T. Duerbaum, and T. Tolle, "A detailed analysis of a resonant gate driver for PWM applications," in *Proceedings on IEEE Applied Power Electronics Conference (APEC)*, pp. 873–878, 2003.

[20] G. Spiazzi, P. Mattavelli, and L. Rossetto, "Effects of parasitic components in high-frequency resonant drivers for synchronous rectification MOSFETs," *IEEE Transactions on Power Electron*, vol. 23, no. 4, pp. 2082–2092, 2008.

[21] L. Balogh, "Design and application guide for high speed MOSFET gate drive circuits," 2001 TI Power Design Seminar, slup169.pdf.

[22] Y. Ren, M. Xu, J. Zhou, and F. C. Lee, "Analytical loss model of power MOSFET," *IEEE Transactions on Power Electronics*, vol. 21, no. 2, pp. 310–319, 2006.

Chapter 3
Continuous current source driver

3.1 Two-channel low-side continuous current source drivers

3.1.1 Operating principle

A two-channel current source driver (CSD) with the continuous current is presented here, which can provide two symmetrical drive signals for driving two metal oxide semiconductor field-effect transistors (MOSFETs). The proposed drive circuit can recover most of the driving energy. It can also reduce the switching loss significantly. The proposed CSD can be used to drive the synchronous rectifier (SR) in a current doubler or full-wave rectifier. It can also be used to drive the primary MOSFETs in push-pull converters.

The proposed CSD is shown in Figure 3.1. The circuit consists of four switches S_1–S_4, which is inherited from dual-channel conventional gate-drive circuit, connecting as a bridge configuration, and an inductor, L_1, connecting across the bridge. Capacitors, C_{g_Q1} and C_{g_Q2}, are the gate capacitors of the power MOSFETs Q_1 and Q_2, respectively. V_c is the voltage source which is applied to gate source of MOSFETs Q_1 and Q_2. In order to simplify the implementation, P-channel MOSFET is used for S_1 and S_2, and N-channel MOSFET is used for S_3 and S_4. It is noted that other implementation methods can also be used to achieve the same objective.

The major objectives of the CSD are: (1) to charge and discharge gate capacitors, C_{g_Q1} and C_{g_Q2}, with minimum energy loss, (2) as fast as possible, and (3) to clamp the voltages across C_{g_Q1} and C_{g_Q2} to either source voltage V_c or zero

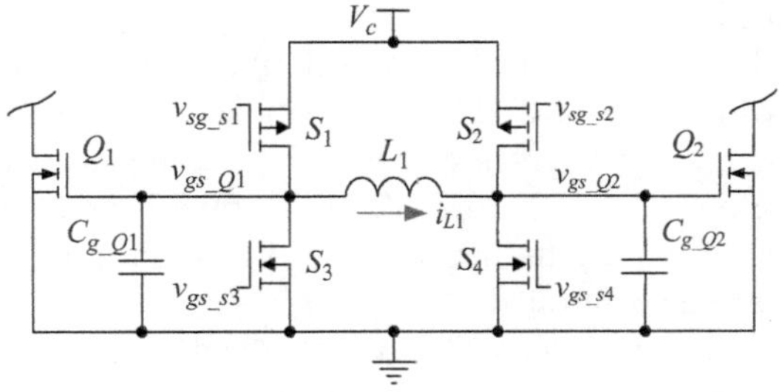

Figure 3.1 Proposed two-channel low-side CSD

via low impedance path. With this CSD, the duty cycles of the voltages across C_{g_Q1} and C_{g_Q2} are the same, which is decided by the gate-drive signals for S_1–S_4. With different gate-drive signals to S_1, S_2, S_3, and S_4; the duty cycle D for Q_1 and Q_2 can be changed from below 0.5, to 0.5, and to above 0.5. The operation is a little different for these three operating conditions. The operation of this circuit under $D > 0.5$ situation is described in detail as shown in Figure 3.2. The typical waveforms are shown in Figure 3.3.

1. Mode 1 [before t_0] (Figure 3.2(a)): S_1, S_4 are turned on and S_2, S_3 are turned off. Q_1 is on and Q_2 is off. Inductor current i_{L1} increases to maximum value at t_0.
2. Mode 2 [t_0, t_1] (Figure 3.2(b)): S_4 is turned off at t_0. Inductor L_1 resonates with the gate capacitor of MOSFET Q_2, C_{g_Q2}. C_{g_Q2} will be charged at this period. The voltage across C_{g_Q2} increases and it will be clamped to the source voltage, V_c, by the body diode of S_2 before t_1. Q_2 is turned on in this time interval. At t_1, S_2 turns on with zero voltage. By controlling the turn-off instant for S_4 (t_0), the turn-on instant of Q_2 can be controlled.
3. Mode 3 [t_1, t_2] (Figure 3.2(c)): S_1, S_2 are on and S_3, S_4 are off. Both Q_1 and Q_2 are on and inductor current i_{L1} is circulating through S_1 and S_2, and remains constant in this interval.
4. Mode 4 [t_2, t_3] (Figure 3.2(d)): S_1 is turned off at t_2 with zero voltage. Inductor L_1 resonates with the gate capacitor of MOSFET Q_1, C_{g_Q1}. C_{g_Q1} is discharged at this period. The voltage across C_{g_Q1} decreases and it will be clamped to zero by the body diode of S_3 before t_3. Q_1 is turned off in this time interval. At t_3, S_3 turns on with zero voltage. By controlling the turn-off instant of S_1 (t_2), the turn-off instant of Q_1 can be controlled.
5. Mode 5 [t_3, t_4] (Figure 3.2(e)): S_2, S_3 are on and S_1, S_4 are off. Q_1 is off and Q_2 is on. The inductor current i_{L1} decreases to zero and it will increase in opposite direction. It reaches the negative maximum at t_4.
6. Mode 6 [t_4, t_5] (Figure 3.2(f)): S_3 is turned off at t_4. Inductor L_1 resonates with capacitor C_{g_Q1}. C_{g_Q1} will be charged. The voltage across C_{g_Q1} increases and it will be clamped to the source voltage V_c by the body diode of S_1 before t_5. Q_1 is turned on in this time interval. At t_5, S_1 turns on with zero voltage. By controlling the turn-off instant of S_3 (t_4), the turn-on instant of Q_1 can be controlled.
7. Mode 7 [t_5, t_6] (Figure 3.2(g)): S_1, S_2 are on and S_3, S_4 are off. Both Q_1 and Q_2 are on and inductor current i_{L1} is circulating through S_1 and S_2, and remains constant in this interval.
8. Mode 8 [t_6, t_7] (Figure 3.2(h)): S_2 is turned off at t_6 with zero voltage. Inductor L_1 resonates with capacitor C_{g_Q2}. C_{g_Q2} will be discharged at this period. The voltage across C_{g_Q2} decreases and it will be clamped to zero by the body diode of S_4 before t_7. Q_2 is turned off in this time interval. At t_7, S_4 turns on with zero voltage. By controlling the turn-off instant for S_2 (t_6), the turn-off instant of Q_2 can be controlled.
9. Mode 9 [t_7, t_8] (Figure 3.2(a)): S_1, S_4 are on and S_2, S_3 are off. Q_1 is on and Q_2 is off. The negative inductor current rises through zero and then further increases. The value of the current i_{L1} will increase to positive maximum at t_8. The next cycle will start at t_8.

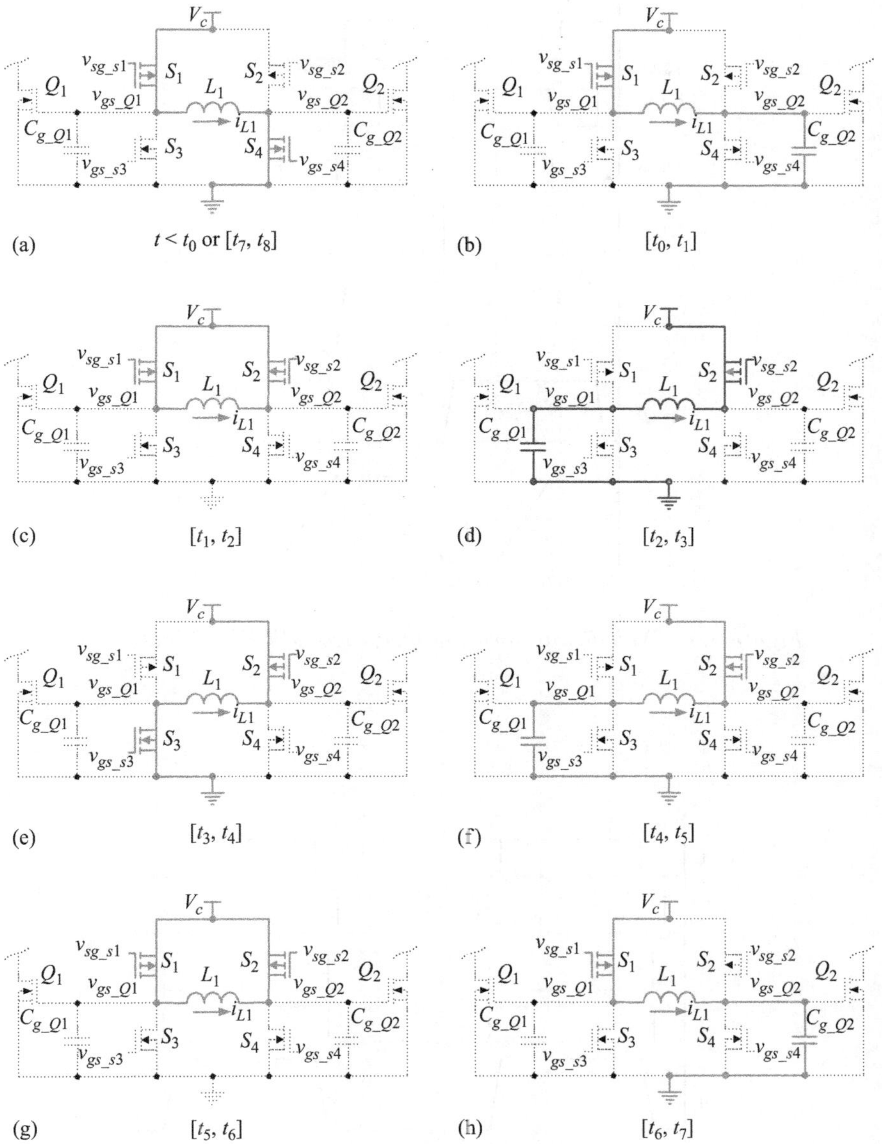

Figure 3.2 Equivalent circuits of eight operation modes in dual low-side symmetrical drive circuit

When the duty cycle D equals 0.5, the operation is similar to that for $D > 0.5$ and the details are not explained here. The key difference is that there is no overlap between the drive signal of S_1 and S_2, so S_1 and S_2 won't conduct simultaneously and there is no current cycling through them. The waveforms are shown in Figure 3.4.

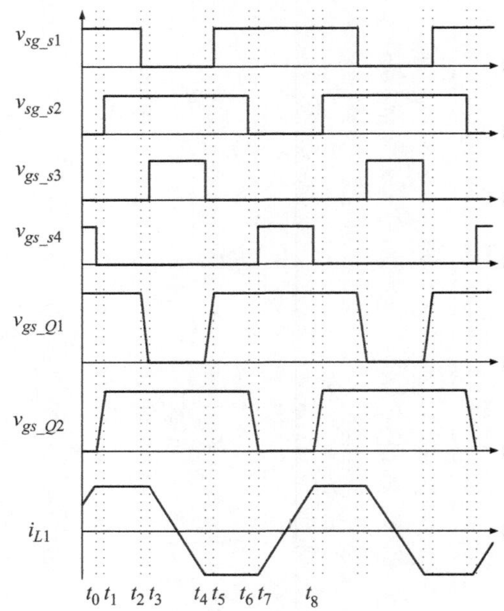

Figure 3.3 Typical waveforms in the proposed CSD with D > 0.5

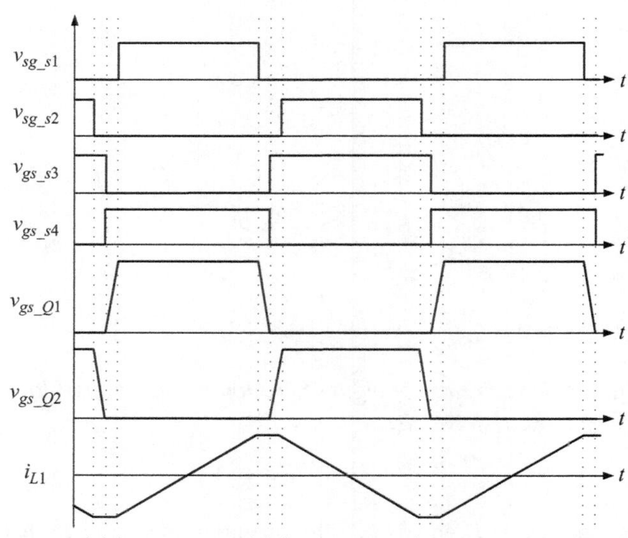

Figure 3.4 Typical waveforms of dual low-side symmetrical drive circuit with D = 0.5

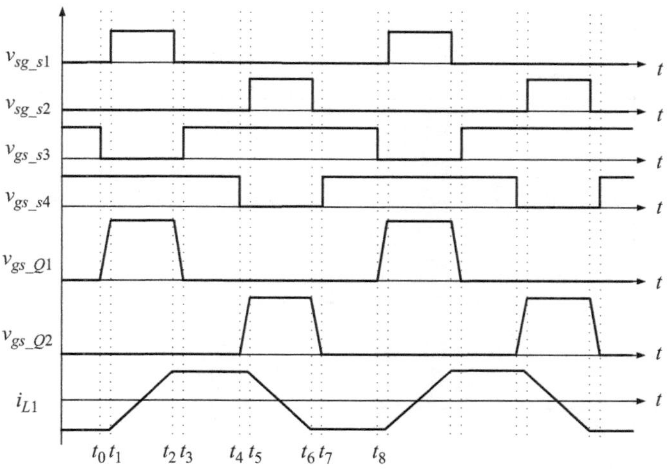

Figure 3.5 Typical waveforms of dual low-side drive circuit with $D < 0.5$

When the duty cycle D is less than 0.5, the operation is similar to that for $D > 0.5$ and the details are not explained here. The key difference is that there is a current cycling though S_3 and S_4 instead of S_1 and S_2. The waveforms are shown in Figure 3.5.

It is noted that when the gate-drive signals for S_1–S_4 are changed, the duty cycle for Q_1 and Q_2 will be changed from below 0.5 to 0.5 and to above 0.5 with smooth transition.

3.1.2 Loss comparison

This section analyzes the loss for the proposed CSD. The key parameter impacting the gate-drive loss is the peak inductor current, I_{Lpeak}, which is used to charge and discharge the gate capacitors of power MOSFETs. Assume the power MOSFETs Q_1 and Q_2 in Figure 3.1 are identical and their desired switching time is t_{sw}, the required peak inductor current, which is also the charge and discharge current, can be given by (3.1):

$$I_{Lpeak} = \frac{Q_g}{t_{sw}} \tag{3.1}$$

where Q_g is the total gate charge of each power MOSFET. It is noted that large I_{Lpeak}, value can reduce the switching time, and therefore, reduce switching loss.

Figure 3.6 redraws the inductor current and the current through S_1–S_4 when $D > 0.5$. The root mean square (RMS) value of the inductor current can be calculated by (3.2).

$$I_{LRMS} = \sqrt{2 \cdot \frac{2D-1}{2} \cdot {I_{Lpeak}}^2 + 2(1-D) \cdot \frac{{I_{Lpeak}}^2}{3}} = I_{Lpeak} \cdot \sqrt{\frac{4D-1}{3}} \tag{3.2}$$

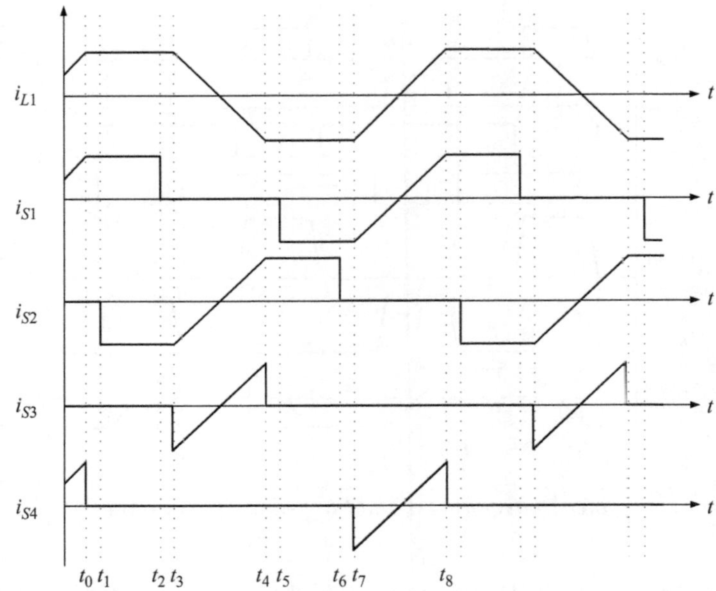

Figure 3.6 Current through inductor and switches S_1–S_4 at $D > 0.5$

At $D = 0.5$, the inductor current becomes triangular and I_{LRMS} equals to $I_{Lpeak}/\sqrt{3}$. The RMS currents through switches S_1 and S_2 are the same, and are given by (3.3):

$$I_{S1RMS} = I_{S2RMS} = \sqrt{2 \cdot \frac{2D-1}{2} \cdot I_{Lpeak}{}^2 + (1-D) \cdot \frac{I_{Lpeak}{}^2}{3}} = I_{Lpeak} \cdot \sqrt{\frac{5D-2}{3}} \tag{3.3}$$

At $D = 0.5$, the RMS current flowing through switches S_1 and S_2 equals to $I_{Lpeak}/\sqrt{6}$. The RMS currents through switches S_3 and S_4 are the same, and given by (3.4):

$$I_{S3RMS} = I_{S4RMS} = I_{Lpeak} \cdot \sqrt{\frac{1-D}{3}} \tag{3.4}$$

At $D = 0.5$, the RMS current equals to $I_{Lpeak}/\sqrt{6}$.

The total power loss of the proposed CSD is the sum of the R_{dson} loss of S_1–S_4, the gate-drive losses of S_1–S_4, the core loss and the copper loss of the resonant inductor L_1, and the gate resistor of the power MOSFET, R_G. There is no switching loss or cross conduction loss as switches S_1–S_4 operate in zero voltage switching (ZVS) condition.

The copper loss of the inductor winding is given by (3.5):

$$P_{copper} = R_{ac} \cdot I_{LRMS}{}^2 \tag{3.5}$$

where R_{ac} is the a.c. resistance of the inductor winding. The inductor core loss, P_{core}, depends upon the inductor design. High-frequency core materials such as 3F5 or PC50 should be used to reduce core loss. Air core inductor may be used when the switching frequency goes above 2 MHz. In that case, core loss is eliminated. The total inductor loss is given by (3.6):

$$P_{ind} = P_{copper} + P_{core} \tag{3.6}$$

Both the charge and discharge currents flow through the internal gate mesh resistance R_G of the power MOSFET and cause resistive loss. The charge and discharge current is the peak value of the inductor current. Thus, the total loss caused by the internal resistance of two power MOSFETs for turn-on and turn-off intervals can be calculated by (3.7):

$$P_{RG} = 2 \cdot 2 \cdot R_G \cdot I_{Lpeak}{}^2 \cdot t_{sw} \cdot f_s \tag{3.7}$$

where t_{sw} is the switching time and f_s is the switching frequency. P_{RG} includes both the charging loss and discharging loss caused by the internal resistance of two power MOSFETs.

The conduction loss caused for S_1 and S_2 can be calculated by (3.8), where $R_{ds1(on)}$ is the on-resistance for S_1 and S_2.

$$P_{top} = 2 \cdot R_{DS1(on)} \cdot I_{Lpeak}{}^2 \cdot \frac{5D-2}{3} \tag{3.8}$$

The conduction loss for S_3 and S_4 can be calculated by (3.9), where $R_{ds3(on)}$ is on-resistance for S_3 and S_4.

$$P_{bott} = 2 \cdot R_{DS3(on)} \cdot I_{Lpeak}{}^2 \cdot \frac{1-D}{3} \tag{3.9}$$

As mentioned before, switches S_3 and S_4 are N-channel MOSFETs, whereas switches S_1 and S_2 can be either N-channel MOSFETs or P-channel MOSFETs depend on design. When the same MOSFET is selected for all switches S_1–S_4, the total conduction loss caused by the switches S_1–S_4 is given by (3.10), where $R_{ds(on)}$ is on-resistance for S_1–S_4.

$$P_{cond} = P_{top} + P_{bott} = 2 \cdot R_{DS(on)} \cdot I_{Lpeak}{}^2 \cdot \frac{4D-1}{3} \tag{3.10}$$

The switching frequency of these switches S_1–S_4 is same as the switching frequency f_s of the power MOSFETs. The total gate-drive loss of all four switches (S_1–S_4) is given by (3.11), where Q_{g_s} is the total gate charge of switches (S_1–S_4), V_{gs_s} is the drive voltage of the switches (S_1–S_4), which is usually 5 V.

$$P_{Gate} = 4 \cdot Q_{g\_s} \cdot V_{gs\_s} \cdot f_s \tag{3.11}$$

Therefore, the total loss of the CSD can be calculated by adding all above losses together and is given by (3.12):

$$P_{DRV} = P_{cond} + P_{RG} + P_{Gate} + P_{ind} \tag{3.12}$$

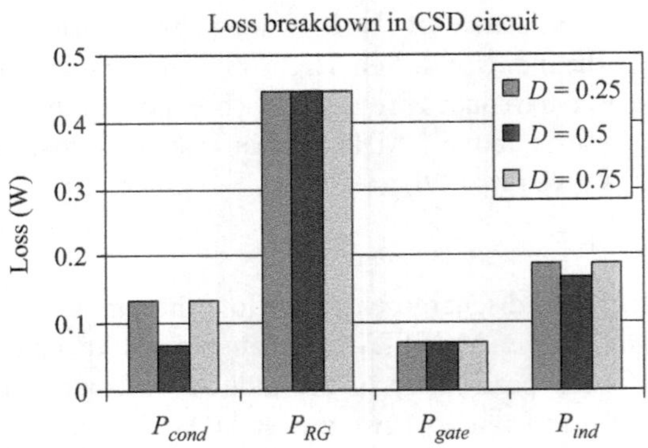

Figure 3.7 Loss breakdown in CSD under different duty cycle

It is noted that the above loss is for two MOSFETs. The above analysis assumes $D \geq 0.5$. When $D < 0.5$, the total loss for CSD can also be calculated in the similar way.

As an example, two IRF6618 are selected as power MOSFET, with total gate charge $Q_g = 93$ nC, when $V_{gs} = 12$ V. Its internal gate resistor R_G is 1 Ω. FDN335N is selected for switches S_1–S_4, with typical $R_{DS(on)} = 0.07$ Ω and total gate charge $Q_{g\_s} = 3.5$ nC. Assume the inductor peak current as 1.2 A. The switching frequency is 1 MHz. DS3316P–2.2 μH is chosen as the resonant inductor. The core loss of the inductor at 1 MHz is 0.147 W and the estimated a.c. winding resistance is 0.044 Ω. The loss breakdown under different duty cycles D is calculated and shown in Figure 3.7.

It is observed from Figure 3.7 that at $D = 0.5$, the total gate-drive loss is 0.75 W. At $D = 0.25$ and 0.75, the total gate-drive loss is 0.88 W. It is noted that with wide duty cycle variation, the total gate-drive loss changes only by 0.13 W.

3.1.3 Advantages of the proposed current source driver

3.1.3.1 Gate-drive loss saving

At switching frequency of 1 MHz and $D = 0.5$, the gate-drive loss of CSD is calculated to be 0.75 W by the analysis in Section 3.1.2.

The gate-drive loss under conventional gate-drive scheme can be calculated to be 2.23 W by (3.13):

$$P_{DRV} = Q_g \cdot V_{gs} \cdot f_s \tag{3.13}$$

There are lots of driver chips based on conventional gate-drive scheme in the market. The loss of the driver chip itself is usually around 300 mW at 1 MHz. Therefore, the total gate-drive-related loss under conventional drive scheme is 2.53 W (2.23 W + 0.3 W). It is noted that the loss will be dissipated in drive chip, which cannot handle such a high power loss. The compromise is to lower the gate

Table 3.1 Calculated gate-drive loss

	Gate charge loss (W)	Driver chip loss (W)	Total drive loss (W)
Conventional driver	2.23	0.3	2.53
Current source driver	0.75	0.04	0.79
Loss saving	1.48	0.26	1.74

voltage to around 5 V. The penalty is increased conduction loss as $R_{ds(on)}$ will be increased at $V_{gs} = 5$ V. In this chapter, $V_{gs} = 12$ V is used for both cases.

In the proposed CSD, some logic circuits and level shift circuits are needed to generate the drive signals of switches S_1–S_4. The loss for these logic circuits is estimated as 40 mW. Therefore, the total gate-drive-related loss for proposed CSD is 0.79 W (0.75 W + 0.04 W). The gate-drive loss saving realized by using CSD is 68.7% (1.74/2.53). Table 3.1 summarizes the above calculation result.

3.1.3.2 Switching loss reduction

It is noted that most of the switching loss happen at Miller plateau. During Miller plateau, the drain current changes from its peak value to zero during turn-off interval or from zero to peak value during turn-on interval. The Miller voltage normally stands around 3 V.

Take the turn-off interval as an example, the theoretical discharge current at Miller plateau is 0.86 A, when R_G (~1 Ω), driver resistance (~1.5 Ω), and optional external resistor (1 Ω) are considered. However, because of the leakage inductance in the discharge loop introduced by printed circuit board (PCB) track and bonding wire inside MOSFET, the actual discharge current is much smaller. Therefore, the turn-off time is longer and turn-off loss is greater. Similarly, in the case of turn-on interval, the charge current at Miller plateau will be even smaller if $V_{gs} = 5$ V is used and therefore, longer turn-on time and greater turn-on loss. This is the problem with conventional gate-drive scheme.

With the proposed CSD, the MOSFET gate is charged and discharged at the peak inductor current during the switching interval (Miller plateau). The current is constant during the switching interval. In addition, the impact of the leakage inductance is removed as it is in series with a much large inductor. Therefore, the turn-on and turn-off times can be significantly reduced and the switching loss can be reduced.

3.1.3.3 Body diode conduction time reduction

It is noted that in order to avoid cross conduction, dead time is introduced for SR MOSFET gate-drive signals. During the dead time, the body diode of the SR field-effect transistor (FET) is on and causes loss. With the proposed CSD circuit, the dead time can be reduced as the SR FET can be turned on and turned off faster as shown in the analysis above. Therefore, the diode conduction loss can be reduced.

3.1.3.4 High noise immunity and alleviation of *dv/dt* effect

With the new CSD, the gate of the power MOSFET is connected to either source or ground via low impedance path, rather than through the external gate resistor and driver's on-resistance (a few ohms) in conventional gate-drive scheme. Therefore, the noise immunity is significantly improved and it is much less likely the synchronous FET will be turned on by *dv/dt* effect.

3.1.3.5 Less impact of parasitic inductance

As the gate is charged/discharged by a constant current source, the negative impact by the parasitic inductance in the gate-drive loop (such as PCB track, lead inside the MOSFET) can also be significantly reduced.

3.1.4 Applications

The proposed CSD can be used in wide range of switching power converters. It can be used to drive the SR in a current doubler circuit, the primary MOSFET in push-pull converters, as well as the so-called bus converter when the duty cycle is 50%. Figure 3.8 shows that the proposed CSD is used to drive the SRs in a current doubler circuit. In this application, the duty cycles for Q_1 and Q_2 are the same and larger than 50%. With the CSD, most of gate-drive energy can be recovered. In addition, the turn-on and turn-off times of Q_1 and Q_2 can also be reduced, which will help reduce their body diode conduction loss.

Figures 3.9 and 3.10 show the current fed push-pull converter and voltage fed push-pull converter with CSD to drive the primary side MOSFET, Q_1 and Q_2.

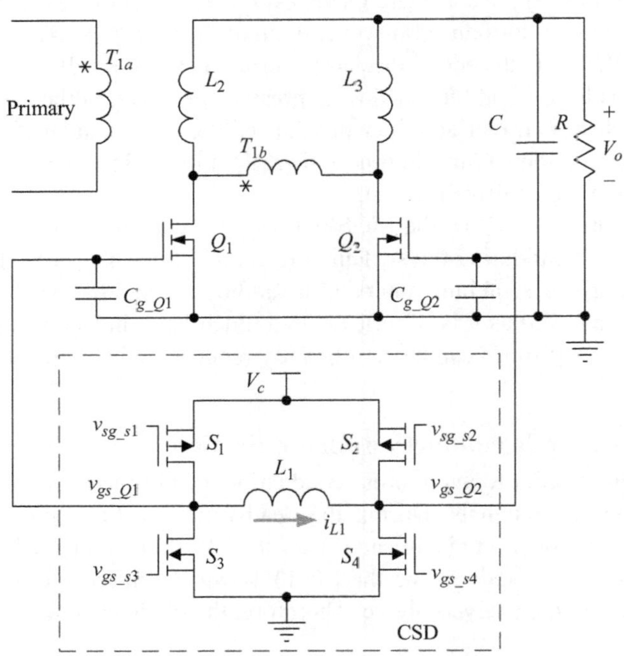

Figure 3.8 Current doubler with CSD

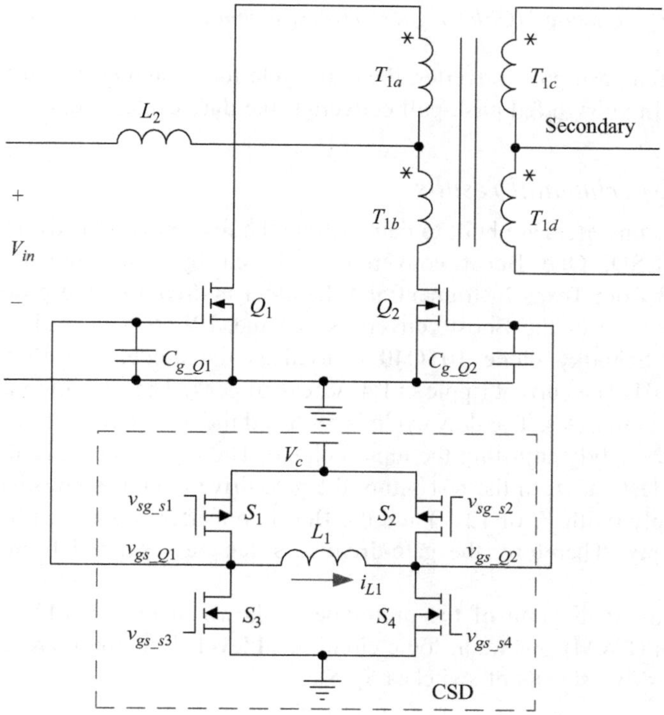

Figure 3.9 Current fed push-pull converter with CSD

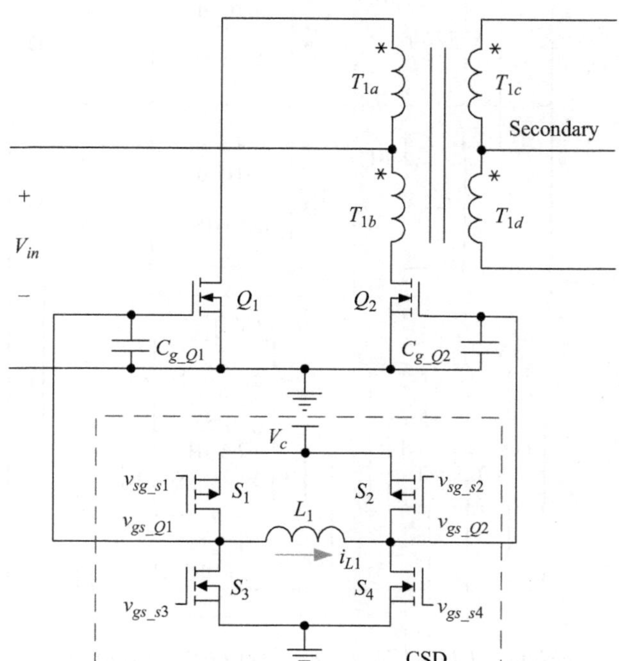

Figure 3.10 Voltage fed push-pull converter with CSD

In current fed push-pull converter, the duty cycle for Q_1 and Q_2 is same and larger than 50%. In voltage fed push-pull converter, the duty cycle is less than 50%.

3.1.5 Experimental results

Two boost converters are built to demonstrate the feasibility and advantages of the proposed CSD. One Boost converter is driven by conventional drive chip UCC27323 from Texas Instrument and the other is driven by the proposed CSD. The power train of the Boost converters is same. IRF6618 is used as the Boost MOSFET. Schottky diode 10TQ40 is used as the diode. The Boost inductor value is 2 μH. The current ripple is 1.4 A peak to peak. The switching frequency is 1 MHz for both cases. The duty cycle is 50% and the output voltage is regulated at 11.35 V by slightly adjusting the input voltage. The input voltage is around 5.7 V for all the test cases. In the test setup, the gate-drive circuit is powered from one power supply (with V_c of 12 V) and the Boost converter is powered from another power supply. Therefore, the gate-drive loss can be separated from the power circuit loss.

The circuit diagram of the prototype is shown in Figure 3.11. Pulse width modulation (PWM) generator, logic circuit, and level shift circuit are designed for generating drive signals of switches S_1–S_4.

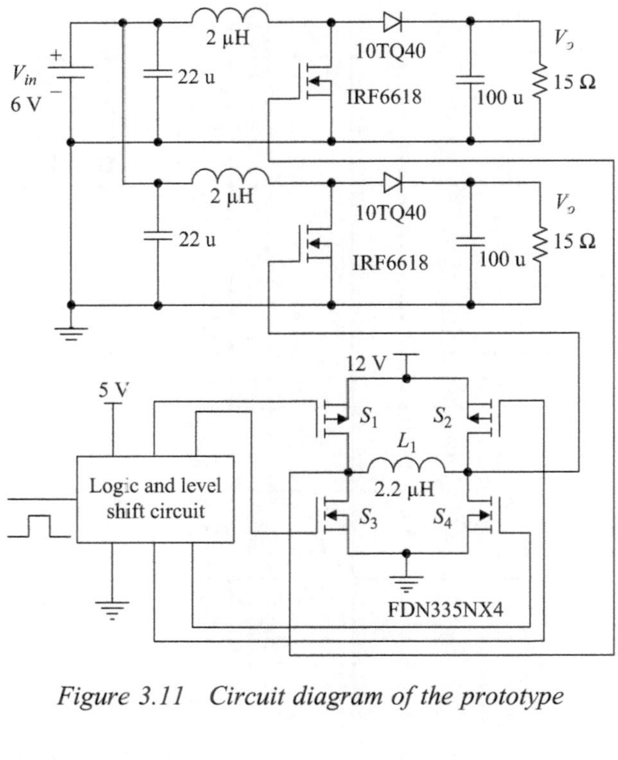

Figure 3.11 Circuit diagram of the prototype

The measured gate-drive loss under conventional drive scheme is 2.61 W (2.53 W from calculation) at 1 MHz operation. The gate-drive loss for proposed current source drive scheme is only 0.864 W (0.75 W from calculation). The loss reduction is 1.746 W, or 67% of the conventional gate-drive loss. The measured gate-drive loss is very close to the calculated loss.

Figure 3.12 shows measured key waveforms for the CSD. Figure 3.12(a) shows the gate signal for S_1, S_2, S_3, and S_4. The dead time is easily achieved. Figure 3.12(b) shows the gate voltage and the resonant inductor current. It is observed that waveform is very clean and the peak inductor current is used to charge and discharge the gate capacitor. No Miller plateau is observed.

Figure 3.13 shows the comparison of gate voltage and gate current for CSD and conventional gate-drive circuit. It is observed from Figure 3.13(a) that the gate current during switching interval is constant and is about 1.3 A. The miller plateau is not observed, which shows that the Miller charge is removed very quickly. It is also observed from Figure 3.13(b) that for conventional gate-drive circuit, the peak current is similar. However, the gate current at Miller plateau is quite small (around 0.3 A) and significant Miller plateau is observed.

The waveforms in Figure 3.13 also indicate that the switching time is reduced for CSD. In order to illustrate this point clearly, the total power train loss of the Boost converter (excluding gate-drive loss) is measured under different load current (0.4 A and 0.8 A) when output voltage is regulated at 11.35 V. For conventional gate-drive scheme, the loss is measured under four conditions, with gate resistor of 2.5 Ω and 1 Ω, as well as with load current of 0.4 A and 0.8 A. Table 3.2 summarizes the total power train loss measurement (excluding gate-drive loss).

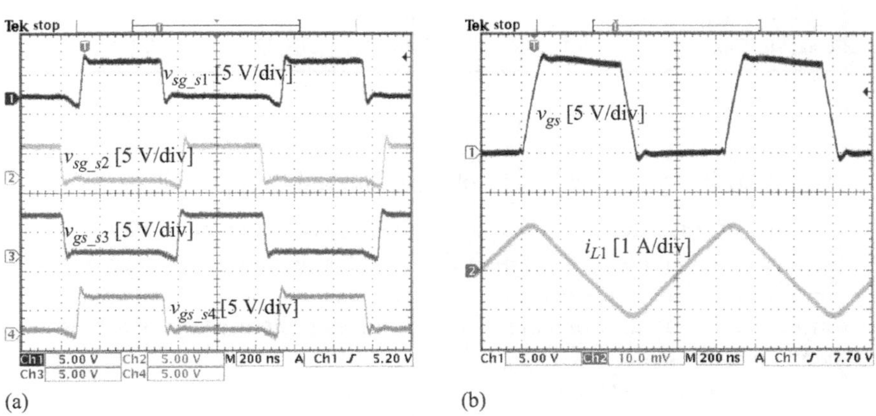

(a) (b)

Figure 3.12 Measured key waveforms of the proposed CSD and conventional gate-drive circuit, time scale 200 ns/div; (a) gate-drive signals of switches S_1–S_4 (5 V/div) and (b) resonant inductor current (1 A/div), and gate-drive voltage (5 V/div)

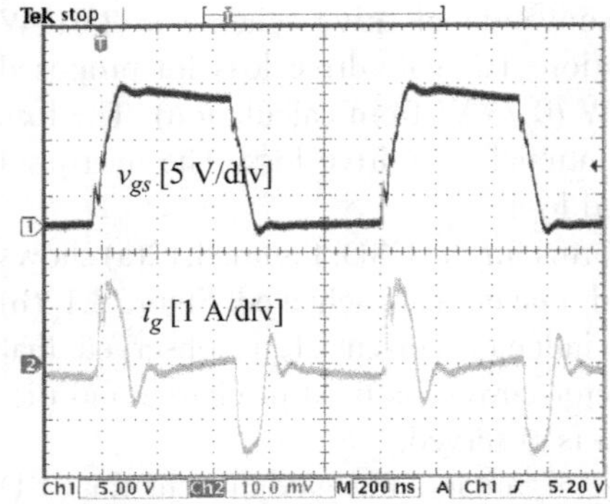

(a)

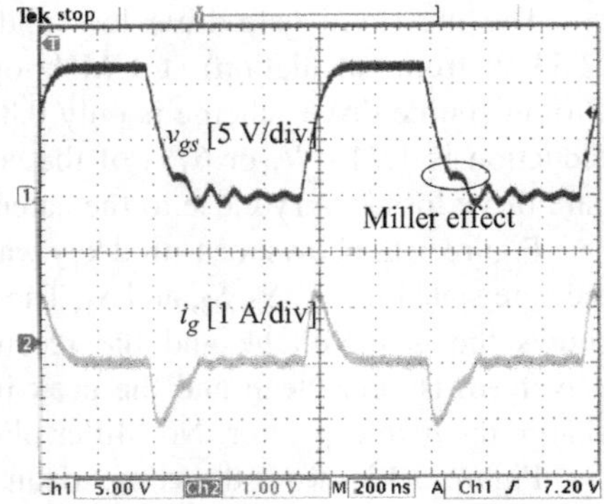

(b)

Figure 3.13 Measured key waveforms of the proposed CSD and conventional gate-drive circuit, time scale 200 ns/div; (a) gate voltage (top, 5 V/div) and gate current (1 A/div) for CSD and (b) gate voltage (top, 5 V/div) and gate current (bottom, 1 A/div) for conventional gate driver

Table 3.2 Total power train loss comparison, $V_{gs} = 12$ V, $V_c = 11.35$ V, $f_s = 1$ MHz

	Load current (A)	**Conventional driver**		**CSD**	**Loss saving (W)**
		R_{g_ext} **(Ω)**	$P_{\_loss}$ **(W)**	$P_{\_loss}$ **(W)**	
Case 1	0.4	2.5	2.07	1.92	0.15
Case 2	0.8	2.5	2.78	2.32	0.46
Case 3	0.4	1	1.98	1.92	0.06
Case 4	0.8	1	2.50	2.32	0.18

The following two points can be observed from the above table:

1. External gate resistor impact switching loss for conventional driver
   For conventional drive, comparing case 1 to case 3, and case 2 to case 4, it can be observed that with smaller gate resistor, the total loss is smaller. The loss difference is due to switching loss difference in these cases as all the other conditions are same.
2. CSD can reduce switching loss
   As can be observed from case 3, with the proposed CSD, the total loss is 1.92 W, while the total loss with conventional driver is 1.98 W. The difference of the total power loss is the switching loss reduction as all the other conditions are same. Therefore, the switching loss reduction is 0.06 W. In case 4, where the current is doubled, the total loss is also increased. The switching loss

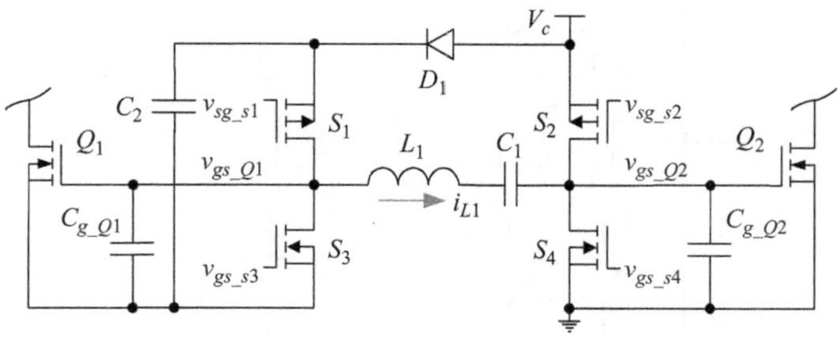

Figure 3.14 Dual-channel high-side and low-side CSD

reduction with CSD is 0.18 W (2.50–2.32 W). It is noted that the power circuit is same for both cases. When the external gate resistor is 2.5 Ω, the switching loss reduction is even further.

The above observation indicates that proposed CSD can reduce the switching loss by reducing the switching time, which is achieved by charging and discharging the gate capacitor at high current level. It is expected that the switching loss reduction will become more significant when the load current is further increased.

3.2 High-side and low-side continuous current source drivers

3.2.1 Operating principle of proposed current source drivers

Different from two-channel low-side CSDs, this section presents high-side and low-side CSD, which is able to drive two MOSFETs in a synchronous buck converter. Similarly, it uses a constant current to charge and discharge MOSFET gate capacitor. Other benefits include simple circuit configuration, better *dv/dt* turn-on immunity, less impact by parasitic inductance.

The proposed CSD is shown in Figure 3.14. The circuit consists of four switches S_1–S_4, which is inherited from dual-channel conventional gate-drive circuit, connecting as a bridge configuration, an inductor, L_1 and a capacitor C_1 connecting across the bridge. Diode D_1 and capacitor C_2 consist of a charge pump circuit for high-side drive. Capacitors, C_{g_Q1} and C_{g_Q2}, are the gate capacitors of the power MOSFETs Q_1 and Q_2, respectively. Q_1 is the high-side switch and Q_2 is the lower side switch. V_c is the gate-drive voltage source. In order to simplify the implementation, P-channel MOSFET is used for S_1 and S_2 and N-channel MOSFET is used for S_3 and S_4. It is noted that other implementation methods can also be used to achieve the same objective.

The proposed CSD shown in Figure 3.14 can operate in both complementary and symmetrical mode. When it operates in complementary mode, it provides two drive signals with duty cycle D and $1 - D$, respectively. This mode is suitable for driving two MOSFETs in a synchronous buck converter or primary MOSFETs in an asymmetrical half-bridge (HB) converter. When operating in symmetrical mode, the proposed circuit provides two drive signals with same duty cycle D and with

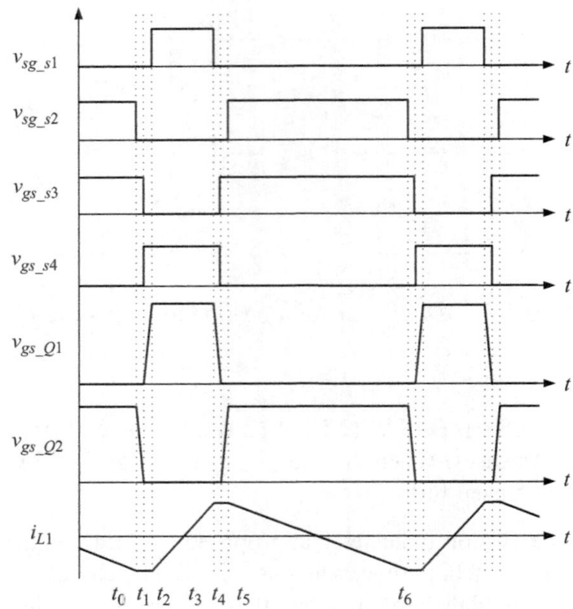

Figure 3.15 Typical waveforms of dual-channel high-side and low-side CSD with complementary drive signal, $D_{Q1} = D$ and $D_{Q2} = 1 - D$

180° out of phase. This mode is suitable for driving HB and full-bridge (FB) configurations.

The following provides the detailed operation of dual high-side and low-side CSD operating at complementary mode. The duty cycle for Q_1 is D and duty cycle for Q_2 is $1 - D$. In high-side and low-side configurations, the current of the resonant inductor will flow through the power train of the converter in addition to the switches S_1–S_4. A synchronous buck converter is used for the operation principle depiction. The key waveforms are shown in Figure 3.15 and a synchronous buck converter with the proposed CSD is shown in Figure 3.16.

In Figure 3.15, v_{sg_S1}, v_{sg_S2}, v_{gs_S3}, and v_{gs_S4} are gate-drive signals for S_1–S_4. v_{gs_Q1} and v_{gs_Q2} are the voltage across C_{g_Q1} and C_{g_Q2}. The rising edge and falling edge of v_{gs_Q1} and v_{gs_Q2} are shown to illustrate the details of charging and discharging interval. i_{L1} is the current waveform through inductor L_1. In the analysis, it is assumed that capacitor C_1 is large and the voltage across C_1 is a d.c. value. If the capacitor value C_1 is small, the operation of the circuit does not change.

The operation of this circuit can be briefly described as following by six operation modes as shown in Figure 3.17:

1. Mode 1 [before t_0] (Figure 3.17(a)): S_1 and S_4 are off while S_2 and S_3 are on. Q_1 is off and Q_2 is on. The inductor current, i_{L1}, increases to negative maximum value. The inductor current i_{L1} flows in the loop consisted of S_2, C_1, L_1, S_3, Q_2, and V_c. The voltage across C_2 is charged to the level of V_c via D_1 and Q_2.

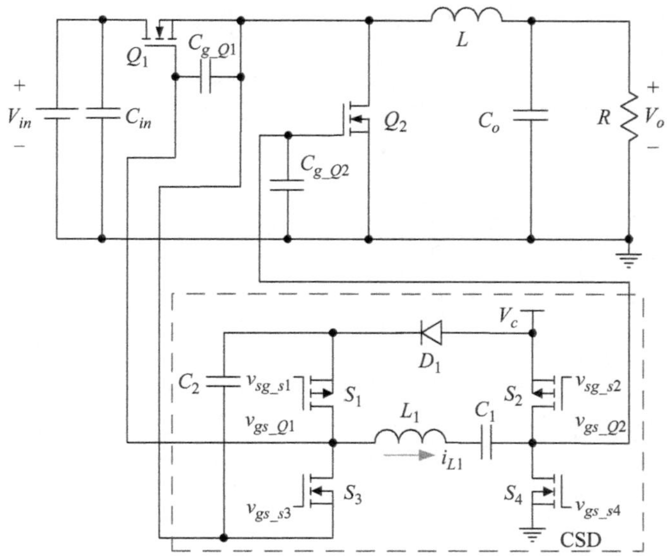

Figure 3.16 Synchronous buck converter with dual high-side and low-side CSD

2. Mode 2 $[t_0, t_1]$ (Figure 3.17(b)): S_2 is turned off at t_0. Inductor L_1 resonates with the input capacitor of MOSFET Q_2, C_{g_Q2}. C_{g_Q2} will be discharged at this period. The voltage across C_{g_Q2} decreases and it will be clamped to zero by the body diode of S_4 before t_1. Q_2 is turned off in this time interval. The current i_{L1} flows in the loop consisted of C_{g_Q2}, C_1, L_1, S_3, and Q_2. Then at t_1, S_3 is turned off and S_4 is turned on with zero voltage simultaneously. By controlling the turn-off instant for S_2 (t_0), the turn-off instant of Q_2 can be controlled.
3. Mode 3 $[t_1, t_2]$ (Figure 3.17(c)): S_3 is turned off and S_4 is turned on with zero voltage at t_1 simultaneously. Inductor L_1 resonates with the input capacitor C_{g_Q1} of MOSFET Q_1. C_{g_Q1} will be charged at this period. The voltage across C_{g_Q1} increases and it will be clamped to the level of V_c by the body diode of S_1 before t_2. Q_1 is turned on in this interval. The current i_{L1} flows in the loop consisted of S_4, C_1, L_1, C_{g_Q1}, Q_1, and V_{in} after Q_1 is turned on. Then at t_2, S_1 is turned on with zero voltage. By controlling the turn-off instant for S_3 (t_1), the turn-on instant for Q_1 can be controlled.
4. Mode 4 $[t_2, t_3]$ (Figure 3.17(d)): S_1 and S_4 are on while S_2 and S_3 are off. Q_1 is on and Q_2 is off. The negative inductor current i_{L1} rises to zero and further increases. The value of the current i_{L1} will increase to positive maximum at t_3. The current i_{L1} flows in the loop consisted of C_2, S_1, L_1, C_1, S_4, V_{in}, and Q_1. Then at t_3, S_1 is turned off with zero voltage.
5. Mode 5 $[t_3, t_4]$ (Figure 3.17(e)): S_1 is turned off at t_3. Inductor L_1 resonates with capacitor C_{g_Q1}. C_{g_Q1} will be discharged at this period. The voltage across C_{g_Q1} decreases and it will be clamped to zero by the body diode of S_3 before t_4.

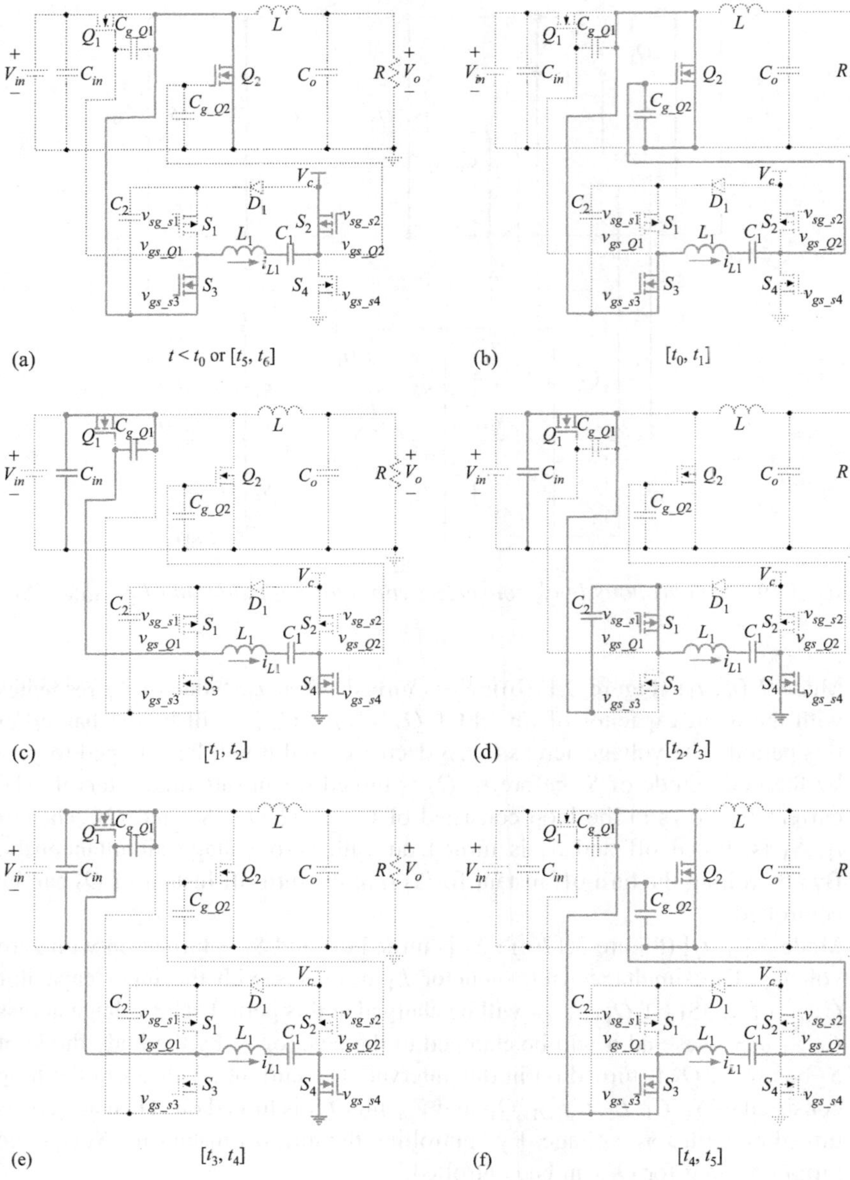

Figure 3.17 Equivalent circuits of six operation modes of the proposed CSD operating in complementary mode

Q_1 is turned off in this time interval. The current i_{L1} flows in the loop consisted of C_{g_Q1}, L_1, C_1, S_4, V_{in} and Q_1. At t_4, S_3 is turned on and S_4 is turned off with zero voltage simultaneously. By controlling the turn-off instant for S_1 (t_3), the turn-off instant of Q_1 can be controlled.

6. Mode 6 $[t_4, t_5]$ (Figure 3.17(f)): S_3 is turned on and S_4 is turned off with zero voltage at t_4 simultaneously. Inductor L_1 resonates with capacitor C_{g_Q2}. C_{g_Q2} will be charged at this period. The voltage across C_{g_Q2} increases and it will be clamped to the source voltage V_c by the body diode of S_2 before t_5. Q_2 is turned on in this time interval. The current i_{L1} flows in the loop consisted of S_3, L_1, C_1, C_{g_Q2}, and Q_2. At t_5, S_2 is turned on with zero voltage. By controlling the turn-on instant for S_3 (t_4), the turn-on instant for Q_2 can be controlled.
7. Mode 7 $[t_5, t_6]$ (Figure 3.17(a)): S_1 and S_4 are off while S_2 and S_3 are on, while Q_1 is off and Q_2 is on. The inductor current i_{L1} decreases to zero and then it will increase in opposite direction. The value of the current i_{L1} will increase to negative maximum at t_6. The current i_{L1} flows in the loop consisted of S_2, C_1, L_1, S_3, Q_2, and V_c. The next cycle will start at t_6.

In a synchronous buck converter shown in Figure 3.16, the d.c. voltage across capacitor C_1 is given by (3.14):

$$V_{C1} = D \cdot V_{in} + (2D - 1) \cdot V_C \tag{3.14}$$

The peak resonant inductor current is given by (3.15):

$$I_{Lpeak} = \frac{(V_{in} + 2V_C) \cdot D \cdot (1 - D) \cdot T_s}{2L} \tag{3.15}$$

Where V_{in} is the input voltage of the converter. V_c is the gate-drive voltage, which could be the input voltage of the converter. T_s is switching period. L is the resonant inductor value. D is the duty cycle of the control MOSFET.

The proposed dual channel high-side and low-side CSD shown in Figure 3.14 can also operate in symmetrical duty cycle mode. In this case, it provides two drive signals with the same duty cycle and with 180° out of phase. A HB converter and a FB converter require this operation mode. The operation is similar to the complementary duty cycle mode and detailed description is not provided here. The key waveforms of this operation mode are shown in Figure 3.18. Figure 3.19 shows a HB converter with dual-channel high-side and low-side CSD. In this converter, the CSD operates in symmetrical mode. The duty cycle of MOSFETs Q_1 and Q_2 are both D.

3.2.2 Advantages

One key point for the proposed CSD is that during MOSFET turn-on transition, its gate capacitor is charged by the peak inductor current, I_{Lpeak}, and during MOSFET turn-off transition, the gate capacitor is discharged by the peak inductor current as well. During charging and discharging period, the inductor current is constant. This fact brings significant advantages to the operation of synchronous buck converter as summarized below.

(a) Smaller gate-drive loss
The gate charge loss can be significantly reduced. The analysis in Section 3.1.1 shows that more than 50% of the gate-drive loss can be recovered at 1 MHz switching frequency.

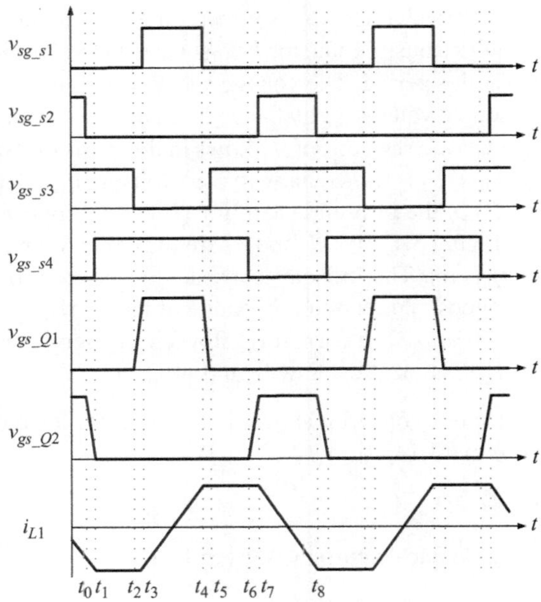

Figure 3.18 Typical waveforms of dual-channel high-side and low-side CSD with symmetrical drive signal, $D_{Q1} = D_{Q2} = D$

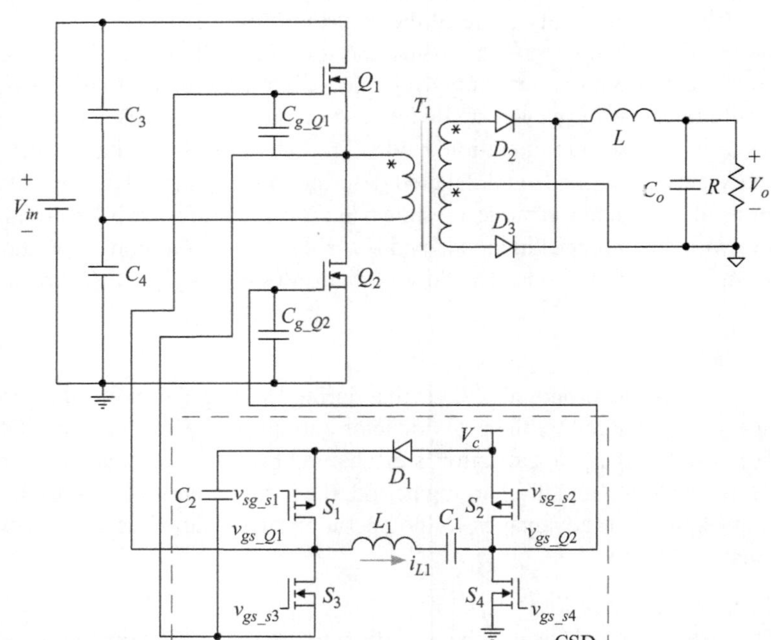

Figure 3.19 Half-bridge converter with dual high-side and low-side drive circuit

(b) Reduced switching loss and body diode conduction loss
The charge current and discharge current is constant and is the peak value of the inductor current. This ensures quick discharge of the Miller capacitor, which is critical to reduce the switching loss, especially the turn-off loss when the turn-off current is higher. An additional benefit from this fact is that the dead time required to avoid cross conduction between the control FET and synchronous FET can also be reduced. Thus, the diode conduction loss can be reduced.

(c) High noise immunity and alleviation of *dv*/*dt* effect
With the proposed CSD, as shown in Figure 3.16, the gates of main MOSFETs, Q_1 and Q_2, are connected to either the voltage source or ground via low impedance path (S_1 to S_4). With the conventional gate-drive scheme, the gate is connected to ground or to V_{cc} through external gate resistor (around 1 Ω) and driver's on-resistance (a few ohms). Therefore, the noise immunity is improved and it is much less likely the synchronous FET will be turned on by *dv*/*dt* effect.

(d) Less impact of parasitic inductance
As the gate is charged/discharged by a constant current source, the impact caused by the parasitic inductance in the gate-drive loop (such as PCB track, lead inside MOSFET) is much less. It is noted that in the conventional gate-drive scheme, the parasitic inductance will reduce the charge and discharge current, and therefore increase switching loss.

3.2.3 Loss analysis

As discussed above, the proposed CSD can reduce gate-drive loss. It can also reduce the switching loss (turn-on and turn-off). This section compares the gate-drive loss and switching loss for a buck converter based on the proposed CSD and the conventional gate-drive circuit.

(a) Gate-drive loss comparison

Two examples are taken to calculate the total gate-drive loss for the proposed CSD and for conventional gate-drive circuit. One example is gate voltage of 5 V for main FET and the other example is gate voltage of 12 V for main FET. MOSFET IRF7821 from International Rectifier is selected as control MOSFET Q_1 and FDS7088N7 from Fairchild is selected as synchronous MOSFET Q_2. FDN335N is selected for switches S_1–S_4, with typical $R_{DS(on)} = 0.045\ \Omega$ at 5 V gate voltage. Its gate charge at ZVS condition is $Q_{g\_s} = 3.5$ nC. The inductor peak current is designed as 1.2 A. The switching frequency is 1 MHz. DS3316P–1.0 μH is used as the inductor at 5 V and DS3316P–2.2 μH is used as the inductor at 12 V. Table 3.3 summarizes the detailed loss breakdown for CSD and the conventional gate-drive circuit.

P_{cond}, P_{RG}, P_{gate}, and P_{ind} (for CSD) are given in Table 3.3. $P_{\_logic}$ indicates the loss for logic circuit required for convert the input PWM signal to the drive signal for S_1–S_4. Its value is similar to the power loss of a conventional gate-drive

Table 3.3 Loss comparison between current source driver and conventional gate driver

	Current source driver (W)						**Conventional gate driver (W)**		
	P_{cond}	P_{RG}	P_{gate}	P_{ind}	$P_{\_logic}$	P_{tot_csd}	P_{g_chg}	$P_{\_driver}$	P_{tot_conv}
$V_{gs}=5$ V	0.043	0.149	0.080	0.015	0.04	**0.327**	0.31	0.15	**0.46**
$V_{gs}=12$ V	0.043	0.321	0.080	0.021	0.04	**0.506**	1.607	0.3	**1.907**

chip operating at no load. The loss is around 0.04 W. P_{g_chg} in Table 3.3 represents the gate charge loss. $P_{\_driver}$ represents the driver chip loss, which is estimated at around 0.3 W for switching frequency of 1 MHz and V_{gs} of 12 V [1]. For 5 V operation, $P_{\_driver}$ is estimated at around 0.15 W.

It can be observed from Table 3.3 that at $V_{gs}=5$ V, the CSD can reduce the total gate-drive loss from 0.46 to 0.327 W, a reduction of 0.133 W, or 29%. At $V_{gs}=12$ V, the CSD can reduce the total gate loss from 1.907 to 0.506 W, a reduction of 1.401 W, or 73%.

It should be noted that for conventional gate-drive circuit, the gate-drive loss at V_{gs} of 12 V is very high and the driver chip will have a high temperature rise. Therefore, V_{gs} of 12 V is not used for 1 MHz switching frequency operation. V_{gs} of 5 V is normally used. The consequence of 5 V V_{gs} is higher conduction loss for the MOSFET. Design compromise has to be made between higher gate-drive loss and lower conduction loss.

With the proposed CSD, the total gate-drive loss at V_{gs} of 12 V is 0.506 W, marginally increased from 0.327 W at $V_{gs}=5$ V and almost same as the gate loss for conventional gate driver at 5 V gate voltage. Therefore, the design compromise between higher gate loss and lower switching loss and conduction loss is not needed. In the simulation and experimental prototype, V_{gs} of 12 V is used for CSD. For conventional gate-drive circuit, V_{gs} of 6 V and 12 V are both used.

(b) Switching loss analysis

Turn-off behavior of the driven MOSFETs under the proposed CSD is much different from that of conventional voltage gate-drive scheme. Switching loss can be reduced by reducing switching time and therefore, significant switching loss reduction is achieved.

Normally, $V_{gs(th)}$ of a power MOSFET is around 2 V, and the plateau voltage V_p is around 3 V. Under conventional gate-drive scheme, assume gate-drive voltage V_{gs} is 5 V and the total gate resistance (including the gate resistor inside the MOSFET, which is around 1–2 Ω, the on resistor of the driver, which is around 1–2 Ω, as well as external gate resistor to damp the gate capacitor, which is usually around 1 Ω) is around 3–5 Ω. Then the average charge current in turn-on transition (charging the Miller plateau) is (5–3 V)/3 Ω = 0.7 A, whereas the average discharge current in turn-off transition is determined by the threshold voltage and plateau voltage and is only 3 V/3 Ω = 1 A. It is also noted that because of the parasitic inductance introduced by PCB track and wire inside the MOSFET, the actual charge current and discharge current is much smaller and the switching loss is much larger [2,3].

An additional benefit for synchronous buck converter with the proposed CSD is that the dead time between the control FET and SR FET can be reduced and therefore, the body diode conduction loss can be reduced.

3.2.4 Experimental results

Both the conventional gate-drive scheme and the proposed CSD scheme are simulated. A totem pole drive topology is used to represent the conventional voltage-drive scheme. A resistor is used to limit the drive current so that the peak current in turn-on transition is 2 A. An experimental prototype is also built to verify the feasibility and advantages of the proposed CSD. As a comparison, a buck converter with conventional gate-drive circuit is also built.

In order to determine the parameters in the CSD, the peak inductor current needs to be decided first. 2 A peak-drive current is chosen for the conventional gate-drive scheme, for the same switching time, only half of the peak inductor current is needed by using the proposed resonant-drive scheme. Once the drive current and duty cycle of the drive signal is determined, the resonant inductor value can be calculated by using (3.15).

Two buck converters are built. One with CSD and the other with the conventional gate-drive circuit. The buck converter parameters are same. The input voltage of the prototype is 12 V. The output is 1.5 V/15 A. The switching frequency is 1 MHz. For buck converter, the control FET is IRF7821 and the synchronous FET is FDS7088N7. The buck inductor value is selected as 0.5 μH with ripple current of 2.6 A peak to peak. For CSD, FDN335N is selected for switches S_1–S_4. The inductor is DS3316P–2.2 μH from Coil Craft. The peak inductor current is 1.2 A.

For conventional gate-drive scheme, the driver chip is TPS2832 from Texas Instrument, which uses adaptive dead-time control circuit to eliminate shoot-through currents through the control FET and SR FET during switching transition. The external gate resistor is 1 Ω. It is noted that when the external gate resistor is set to zero, the measurement result does not change.

For CSD, the gate-source voltage is 12 V. For conventional gate-drive circuit, gate voltage of 12 V and 6 V are both used. It is noted that with V_{gs} of 12 V, the driver chip is very hot and it cannot operate continuously for more than 10 minutes.

The total power loss for two converters under different load current is measured and summarized in Figure 3.20.

It is observed from the measurement that significant power loss reduction is achieved. It is observed from the above figure that at 1.5 V/15 A output, the power loss reduction is 2.15 W at 15 A output current. It is almost 10% of the output power (22.5 W).

A more detailed look at the measurement data shows that when the load current increases, the total power loss reduction achieved by the CSD is also increased. At 5 A load current, the total loss reduction is 1.03 W and at 15 A, the total loss

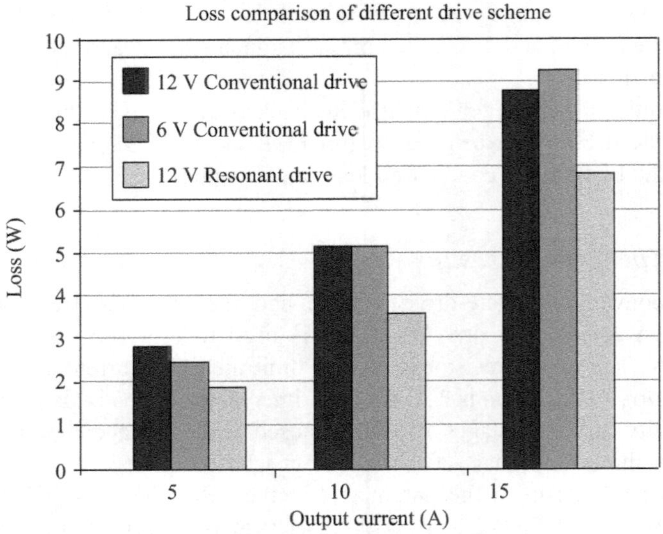

Figure 3.20 Loss comparison of conventional drive scheme and resonant drive scheme

reduction is increased to 2.15 W. As the gate-drive loss is independent of load current and the conduction loss is same for both gate-drive schemes, the result shows that the switching loss is reduced by the CSD scheme. The power loss reduction due to switching loss reduction is very significant.

The proposed CSD also beats the conventional gate-drive scheme with 6 V drive voltage. It is observed that less energy is saved under light load condition by using CSD as the driving loss under 6 V conventional drive scheme is reduced dramatically. Whereas under heavy load condition, more energy is saved by using CSD, since the drive loss reduction benefit is overwhelmed by the conduction loss increasing.

Figure 3.21 shows the measured waveforms with CSD. As shown in Figure 3.21(a), the gate-drive signal for v_{gs1} (control FET) is very clean and smooth. No Miller plateau is observed as the Miller charge is removed very quickly. The total rise time and fall time (between 0 V and 12 V) of v_{gs1} is less than 20 ns. This indicates that the switching time is much less than that, and therefore, very fast switching is achieved. The ringing in v_{gs2} is caused by noise picked by the probe. Figure 3.21(b) shows the drain to source voltage of SR. Figure 3.21(c) shows the resonant inductor current waveform. The peak inductor current, which is also the charge and discharge current, is about 1.2 A.

The experimental results presented in this section demonstrate that the proposed CSD can significantly reduce the power loss, especially the switching loss, and, therefore, improve the efficiency. With CSD, the loss saving of 2.15 W is achieved for 22.5 W (1.5 V/15 A) output, or almost 10%.

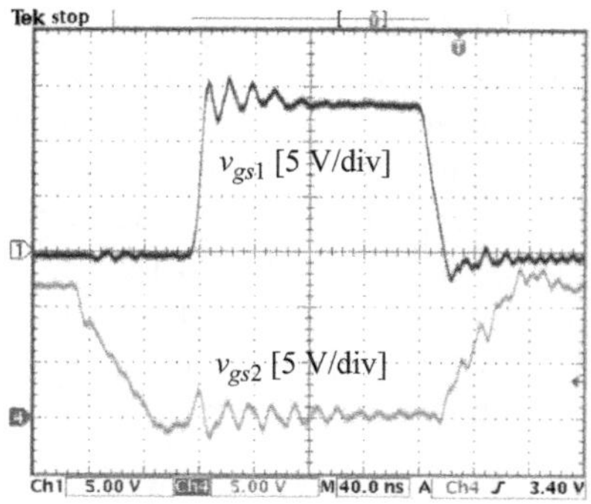

(a)

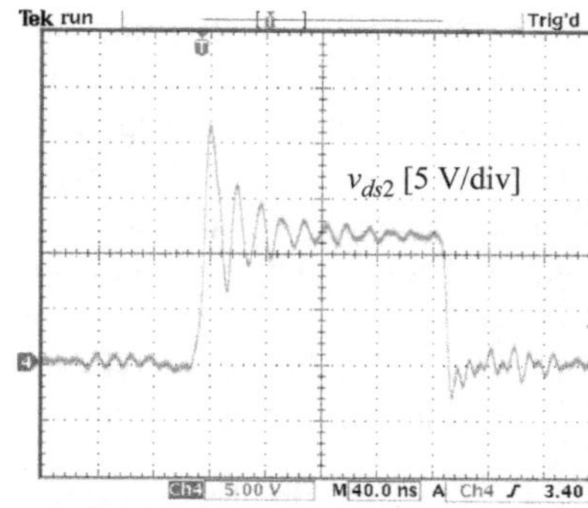

(b)

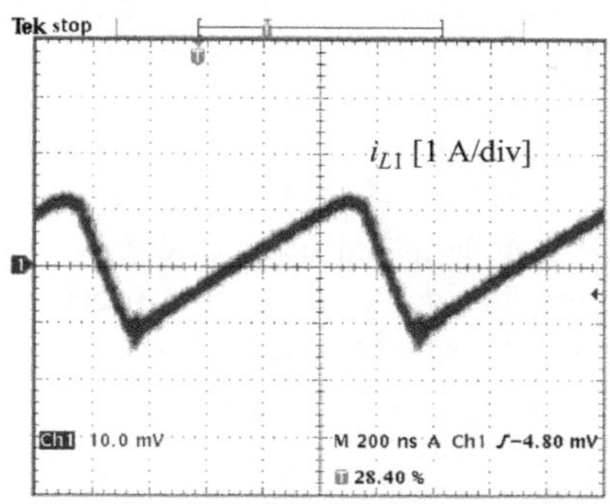

(c)

Figure 3.21 Measured key waveforms of the proposed CSD; (a) top trace: v_{gs1} (control FET). Bottom trace: v_{gs2} (SR FET), scale: 5 V/div, 40 ns/div, (b) drain-to-source voltage for SR FET scale: 5 V/div, 40 ns/div and (c) resonant inductor current 1 A/div (10 mV/div), 200 ns/div

3.3 Accurate switching loss model with current source drivers

3.3.1 Proposed MOSFET loss model with current source resonant driver

This section develops an analytical loss model of the power MOSFET driven by the CSD. Closed-formed analytical equations are derived to investigate switching characteristics due to the parasitic inductance. The comparison between a voltage driver and a resonant driver is presented concentrating on the common source inductance in detail. The modeling and simulation results prove that a current source resonant driver can significantly reduce the propagation impact of the common source inductance on the switching transition, which leads to a significant reduction of the switching transition time and the switching loss.

The MOSFET switching loss models can be classified into: (1) a physical-based model using physical parameters of the device, (2) a behavior model provided by device vendor supplies, and (3) an analytical model (also called a mathematical model). The physical-based model and the behavior model are convenient using simulation software, but simulation cannot be used for the purpose of design and optimization directly. It should be stressed that the piecewise loss model by linearizing the switching waveforms is no longer valid due to the parasitic inductance at high frequency. Therefore, we need a new analytical loss model to predict the optimal design solution for a resonant driver.

3.3.1.1 Circuit and basic assumption

Figure 3.22 shows a basic converter circuit including a MOSFET in series with a diode D_1, with d.c. input voltage V_D and an inductive load. The simplified equivalent circuit for the switching transition is shown in Figure 3.23, where MOSFETs M_1 is represented with a typical capacitance model, the clamped inductive load is replaced by a constant current source and the current source gate driver is simplified as a current source (I_G) since the charge and discharge current is kept constant during the switching transition. L_D is the switching loop inductance including the packaging inductance and any unclamped portion of the load inductance. L_S is the common source inductance, which is shared by the main current path and the gate diver loop. The critical MOSFET parameters are as follows: (1) the gate-to-source capacitance C_{GS}, the gate-to-drain capacitance C_{GD}, the drain-to-source capacitance C_{DS}; (2) the gate equivalent series resistance (ESR) R_G (external and internal); (3) the threshold voltage V_{th}; and (4) transconductance g_{fs}.

For purpose of transient analysis, we make the following simplifying assumptions:

1. $i_D = g_{fs}(v_{GS} - V_{th})$ and MOSFET is ACTIVE, provided $v_{GS} > V_{th}$ and $v_{DS} > i_D R_{DS(on)}$;

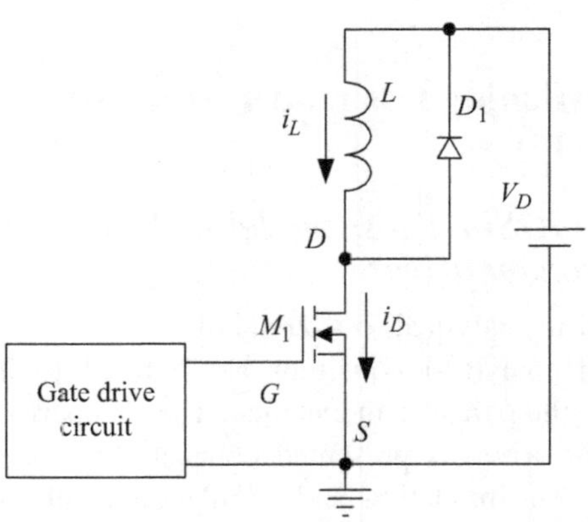

Figure 3.22 Circuit with a clamped inductive load

2. For $v_{GS} < V_{th}, i_D = 0$, and MOSFET is OFF;
3. When $g_{fs}(v_{GS} - V_{th}) > v_{DS}/R_{DS(on)}$, the MOSFET is fully ON.

3.3.1.2 Analytical modeling of main switching transition

During the main switching transition period, the MOSFET enters its active state and the linear transfer characteristics is assumed as given in (3.16) [4], where $i_D(t)$ is the instantaneous switching current of the MOSFET and $v_{GS}(t)$ is the instantaneous gate-to-source voltage of the MOSFET:

$$i_D(t) = g_{fs}(v_{GS}(t) - V_{th}) \tag{3.16}$$

According to the equivalent circuit in Figure 3.23, the circuit equations take the form:

$$I_G = C_{GD}\frac{dv_{GD}}{dt} + C_{GS}\frac{dv_{GS}}{dt} \tag{3.17}$$

and

$$v_{GD} = v_{GS} - v_{DS} \tag{3.18}$$

So

$$I_G = (C_{GS} + C_{GD})\frac{dv_{GS}}{dt} - C_{GD}\frac{dv_{DS}}{dt} \tag{3.19}$$

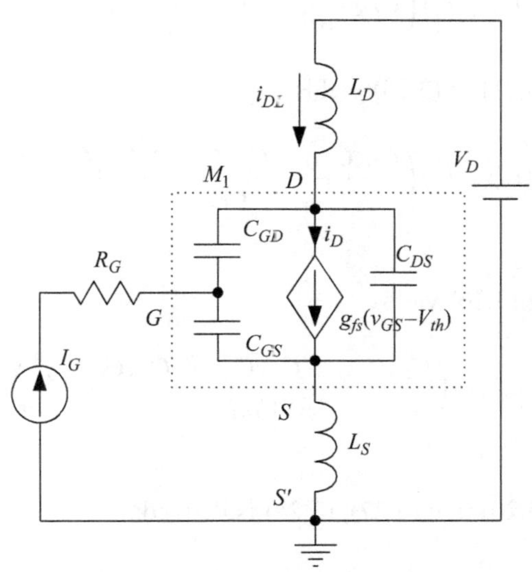

Figure 3.23 Equivalent circuit of MOSFET switching transition

From (3.19), dv_{DS}/dt is solved as

$$\frac{dv_{DS}}{dt} = \frac{C_{GS} + C_{GD}}{C_{GD}} \cdot \frac{dv_{GS}}{dt} - \frac{I_G}{C_{GD}} \tag{3.20}$$

So d^2v_{DS}/dt^2 and d^3v_{DS}/dt^3 are respectively

$$\frac{d^2v_{DS}}{dt^2} = \frac{C_{GS} + C_{GD}}{C_{GD}} \cdot \frac{d^2v_{GS}}{dt^2} \tag{3.21}$$

$$\frac{d^3v_{DS}}{dt^3} = \frac{C_{GS} + C_{GD}}{C_{GD}} \cdot \frac{d^3v_{GS}}{dt^3} \tag{3.22}$$

During the switching interval, the change of the switching loop current i_{DL} induces a voltage across the parasitic inductance. The drain-to-source voltage v_{DS} is given as

$$v_{DS} = V_D - L_D\frac{di_{DL}}{dt} - L_s\frac{d(i_{DL} + I_G)}{dt} = V_D - (L_D + L_s)\frac{di_{DL}}{dt} \tag{3.23}$$

And

$$i_{DL} = C_{GS}\frac{dv_{GS}}{dt} + C_{DS}\frac{dv_{DS}}{dt} + g_{fs}(v_{GS} - V_{th}) - I_G \tag{3.24}$$

Substituting (3.24) to (3.23) yields

$$v_{DS} = V_D - (L_D + L_s)\left(C_{GS}\frac{d^2v_{GS}}{dt^2} + C_{DS}\frac{d^2v_{DS}}{dt^2} + g_{fs}\frac{dv_{GS}}{dt}\right) \tag{3.25}$$

Substituting (3.21) to (3.25) yields

$$v_{DS} = V_D - (L_D + L_s)\left(\frac{C_{GS}C_{GD} + C_{DS}C_{GD} + C_{DS}C_{GS}}{C_{GD}} \cdot \frac{d^2v_{GS}}{dt^2} + g_{fs}\frac{dv_{GS}}{dt}\right) \tag{3.26}$$

Differentiating (3.26) yields

$$\frac{dv_{DS}}{dt} = -(L_D + L_s)\left(\frac{C_{GS}C_{GD} + C_{DS}C_{GD} + C_{DS}C_{GS}}{C_{GD}} \cdot \frac{d^3v_{GS}}{dt^3} + g_{fs}\frac{d^2v_{GS}}{dt^2}\right) \tag{3.27}$$

Substituting (3.20) into (3.27), (3.28) is derived:

$$A\frac{d^3v_{GS}(t)}{dt^3} + B\frac{d^2v_{GS}(t)}{dt^2} + C\frac{dv_{GS}(t)}{dt} = I_G \tag{3.28}$$

where parameters A, B and C are represented in terms of the device parameters (C_{GS}, C_{GD}, C_{DS}, g_{fs}, and R_G) and the equivalent circuit parameters (L_D and L_S) as $A = (L_D + L_s)(C_{GS}C_{GD} + C_{DS}C_{GD} + C_{DS}C_{GS})$, $B = g_{fs}(L_D + L_s)C_{GD}$ and $C = C_{GS} + C_{GD}$.

For turn-on transition, the initial condition for (3.28) is $v_{GS}(0) = V_{th}$. Then (3.28) is solved to give either sinusoidal or exponential solutions, depending on the relative magnitudes of B^2 and AC.

When $B^2 - 4AC < 0$, sinusoidal solution occurs and $v_{GS}(t)$ takes the form:

$$v_{GS}(t) = \frac{B}{C^2} \cdot I_G \cdot \exp\left(-\frac{t}{T_2}\right) \cdot \cos(\omega_1 t) + \left(\frac{B^2}{2C^2 \cdot \sqrt{4AC - B^2}} - \frac{\sqrt{4AC - B^2}}{2C^2}\right) \cdot I_G \cdot \exp\left(-\frac{t}{T_1}\right) \cdot \sin(\omega_1 t) + \frac{I_G \cdot t}{C} - \frac{B}{C^2} \cdot I_G + V_{th} \tag{3.29}$$

where $T_1 = \frac{2A}{B}$, $\omega_1 = \frac{\sqrt{4AC - B^2}}{2A}$.

When $B^2 - 4AC > 0$, exponential solution occurs. Then $v_{GS}(t)$ takes the form:

$$v_{GS}(t) = -\frac{\left(\sqrt{B^2 - 4AC} + B\right) \cdot I_G \cdot A \cdot \exp\left(-\frac{t}{T_2}\right)}{\left(\sqrt{B^2 - 4AC} - B\right) \cdot C \cdot \sqrt{B^2 - 4AC}} + \frac{\left(\sqrt{B^2 - 4AC} - B\right) \cdot I_G \cdot A \cdot \exp\left(-\frac{t}{T_3}\right)}{\left(\sqrt{B^2 - 4AC} + B\right) \cdot C \cdot \sqrt{B^2 - 4AC}} + \frac{I_G \cdot t}{C} - \frac{B \cdot I_G}{C^2} + V_{th}, \tag{3.30}$$

where $T_2 = \frac{2A}{B - \sqrt{B^2 - 4AC}}$, $T_3 = \frac{2A}{B + \sqrt{B^2 - 4AC}}$.

Then, by substituting $v_{GS}(t)$ to (3.16) and (3.26), $i_D(t)$ and $v_{DS}(t)$ of the MOSFET can be calculated respectively. The turn-off transition is similar to the turn-on transition except for the initial condition becomes $v_{GS}(0) = V_{th} + \frac{I_L}{g_{fs}}$.

3.3.2 Analytical modeling and simulation results

The modeling results in Section 3.1.1 are presented in this section. The turn-on and turn-off transients are divided into several intervals, during which the gate-to-source voltage $v_{GS}(t)$, the drain current $i_D(t)$ and the drain voltage $v_{DS}(t)$ can be calculated analytically with corresponding boundary conditions and constraints. Once the instantaneous waveforms of $v_{GS}(t)$, $i_D(t)$, and $v_{DS}(t)$ are solved, the switching transition time and the switching loss can be easily obtained.

In the experimental prototype, MOSFET Si7860 from Vishay is used and the circuit specifications and the device parameters are listed in Table 3.4. The estimated value of the parasitic inductances are $L_s = 1$ nH and $L_D = 2$ nH by Maxwell

Table 3.4 Circuit specifications and device parameters in analytical modeling

$V_D = 12$ V, $I_L = 20$ A, $f_s = 1$ MHz						
C_{GS} (pF)	C_{GD} (pF)	C_{DS} (pF)	V_{th} (V)	g_{fs} (S)	R_G: current driven (Ω)	R_G: voltage driven (Ω) (external and internal)
1600	200	500	1.8	60	1	1.5

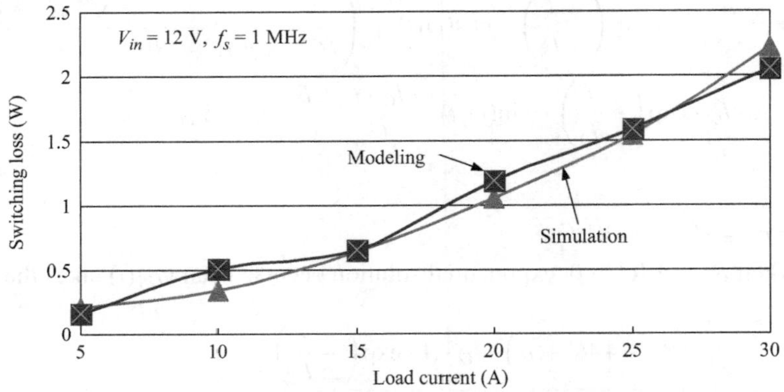

Figure 3.24 Switching loss comparison between modeling and simulation

simulation software. In this case, since B^2 is more than $4AC$ depending on the above parameters, the exponential solution occurs as (3.30).

Figure 3.24 shows the switching loss comparison between the above model using Mathcad software and the simulation results based on Si7860AD SPICE model provided by Vishay. It is noted that the modeling results are in good agreement with the simulation results.

However, as for a current source resonant driver, the great advantage is that the common source inductance is absorbed by the resonant inductor to ensure the constant drive current during the switching transition. Therefore, the propagation impact of the common source inductance on the switching transition is eliminated, which leads to a significant reduction of the switching time and the switching loss.

Figure 3.25 gives the comparison of the turn-on and turn-off losses as a function of the common source inductance on the basis of above analytical loss model. With today's MOSFET packaging and compact PCB layout, the parasitic inductance is usually less than 2 nH and the parasitic values can be extracted by Maxwell simulation software. It is observed that with the resonant driver, the common source inductance L_s does not increase the switching loss of the MOSFET. Therefore, the resonant driver significantly reduces the switching loss compared to a conventional voltage source driver due to the common source inductance.

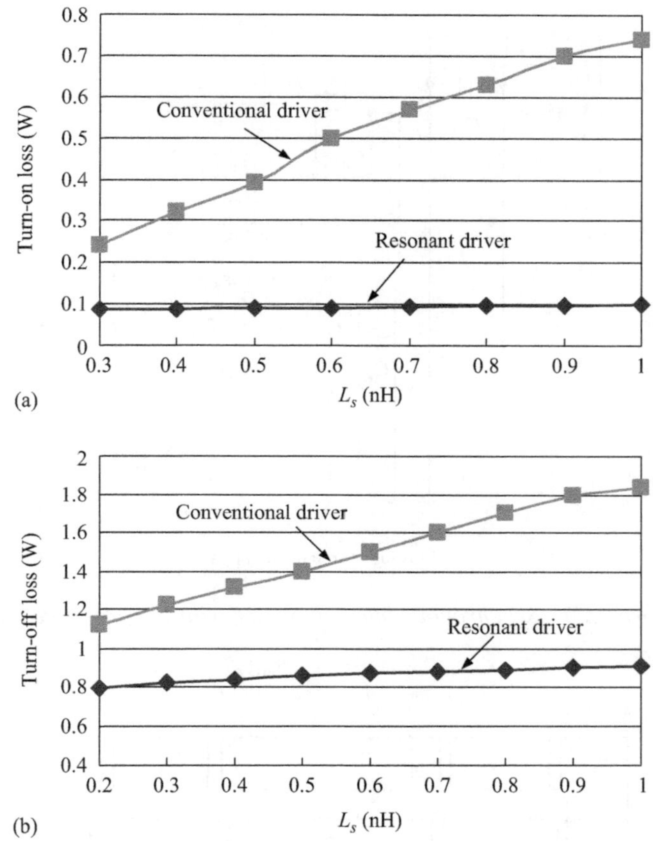

Figure 3.25 Switching loss comparison as function with different common source inductances ($V_{in} = 12$ V, $I_o = 20$ A, $f_s = 1$ MHz, $L_D = 2$ nH); (a) turn-on loss and (b) turn-off loss

3.3.3 Proposed optimal design with the accurate switching loss model

Figure 3.26 shows the resonant driver under investigation. In next section, this resonant driver is optimized for a buck converter operating at 1 MHz switching frequency and 12 V input.

Figure 3.27 shows the key waveforms. C_{g1} and C_{g2} are the gate-source capacitors of the control FET Q_1 and the synchronous FET Q_2, respectively, and they are not external capacitors. In Figure 3.27, the gate-to-source signals of Q_1 and Q_2 have a small crossover level, which is less than the threshold voltage of the devices, to minimize the dead time and reduce the conduction loss and reverse recovery of the body diode. It should also be pointed that an additional margin should be set for this crossover level according to the threshold voltage at high temperature, which reduces with temperature increasing.

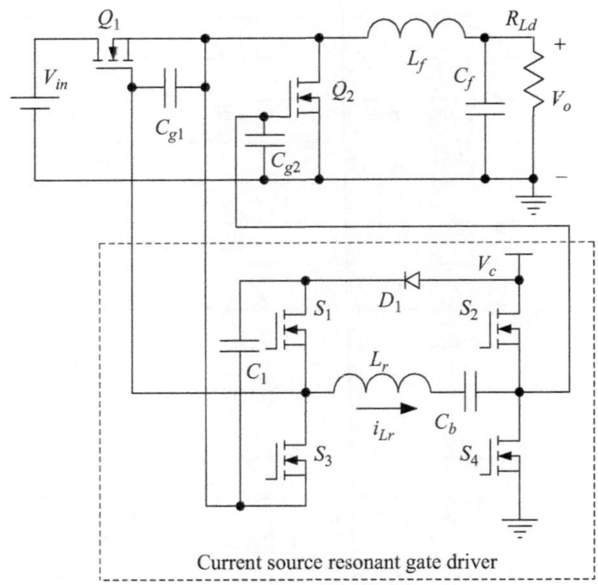

Figure 3.26 Synchronous buck converter with the proposed CSD

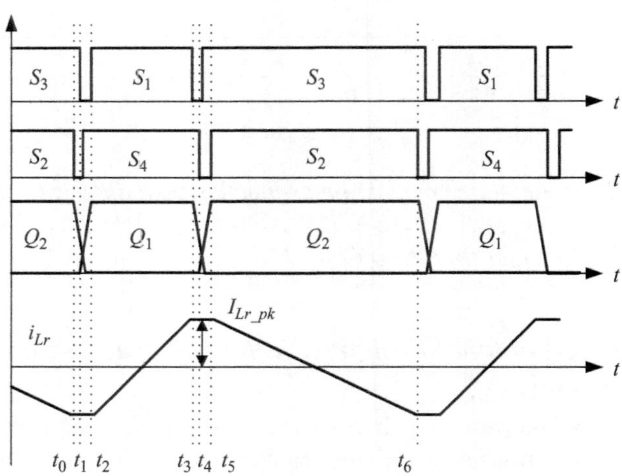

Figure 3.27 Key waveforms

In order to achieve design optimization, the loss analysis of the CSD is given. The total power loss of the proposed resonant driver includes: (1) the resistive loss and the gate-drive loss of switches S_1–S_4; (2) the loss of the resonant inductor; (3) the resistive loss caused by the internal gate mesh resistance of the power MOSFETs. The detail loss analysis was presented in [5].

As seen from the principle of operation in Figure 3.27, the peak current I_{Lr_pk} of the resonant inductor L_r is regarded as the current source magnitude I_G. So the higher I_{Lr_pk} is, the shorter of switching transition is, thus more switching loss can be saved. On the other hand, higher I_{Lr_pk} will result in a larger RMS value of the inductor circulating current i_{Lr} since the waveform of i_{Lr} is triangular, which increases the resistive circulating loss in the drive circuit and decreases the gate energy recovery efficiency. Therefore it is critical to decide I_{Lr_pk} (i.e., I_G) properly so that the maximum loss saving can be achieved.

The general method proposed here is to find the optimal solution on the basis of the object function that adds the switching loss and the resonant drive circuit loss together. The object function should be a U-shaped curve as function of the drive current I_G, and the optimization solution is simply located at the lowest point of the curve. It is noted that the analytical loss model proposed in Section 3.3.1 is used to calculate the switching loss since the piecewise loss model is no longer valid due to the parasitic inductance at high frequency. The demonstration of the optimization methodology is employed to the new CSD. The specifications are: $V_{in} = 12$ V; $V_o = 1.5$ V; $I_o = 30$ A; $V_c = 8$ V; $f_s = 1$ MHz; Q_1: Si7860DP; Q_2: Si7336ADP.

First, the switching loss of the control FET as function of drive current I_G is given in (3.31):

$$P_{Q1} = \int_0^{t_{sw(on)\_Q1}} v_{ds(on)\_Q1} \cdot i_{d(on)\_Q1} dt \cdot f_s + \int_0^{t_{sw(off)\_Q1}} v_{ds(off)\_Q1} \cdot i_{d(off)\_Q1} dt \cdot f_s \tag{3.31}$$

where $v_{ds(on)\_Q1}$ and $v_{ds(off)\_Q1}$ are the drain-to-source voltages during turn-on interval and turn-off interval, respectively; $i_{(on)\_Q1}$ and $i_{d(off)\_Q1}$ are the drain currents at turn-on interval and turn-off interval, respectively; $t_{sw(on)\_Q1}$ is the turn-on switching transition time and $t_{sw(off)\_Q1}$ is the turn-off switching transition time. In Figure 3.28, the switching loss $P_{switching}$ as function of driven current I_G is given, illustrating that the switching loss reduces when I_G increases.

Secondly, the total loss of the CSD as function of drive current I_G is calculated. In Figure 3.28, the loss of the total gate-drive circuit $P_{circuit}(I_G)$ as function of drive current I_G is given, illustrating that the resonant drive circuit loss increases when I_G increases.

Thirdly, in order to find the optimized gate-drive current, the objective function is established by adding the switching loss and the resonant driver circuit loss together as

$$F(I_G) = P_{circuit}(I_G) + P_{switching}(I_G) \tag{3.32}$$

In Figure 3.28, the objective function $F(I_G)$ with the drive current I_G is given, which is a U-shaped curve. Therefore, the optimization solution can be found at the lowest point of the curve, and accordingly, the gate-drive current I_G is chosen as 1.2 A.

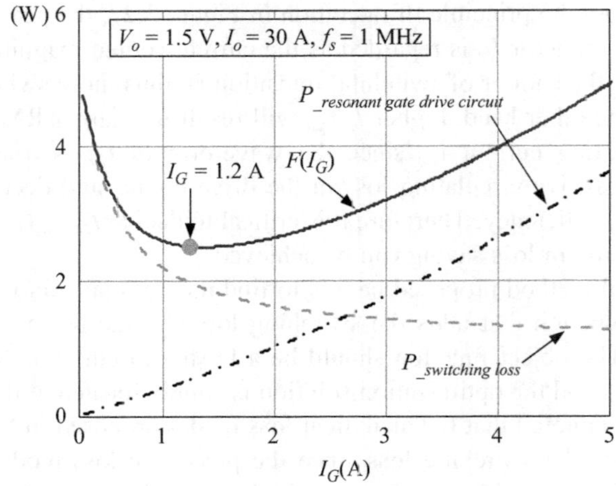

Figure 3.28 Objective function $F(I_G)$ as function of current I_G

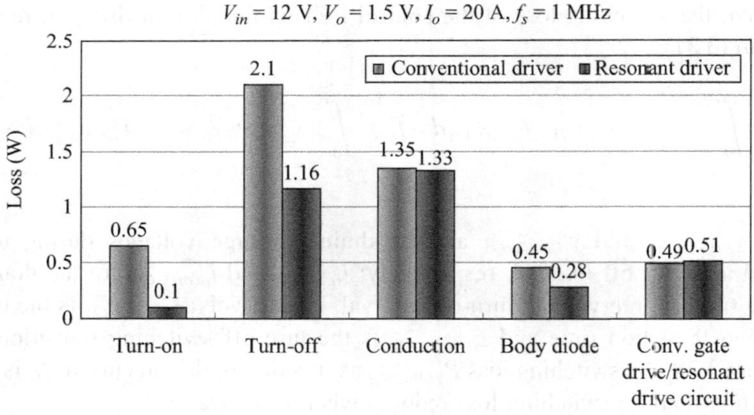

Figure 3.29 Loss breakdown between the CSD and the conventional gate driver (V_{in} = 12 V, V_o = 1.5 V and I_o = 20 A)

Finally, from the selected gate-drive current, the calculated resonant inductor value from (3.33) is 1.5 μH:

$$L_r = \frac{(V_{in} + 2V_c) \cdot D \cdot (1 - D)}{2 \cdot I_G \cdot f_s} \tag{3.33}$$

where V_{in} = 12 V, f_s = 1 MHz, V_o = 1.5 V, V_c = 8 V and I_G = 1.2 A.

Figure 3.29 illustrates the loss breakdown comparison based on the analytical loss model between the above optimized resonant driver and the conventional diver. At V_o = 1.5 V and I_o = 20 A, the turn-on loss is reduced by 0.55 W and the

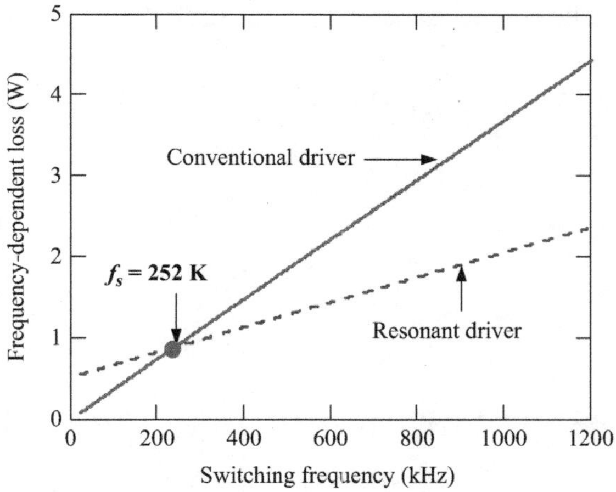

Figure 3.30 Loss comparison between resonant driver and conventional driver with different switching frequencies (V_{in} = 12 V, V_o = 1.5 V and I_o = 20 A)

turn-off loss is reduced by 0.94 W. The total loss reduction is 1.66 W, which is 5.5% (1.66 W/1.5 V/20 A) of the output power.

Figure 3.30 shows the loss comparison between the resonant driver and the conventional driver as function of the switching frequency based on the above circuit parameters. It is interesting to observe that as long as the switching frequency is above 252 kHz, the loss of the resonant driver is always lower than that of the conventional driver, which means that the resonant driver can always achieve loss saving over the conventional driver. The higher switching frequency, the more loss saving we have. However, when the switching frequency is lower than 252 kHz, the loss of the resonant driver is higher than that of the conventional driver. The reason is that though the frequency-dependent loss is reduced, it is offset by the circulating loss of the resonant driver circuit itself, which means the resonant drive cannot achieve the loss saving when the switching frequency is below 252 kHz.

3.3.4 Experimental verification and discussion

A 1 MHz synchronous buck converter with the new resonant driver was built to verify the modeling results and demonstrate the advantages of the proposed resonant driver. The specifications are as follows: input voltage V_{in} = 12 V; output voltage V_o = 1.5 V; output current I_o = 30 A; switching frequency f_s = 1 MHz; resonant driver voltage V_c = 8 V. The PCB uses six-layer 2 oz copper. The components used in the circuit are listed as follows: Control FET Q_1: Si7860DP (30 V N-channel, $R_{DS(on)}$ = 11 mΩ@V_{GS} = 4.5 V, Vishay), Synchronous FET Q_2: Si7336ADP (30 V N-channel, $R_{DS(on)}$ = 4 mΩ@V_{GS} = 4.5 V, Vishay), Drive switches S_1–S_4: FDN335N (20 V N-channel, $R_{DS(on)}$ = 70 mΩ@V_{GS} = 4.5 V,

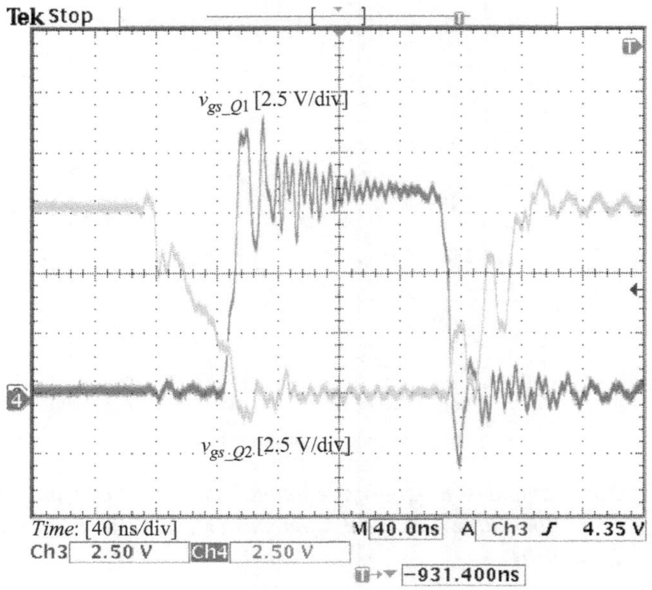

Figure 3.31 Waveforms of the gate signals v_{gs_Q1} (control FET) and v_{gs_Q2} (synchronous FET)

Fairchild), Output filter inductance: $L_f = 330$ nH ($R = 1.3$ mΩ, IHLP-5050CE-01, Vishay), Resonant inductor: $L_r = 1.5$ μH.

Figure 3.31 shows the gate-drive signal v_{gs_Q1} of the control FET Q_1 and v_{gs_Q2} of the synchronous FET Q_2. The crossover level of these two gate signals is less than the threshold voltage of the switches so that the dead time can be minimized significantly and the shoot-through can also be avoided. It is observed that v_{gs_Q1} is smooth and the miller plateau is less than 5 ns due to the constant charging current. Moreover, the rise time and fall time of v_{gs_Q1} is less than 15 ns, which indicates the gate charging time is significantly reduced compare to a conventional driver. Figure 3.32 shows the drain-source voltage v_{ds_Q2} and the gate signal v_{gs_Q2} of the synchronous FET. It can be seen from v_{ds_Q2} that the body diode conduction time is small, which reduces the conduction loss and the reverse recovery loss of the body diode significantly.

Figure 3.33 shows the gate-drive current i_g and the gate signal v_{gs_Q2} of the synchronous FET. It is noted that the gate-drive current keeps constant during the switching interval disregarding the current oscillations. Because the extra wire length is used to allow the insertion of a current probe to measure the current waveforms, this introduces higher stray inductances, which causes the parasitic oscillations of the current waveform at high frequency. Figure 3.34 shows the resonant inductor current i_{Lr} and its peak current value is 1.2 A, which is the optimized value of the drive current.

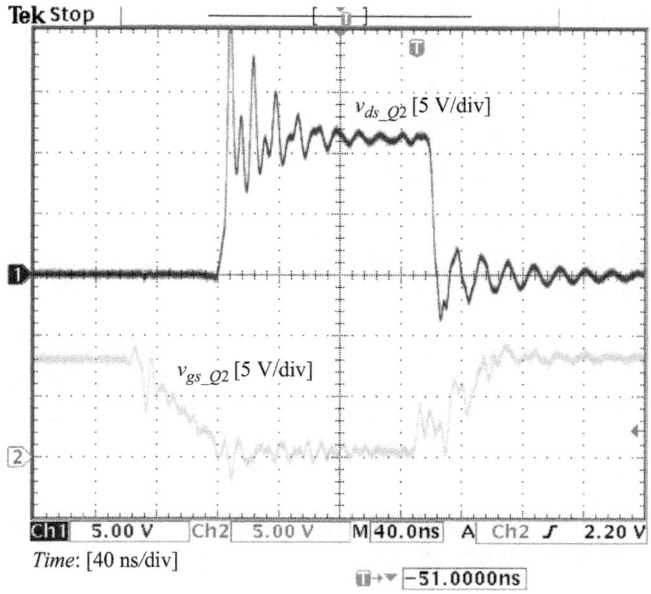

Figure 3.32 Waveforms of the drain-source voltage v_{ds_Q2} and the gate signal v_{gs_Q2} (synchronous FET)

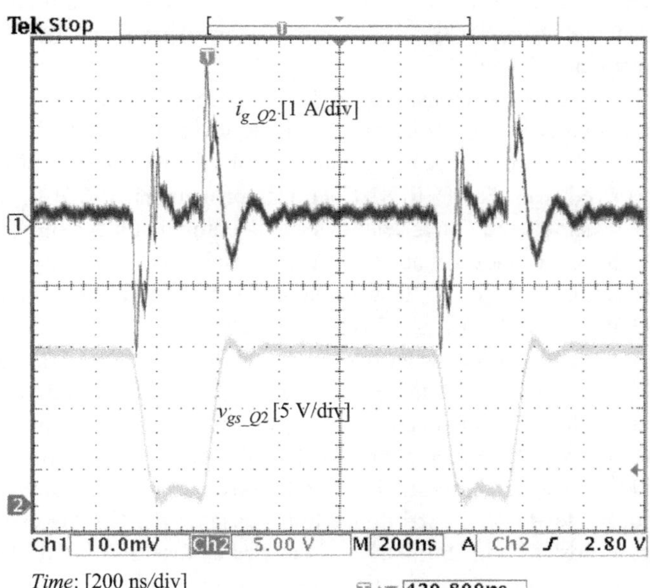

Figure 3.33 Waveforms of the gate drive current i_g and the gate signal v_{gs_Q2} (synchronous FET)

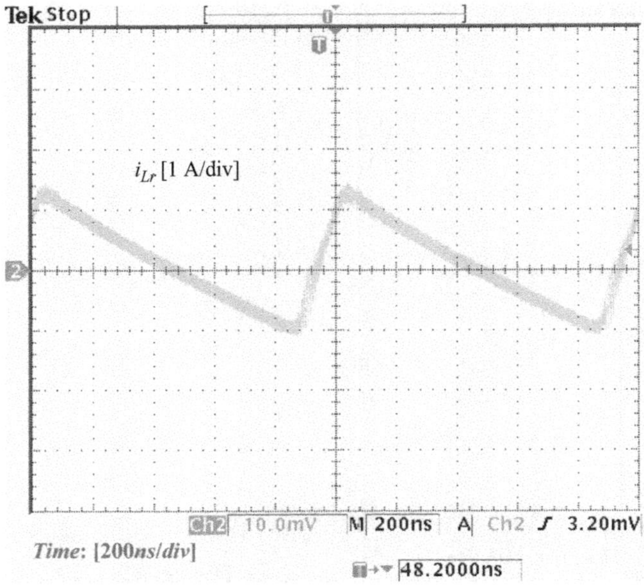

Figure 3.34 Waveform of the resonant inductor current i_{Lr}

In order to illustrate efficiency improvement by the resonant driver, a benchmark of a synchronous buck converter with the conventional gate driver was built. A Predictive Gate Drive UCC 27222 from Texas Instruments was used as the conventional voltage driver.

Figure 3.35 shows the measured efficiency comparison for the CSD and the conventional gate driver at 1.5 V output. It is observed that at 20 A, the efficiency is improved from 82.7% to 86.6% (an improvement of 3.9%) and at 30 A, the efficiency is improved from 76.9% to 83.6% (an improvement of 6.7%).

Figure 3.36 shows the converter power loss comparison for the CSD and the conventional driver. It shows that at 20 A, the total loss is reduced from 6.28 to 4.64 W, a reduction of 26%. At 30 A output current, the total loss is reduced from 13.52 to 8.83 W, a reduction of 34.7%.

Another interesting observation from Figure 3.36 is that for same power loss of 6.4 W, the buck converter with the conventional gate drive can only provide 20 A output current, while the buck converter with the CSD can provide 25 A (an improvement of 25%). In other words, if the total voltage regulator (VR) output current is 100 A, we need five phases buck converters if the conventional gate driver is used, while we only need to use four phases if the CSD is used. This will be a significant cost saving. Similarly, if the power loss is limited at 8.8 W, the buck converter with the conventional driver can only provide about 24 A while with the CSD the converter will be able to provide 30 A (an improvement of 25%). In other words, if the total load current is 150 A, we will need six (150/24 = 6.25)

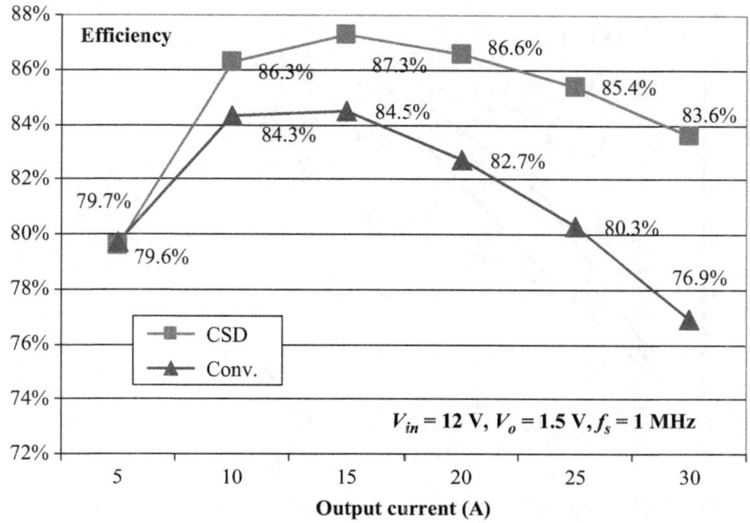

Figure 3.35 Efficiency comparison at 1.5 V/30 A condition: top, current source gate driver (CSD); bottom: conventional driver (Conv.)

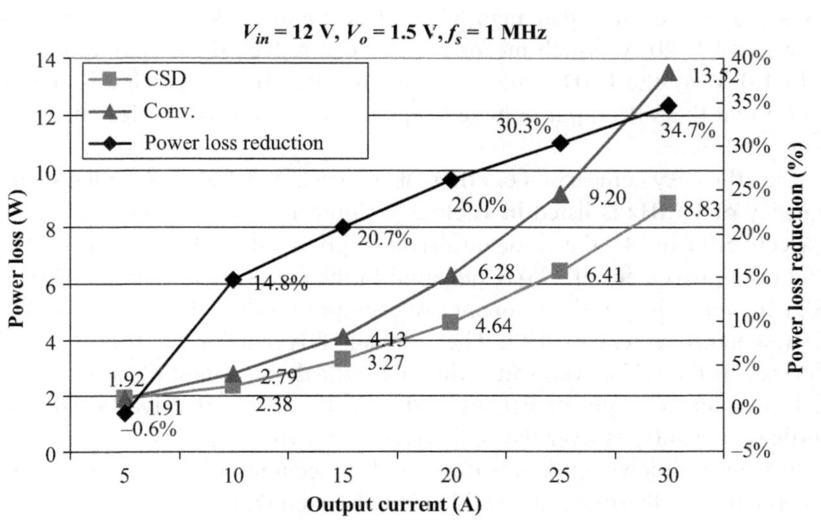

Figure 3.36 Power loss comparison at 1.5 V/30 A condition: top, current source gate driver (CSD), bottom, conventional driver (Conv.)

phases for the conventional gate driver while we only need five (150/30 = 5) phases for the CSD.

Figure 3.37 shows the measured efficiency for the CSD at different output voltages and load currents. It is observed that at 1.0 V/20 A, the efficiency is

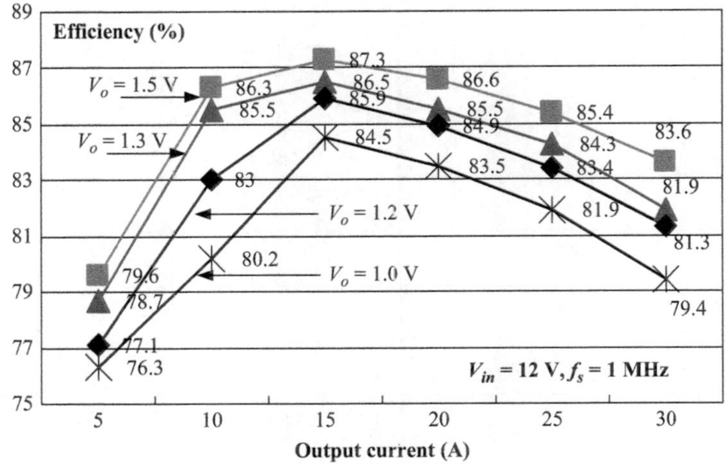

Figure 3.37 Efficiency with different output voltages and currents with the resonant driver

83.5% and it is even higher than 82.7% (see Figure 3.35) of the conventional driver at 1.5 V/20 A, which means that we can reduce the output voltage from 1.5 to 1.0 V by the CSD without penalizing the efficiency. This is important because that the VR output voltage keeps reducing and will be less than 1 V in the near future.

The efficiency comparison of different approaches of 12 V VRs at the switching frequency of 1 MHz is listed in Table 3.5. Compared to the tapped-inductor (TI) buck converter in [4], the resonant driver improves the efficiency from 84% to 87% (an improvement of 3%). Compared to the soft-switching phase-shift buck (PSB) converter in [6], the resonant driver improves the efficiency from 82% to 86% (an improvement of 4%). The resonant driver achieves almost the same efficiency as the self-driven soft-switching buck-derived multiphase converter in [7]. But in terms of power density and cost, the resonant driver approach has significant advantages over the self-driven soft-switching buck converter which requires an additional bulk transformer that occupies plenty of space on the motherboard. Furthermore, it should be emphasized that the resonant driver does not change the multiphase buck architecture of today's VRs featuring lower cost and simple control while improving the efficiency in a cost-effective manner. However, other buck derived approaches requiring additional magnetic components not only reduce the power density significantly, but also increase the cost and the complexity of the circuits and control scheme. In particular, it should also be mentioned that an efficiency improvement of 1.6% at 1.5 V/20 A is also achieved over the Toshiba synchronous buck Multi Chip Module using semiconductor integration approach to minimize the parasitic inductances [8].

Table 3.5 Efficiency comparison between the resonant driver and different state of the art VR approaches

Input voltage: 12 V and switching frequency: 1 MHz			
VR topologies	**Conversion efficiency (%)**	**Output current/phase (A)**	**Output voltage (V)**
Tapped-inductor (TI) buck converter [4]	84	12.5	1.5
Toshiba synchronous buck Multi Chip Module (TB7001FL) [8]	85	20	1.5
Proposed resonant driver (Figure 3.37)	87	12.5	1.5
	86.6	20	1.5
Soft-switching phase-shift buck (PSB) converter [6]	82	17.5	1.3
Self-driven soft-switching buck-derived multiphase converter [7]	84.7	25	1.3
Proposed resonant driver (Figure 3.37)	86	17.5	1.3
	84.3	25	1.3

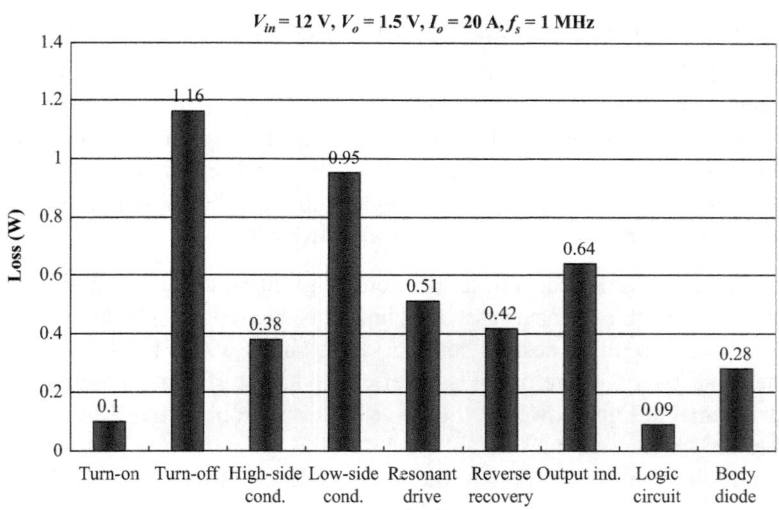

Figure 3.38 Loss breakdown

Figure 3.38 gives the total loss breakdown of the buck VR with the resonant driver. Figure 3.39 shows the measured efficiency and the analytical efficiency based on the loss model. It can be seen that the modeling results matches the experimental result well.

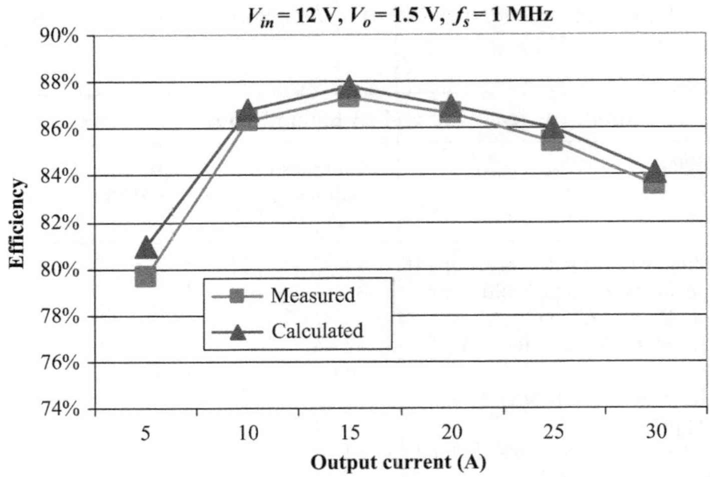

Figure 3.39 Loss model verification with experimental results

3.4 High-side and low-side current source drivers

3.4.1 Problem of high-side and low-side current source drivers

In particular, the current-source driver as shown in Figure 3.26 for a buck converter significantly reduces the switching loss since the impact of the parasitic inductances, especially common source inductance, can be significantly reduced. This gate-drive circuit features simplicity, low cost and efficiency improvement. However, carefully investigating the equivalent circuits of operation, it is noted that this driver circuit in Figure 3.26 has several drawbacks:

1. In a buck converter for VR applications, high-drive current is desired for the control MOSFET to ensure fast switching speed to reduce switching loss, while low-drive current is desired for the synchronous MOSFET to achieve gate energy recovery. Therefore, it is beneficial to have different gate-drive currents for optimal design. However, this drive circuit can only provide identical drive currents for the two buck MOSFETs;
2. Due to the reverse recovery of the body diode, the switching node has severe oscillation as shown in Figure 3.40. This switching node is actually in series with the level-shift capacitor of the drive circuit, which makes the circuit sensitive to the reverse recovery noise in practical applications;
3. The resonant inductor current flows through the control MOSFET and synchronous MOSFET, which causes additional conduction loss;
4. To enhance light load efficiency, the switching frequency can be reduced or diode emulation can be used to turn off the synchronous MOSFET to allow discontinuous conduction mode by detecting when the inductor current reaches zero. However, it is difficult to achieve these advanced features using this drive topology.

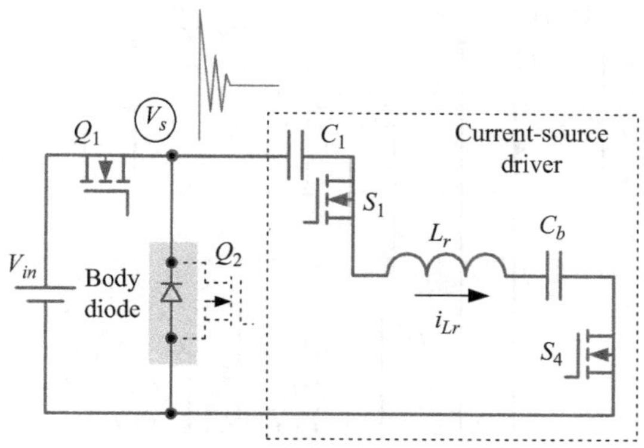

Figure 3.40 Equivalent circuit during switching transition

In order to solve the above problems and improve the driver's performance, a new current-source gate driver is introduced in the next section.

3.4.2 Proposed decoupled high-side and low-side current source drivers

One objective of the proposed current-source driver is to achieve independent drive control for the control and synchronous MOSFETs to achieve optimal performance. For the control MOSFET, the optimal design involves a tradeoff between switching loss and drive circuit loss; while for the SR MOSFET, optimal design involves a tradeoff between body diode conduction loss and drive circuit loss. Moreover, the drive currents should not go through the main power MOSFETs.

All the above features can be achieved by the proposed new drive circuit as shown in the dotted area in Figure 3.41. In Figure 3.41, Q_1 is the control MOSFET and Q_2 is the synchronous MOSFET in a buck converter. Figure 3.42 gives the key waveforms.

3.4.2.1 Principle of operation

In Figure 3.41, there are two sets of the drive circuits (CSD #1 and CSD #2) and each of them has the structure of the HB topology, consisting of drive MOSFETs S_1–S_2 and S_3–S_4, respectively.

It is noted that the drive MOSFETs (S_1–S_4) are similar to those in a conventional driver integrated circuit (driver IC), which means S_1–S_4 can also be easily implemented in a driver IC with standard complementary metal oxide semiconductor (CMOS) technology to achieve low profile and high power density with low footprint packages such as small outline integrated circuit (SOIC) package or Quad Flat No-lead (QFN) package, which are pin-to-pin compatible with conventional driver IC. The new current source gate-driver circuit is analyzed with discrete components and the timing is tuned using digital complex programmable logic device (CPLD). If the drive circuit is integrated into an IC chip, the circuit timing can also be optimized.

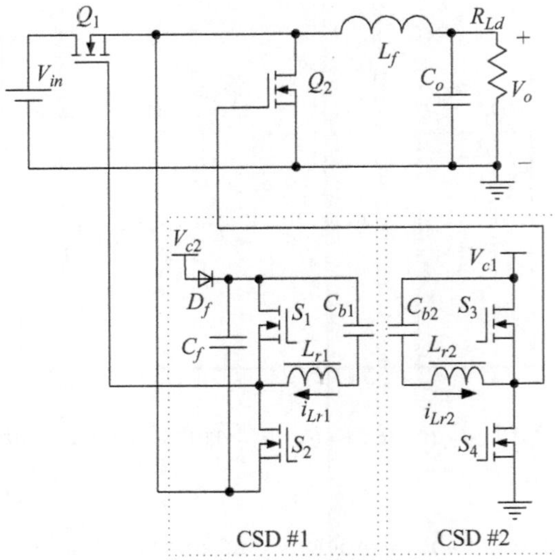

Figure 3.41 Buck converter with proposed current-source driver

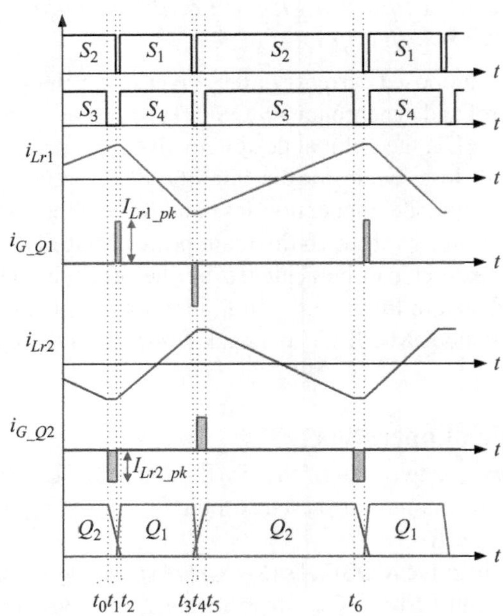

Figure 3.42 Key waveforms of the new drive circuit

As illustrated in Figure 3.42, S_1 and S_2 are switched out of phase with complimentary control to drive Q_1, while S_3 and S_4 are switched out of phase with complimentary control to drive Q_2. With complimentary control, all the drive switches can achieve ZVS. Driver #1 can also be regarded as a level-shift version of driver #2. V_{c1} and V_{c2} are the drive voltages and they could use the same, or independent drive voltages if desired. The diode D_f provides the path to charge C_f to the voltage of the drive voltage V_{c2}. C_{b1} and C_{b2} are the blocking capacitors.

There are six switching modes in a switching period and the equivalent circuits are given in Figure 3.43 accordingly. D_1–D_4 are the body diodes and C_1–C_4 are the intrinsic drain-to-source capacitors of S_1–S_4, respectively. C_{gs1} and C_{gs2} are intrinsic gate-to-source capacitors of Q_1 and Q_2, respectively. The switching transitions of charging and discharging C_{gs1} and C_{gs2} are during the interval of $[t_0, t_2]$ and $[t_3, t_5]$. The peak currents i_{G_Q1} and i_{G_Q2} during $[t_0, t_2]$ and $[t_3, t_5]$ are constant during switching transition, which ensures fast charging and discharging of the MOSFET Q_1 gate capacitor including the Miller capacitor.

1. Mode 1 $[t_0, t_1]$ (Figure 3.43(a)): Prior to t_0, S_2 and S_3 conduct and the inductor current i_{Lr1} increases in the positive direction, while i_{Lr2} increases in the negative direction. Q_2 is on. At t_0, S_3 turns off. i_{Lr2} charges C_3 and discharges C_4 plus C_{gs2} simultaneously. Due to C_3 and C_4, S_3 achieves zero-voltage turn-off. The voltage of C_3 rises linearly and the voltage of C_4 decays linearly.
2. Mode 2 $[t_1, t_2]$ (Figure 3.43(b)): At t_1, v_{c3} rises to V_{c1} and v_{c4} decays to zero. The body diode D_4 conducts and S_4 turns on with zero-voltage condition. The gate-to-source voltage of Q_2 is clamped to ground through S_4. At t_1, S_2 turns off. i_{Lr1} charges C_2 plus the input capacitor C_{gs1} and discharges C_1 simultaneously. Due to C_1 and C_2, S_2 is zero-voltage turn-off. The voltage of C_2 rises linearly and the voltage of C_1 decays linearly.
3. Mode 3 $[t_2, t_3]$ (Figure 3.43(c)): At t_2, v_{c2} rises to V_{c2} and v_{c1} decays to zero. The body diode D_1 conducts and S_1 turns on under zero-voltage condition. The gate-to-source voltage of Q_1 is clamped to V_{c2} through S_1. i_{Lr1} and i_{Lr2} both decrease.
4. Mode 4 $[t_3, t_4]$ (Figure 3.43(d)): Before t_3, i_{Lr1} and i_{Lr2} each changed polarity. At t_3, S_4 and S_1 turns off. i_{Lr2} charges C_4 plus C_{gs2} and discharges C_3. Due to C_3 and C_4, S_4 achieves zero-voltage turn-off. The voltage of C_4 rises linearly and the voltage of C_3 decays linearly. i_{Lr1} charges C_1 and discharges C_2 plus C_{gs1} simultaneously. Due to C_1 and C_2, S_1 achieves zero-voltage turn-off.
5. Mode 5 $[t_4, t_5]$ (Figure 3.43(e)): At t_4, v_{c1} rises to V_{c2} and v_{c2} decays to zero. The body diode D_2 conducts and S_2 turns on with zero-voltage. The gate-to-source voltage of Q_1 is clamped to zero through S_2.
6. Mode 6 $[t_5, t_6]$ (Figure 3.43(f)): At t_5, v_{c4} rises to V_{c1} and v_{c3} decays to zero. The body diode D_3 conducts and S_3 turns on with zero-voltage. The gate-to-source voltage of Q_2 is clamped to V_{c1} through S_3.

3.4.2.2 Advantages of proposed current-source gate driver

The advantages of the new drive circuit are highlighted as follows:

1. Significant reduction of the switching transition time and switching loss
During the switching transition $[t_1, t_2]$ and $[t_3, t_4]$ (see Figure 3.42), the proposed current source driver uses the peak portion of the resonant inductor current to drive the control MOSFET and absorbs the common source

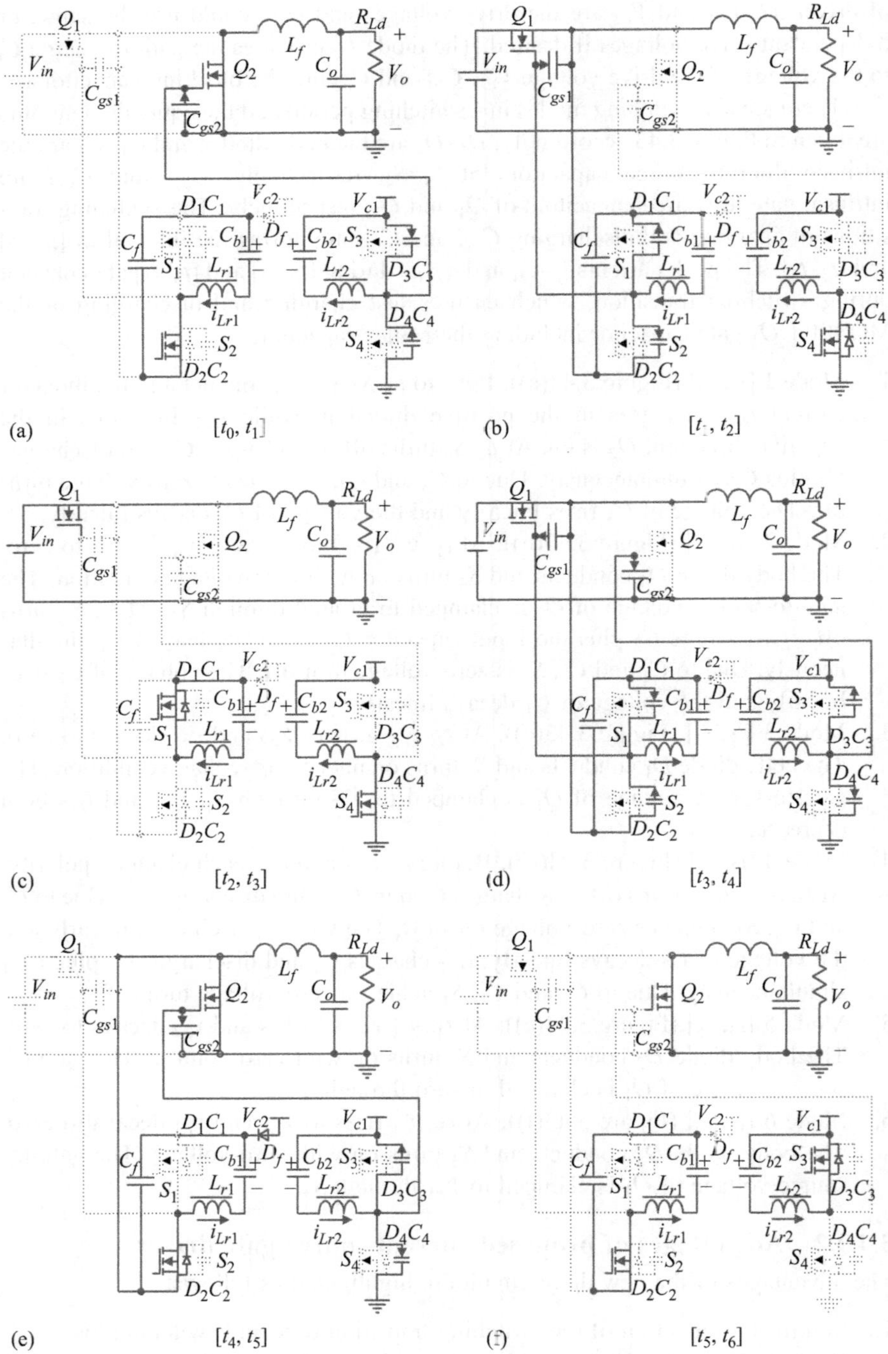

Figure 3.43 Equivalent circuits of operation

inductance. This significantly reduces the propagation impact of the parasitics during the switching transition, which leads to a reduction of the switching transition time and the switching loss.

2. Gate energy recovery
Current source drivers, using an inductor as a current source, store energy in the inductance, which can be recovered to the driver supply voltage rail. In the proposed driver, the inductor returns its stored energy to the line during intervals $[t_0, t_2]$ for the SR MOSFET, Q_2. Energy is also returned to the driver supply during the corresponding time intervals for the control MOSFET, Q_1. Owing to gate energy recovery, high gate-drive voltage can be used to further reduce $R_{DS(on)}$ conduction losses.
3. Reduced SR body diode conduction
Quick switching is also beneficial to minimize the dead time between the control and SR MOSFETs. As the gate-source voltage rises from the plateau voltage to V_{cc}, the $R_{DS(on)}$ of the SR is decreasing. Quick switching enables a fast transition to the minimum value of $R_{DS(on)}$ at $v_{gs} = V_{cc}$. This results in conduction loss savings and reverse recovery loss in the body diode of the SR MOSFET.
4. ZVS of the drive switches
Through the mode analysis, it is noted that all of the drive switches are able to achieve ZVS, which is beneficial for high frequency (i.e., >1 MHz).
5. High noise immunity
With the new drive circuit, the gate terminals of the buck MOSFETs (Q_1 and Q_2) are clamped to either the drive voltage source via a low impedance path (S_1 and S_3 with fairly small $R_{DS(on)}$) or their source terminals (S_2 and S_4). This offers high noise immunity and leads to the alleviation of *dv/dt* effect to prevent $C_{dv/dt}$ induced turn-on of the SR MOSFET.

3.4.2.3 Drive circuit design analysis

The two current-source drivers have similar driver losses, except each operates with different peak inductor currents. With CSD #1, the drive loss includes: (1) the resistive loss and gate-drive loss of drive switches S_1–S_2; (2) the loss of the resonant inductor L_{r1}; and (3) the resistive loss caused by the internal gate mesh resistance of the power MOSFETs.

Using volt-second balance across the resonant inductor, the DC voltage, v_{cb1}, across the blocking capacitor, C_{b1}, is given by

$$v_{Cb1} = (1 - D) \cdot V_{c2} \tag{3.34}$$

where D is the duty cycle of S_1 and V_{c2} is the drive voltage.

The blocking capacitor value is found using:

$$C_{b1} = \frac{I_{Lr1\_pk}}{4 \cdot k \cdot V_{c1} \cdot f_s} \tag{3.35}$$

where k is the percent ripple on C_{b1} and f_s is the switching frequency. For example, for $V_{c1} = 7$ V, $I_{Lr1\_pk} = 1.5$ A, $k = 5\%$ and $f_s = 1$ MHz, then $C_{b1} = 1.0$ μF should be used.

The relationship of the inductor value L_{r1} and the peak inductor current I_{Lr1_pk} is given by (3.36):

$$L_{r1} = \frac{V_{c1} \cdot D \cdot (1 - D)}{2 \cdot I_{Lr1\_pk} \cdot f_s} \tag{3.36}$$

By choosing the proper peak inductor current (i.e., drive current for the power MOSFET), the resonant inductor value can be obtained using (3.36).

3.4.2.4 Optimal design

One advantage of this new drive circuit is that it can drive the control and synchronous MOSFET independently with different gate currents to achieve an optimized design.

For the control MOSFET, Q_1, optimal design involves a tradeoff between the switching loss and the drive circuit loss. Following the optimal design procedure using the analytical loss model in Section 3.3, the gate-drive current can be decided. The sum of the switching loss and drive circuit loss can be plotted as $P_{Q1\_Optimal}$ and the optimal gate current, I_{G_Q1} can be determined from the graph (at the minimum point of $P_{Q1\_Optimal}$). For the given application and parameters in the experimental results section, the curves are given in Figure 3.44. In Figure 3.44, the optimal gate current is 1.5 A and the resonant inductor value is 1.0 μH using (3.36). In this example, $P_{switching} = 2.3$ W and $P_{Drive} = 0.5$ W.

The synchronous MOSFET, Q_2, operates with ZVS since its output capacitance is discharged to zero voltage before it turns on. Therefore, for the synchronous MOSFET, optimal design involves a tradeoff between body diode conduction loss and gate-drive loss.

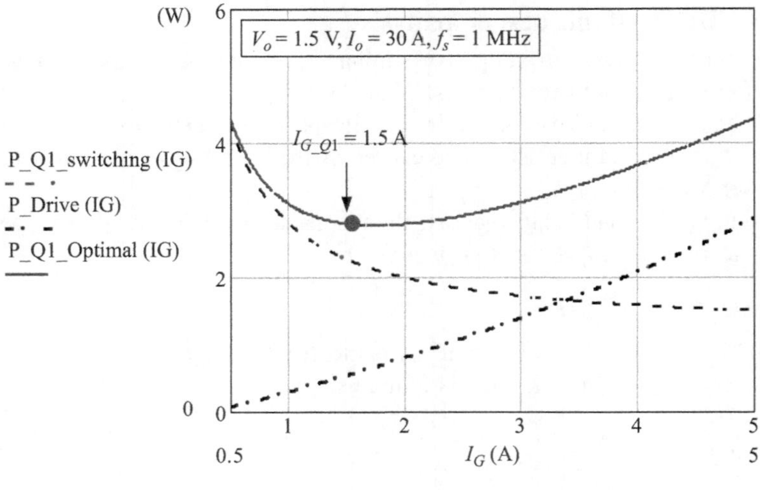

Figure 3.44 Optimization curves for the control MOSFET Q_1: power loss vs. gate current

The body diode conduction loss can be estimated as:

$$P_{body\_Q2} = V_{body\_Q2} \cdot I_o \cdot f_s \cdot t_{body} \tag{3.37}$$

where t_{body} is body diode conduction time, which can be estimated using (3.38). Equation (3.38) assumes that the body diode conducts during the interval when the gate voltage is between the threshold and until the gate voltage is large enough so that the $R_{DS(on)}$ of the synchronous MOSFET is less than about 20 mΩ, which means the voltage drop across the channel is less than the body diode drop. The values for Q_{g_Q2} (V_{20mohm}) and $Q_{g\_Q2}(V_{th\_Q2})$ can be estimated using the MOSFET manufacturer datasheets:

$$t_{body} = 2\left[\frac{Q_{gQ2}(V_{20mohm}) - Q_{gQ2}(V_{thQ2})}{I_{gQ2}}\right] \tag{3.38}$$

Using P_{Drive} and P_{body_Q2}, the sum of the two loss components can be plotted as $P_{Q2\_Optimal}$ and the optimal gate current, I_{G_Q2}, can be determined from the graph (at the minimum point of $P_{Q2\_Optimal}$). For the given application and parameters in the experimental results section, the curves are given in Figure 3.45. In Figure 3.45, the optimal gate current is 1.1 A and the resonant inductor value is 1.2 μH by using (3.36). In this example, $P_{Drive} = 0.33$ W, $P_{body\_Q2} = 0.58$ W, $t_{body} = 27$ ns.

Figure 3.46 illustrates the loss breakdown comparison between the new current-source driver with the optimal design and the conventional diver. At $V_o = 1.5$ V, $I_o = 20$ A and $f_s = 1$ MHz, the most significant loss reduction is the switching loss. The turn-on loss is reduced by 0.5 W and the turn-off loss is reduced by 0.7 W. The conduction loss and the body diode loss are also reduced by 0.05 W and 0.2 W, respectively. The total loss reduction is 1.45 W.

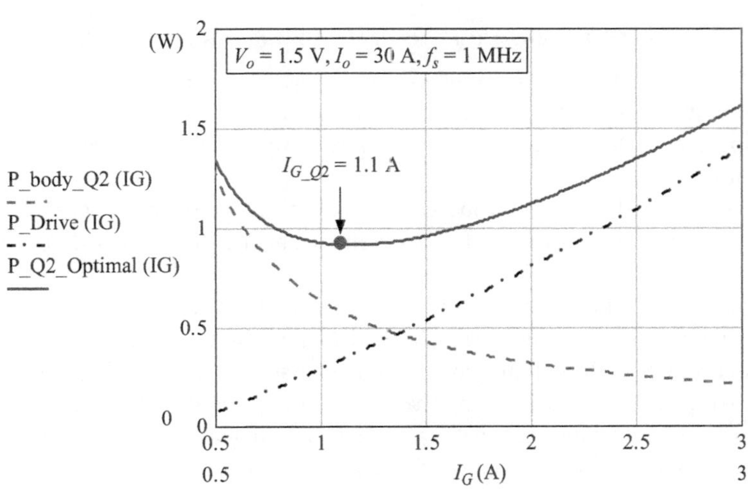

Figure 3.45 Optimization curves for the synchronous MOSFET Q_2: power loss vs. gate current

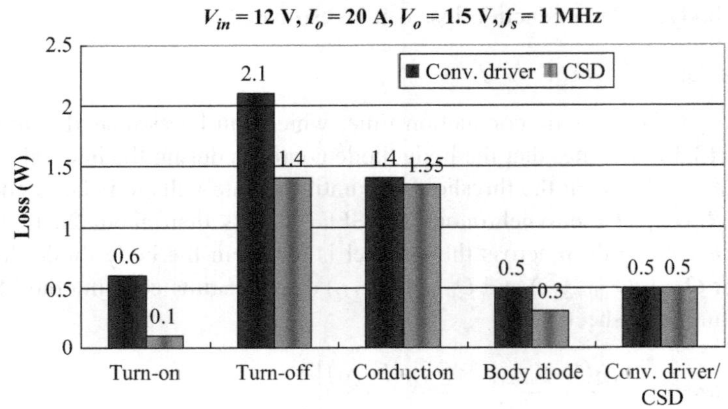

Figure 3.46 Loss breakdown between the current-source gate driver and conventional gate driver

3.4.2.5 Application extension

Half-bridge converter and FB converters can achieve ZVS using complimentary control and phase-shift control respectively. However, turn-off loss still exits and can be high due to the parasitics of the main power MOSFETs. Turn-off loss therefore becomes a concern at MHz switching frequencies.

It is noted that the proposed gate-drive circuit can also be extended to drive two MOSFETs in one leg of the HB or FB topologies to achieve switching loss reduction and gate energy savings at high switching frequencies (>1 MHz), as shown in Figure 3.47.

3.4.3 New current-source gate driver with integrated magnetics

It can be observed the waveforms of two inductor current i_{Lr1} and i_{Lr2} in Figure 3.42 that i_{Lr1} is nearly a mirror image about the time axis of i_{Lr2}. This provides good ripple cancellation effect of the magnetic flux enabling potential inductors integration.

Figure 3.48 shows the proposed current-source gate driver with integrated magnetics. It is noted that the reference voltages of the two drivers do not need to be the same.

Figure 3.49 illustrates an integrated inductor structure for current-source driver. The two resonant inductors (L_{r1} and L_{r2}) are built on the two outer legs of one E-I core with different air gaps, g_1 and g_2, respectively. The fluxes Φ_1 and Φ_2 in the two outer legs generated by the two windings flow through the center leg, which is a low-reluctance magnetic path with no air gap. Though L_{r1} and L_{r2} are built on the same E-I core, there is no interaction between the two flux loops of Φ_1 and Φ_2 and there is no coupling effect between L_{r1} and L_{r2}. Therefore, the principle of operation of the driver circuits with the integrated inductors do not change; however, the core number is reduced from two to one to one.

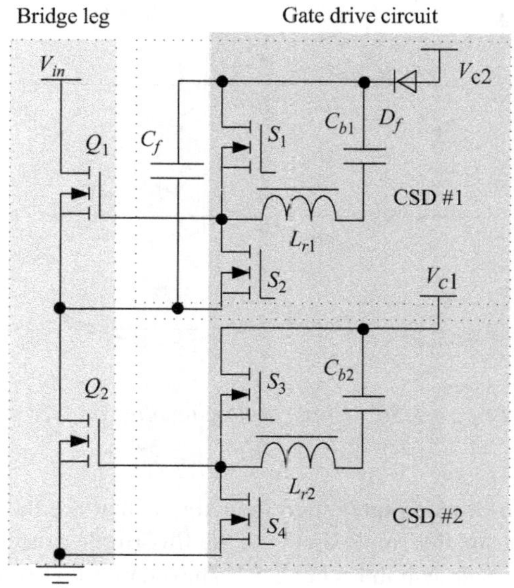

Figure 3.47 Bridge leg with the new current-source driver

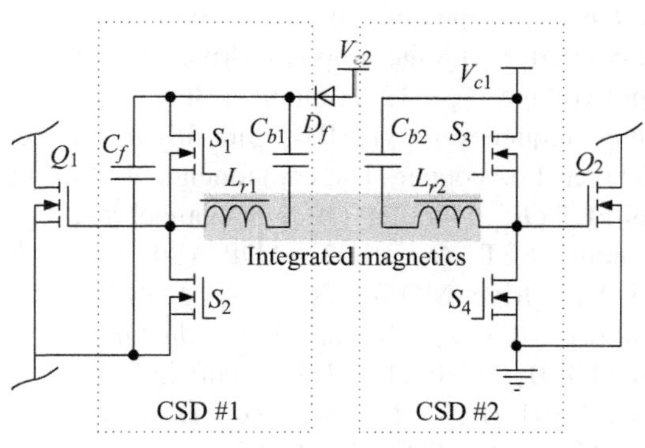

Figure 3.48 Proposed current-source driver with integrated inductor

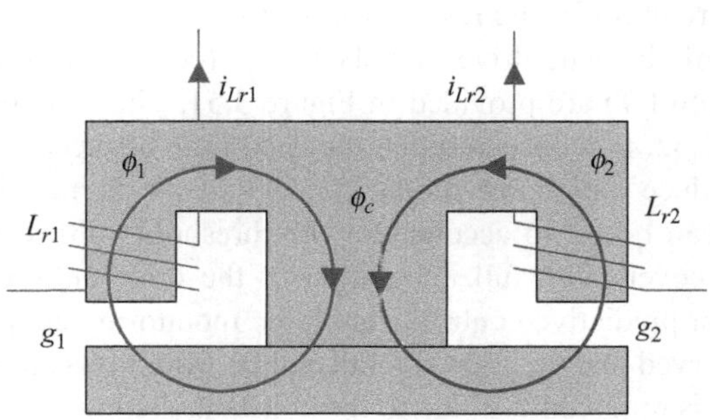

Figure 3.49 Integrated inductor structure

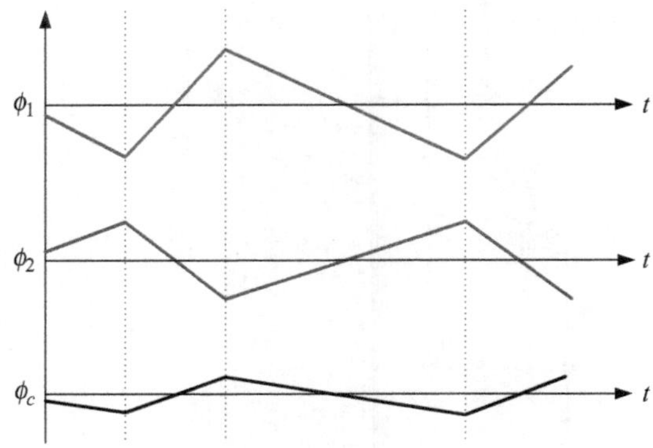

Figure 3.50 Flux ripple cancellation effect

Another benefit of using integrated inductors is that the flux $\mathbf{\Phi}$ ($\mathbf{\Phi} = \mathbf{\Phi}_1 + \mathbf{\Phi}_2$) in the center leg has smaller ripple owing to the flux ripple cancellation effect of the current i_{Lr1} and i_{Lr2} as shown in Figure 3.50. The smaller flux ripple helps to reduce the core losses in the center leg.

3.4.4 Experimental results and discussion

In order to verify the advantages of the new current-source gate-drive circuit, a synchronous buck converter with the proposed driver was built. The specifications are as follows: input voltage $V_{in} = 12$ V; output voltage $V_o = 1.5$ V; output current $I_o = 30$ A; switching frequency $f_s = 1$ MHz; gate-driver voltage $V_{c1} = V_{c2} = 8$ V. The PCB uses six-layer 2 oz copper. The components used in the circuit are listed as follows: Control FET Q_1: Si7860DP (30 V N-channel, $R_{DS(on)} = 11$ mΩ@$V_{GS} =$ 4.5 V), Synchronous FET Q_2: Si7336ADP (30 V N-channel, $R_{DS(on)} =$ 4 mΩ@$V_{GS} = 4.5$ V), Drive MOSFETs S_1–S_4: FDN335N (20 V N-channel, $R_{DS(on)} = 70$ mΩ@$V_{GS} = 4.5$ V), Output filter inductance: $L_f = 330$ nH ($R_{dc} =$ 1.3 mΩ, IHLP-5050CE-01, Vishay) and Resonant inductors: $L_{r1} = 1$ μH ($I_{pk1} =$ 1.5 A) and $L_{r2} = 1.2$ μH ($I_{pk2} = 1.1$ A). The driver was built using discrete components and an Altera Max II EPM240 CPLD was used to generate the driver gating signals. The bootstrap and level-shift circuit were used for the non-ground referenced switches. Surface mount (SMT) power inductors (DO1608C series) from Coil Craft are used for the resonant inductors.

Waveforms of the gate-drive signals v_{gs_Q1} (control MOSFET) and v_{gs_Q2} (synchronous MOSFET) are provided in Figure 3.51. The zoomed waveforms are shown in Figure 3.52. It is observed that the dead time between two gate signals is minimized to reduce the body diode conduction. It is noted that additional fixed dead time can be set to account for the threshold voltage variation at high temperature. Moreover, with full integration of the drive circuit into a drive IC, adaptive control or predictive control using logic monitoring circuitry can be used.

It is also observed that v_{gs_Q1} is smooth and no Miller plateau is observed since the Miller charge is removed fast due to the constant charging/discharging current.

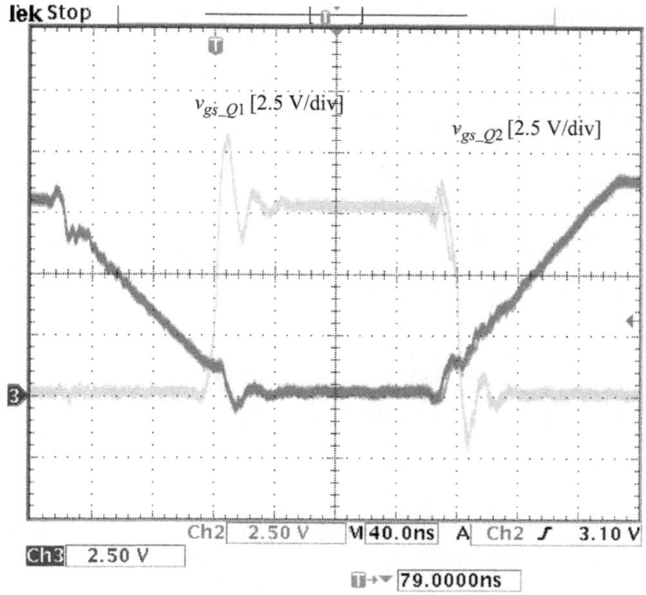

Figure 3.51 The gate signals v_{gs_Q1} (control FET) and v_{gs_Q2} (synchronous FET)

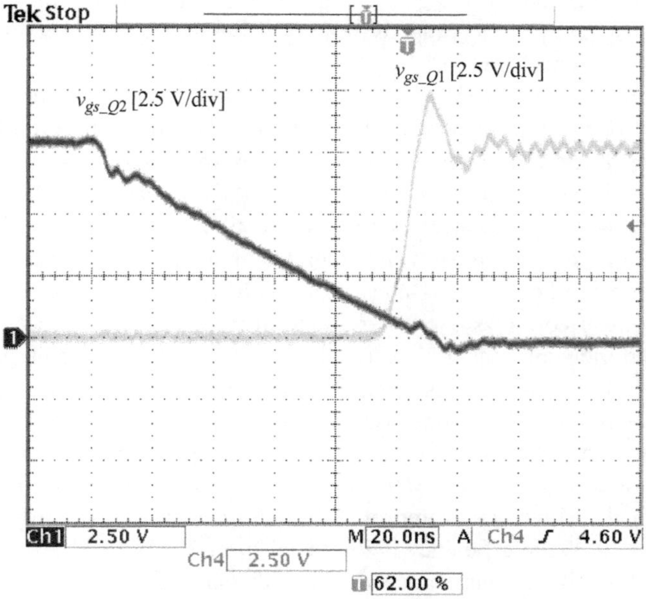

Figure 3.52 Zoomed gate signals v_{gs_Q1} (control FET) and v_{gs_Q2} (synchronous FET)

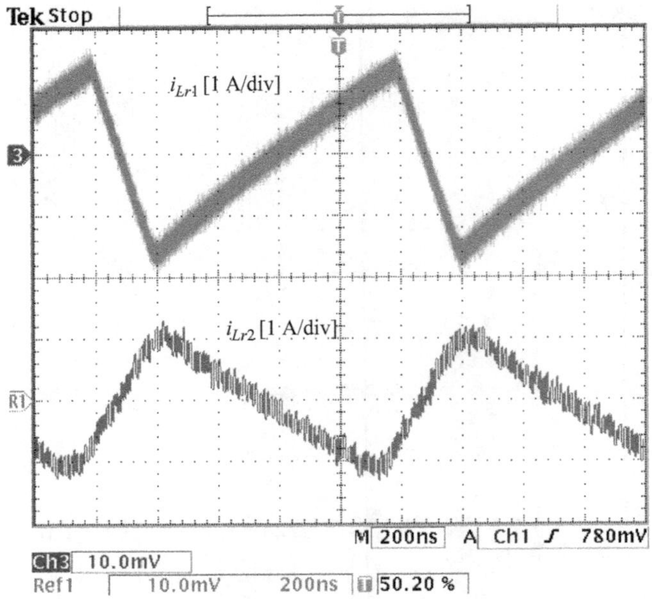

Figure 3.53 Waveforms of resonant inductor currents (discrete resonant inductors)

It is also noted that due to the gate energy recovery, 8 V drive voltage is used to reduce the $R_{DS(on)}$ by 25% compared to 5 V drive voltage. This translates into a reduction of the conduction loss by 25%.

Waveforms of the resonant inductor currents, i_{Lr1} and i_{Lr2}, are illustrated in Figure 3.53 for the prototype with discrete inductors. It is observed that for the control MOSFET, the drive current i_{Lr1} is the optimal value, 1.5 A, while for the synchronous MOSFET, the drive current i_{Lr2} is the optimal value, 1.1 A.

Waveforms of the resonant inductor currents, i_{Lr1} and i_{Lr2}, are illustrated in Figure 3.54 for the prototype with integrated inductors. It is observed that the mirror relationship between i_{Lr1} and i_{Lr2} leads to the feasibility of inductor integration and lower core loss due the magnetic flux cancellation effect. In addition, the integrated inductor does not change the operation of the drive circuits.

A benchmark synchronous buck converter with a UCC27222 conventional gate driver was built. The same parameters are used as the current-source gate driver. Figure 3.55 illustrates the measured efficiency comparison for the current-source gate driver and the conventional gate driver at 1.5 V output. It is observed that at 20 A, the efficiency is improved from 84% to 87.3% (an improvement of 3.3%) and at 30 A, the efficiency is improved from 79.4% to 82.8% (an improvement of 3.4%). The current-source driver with one integrated inductor achieves similar efficiency as the driver with two discrete inductors.

Figure 3.56 shows the converter power loss comparison for the CSD and the conventional driver. It is noted that 30 A load, the proposed current source gate

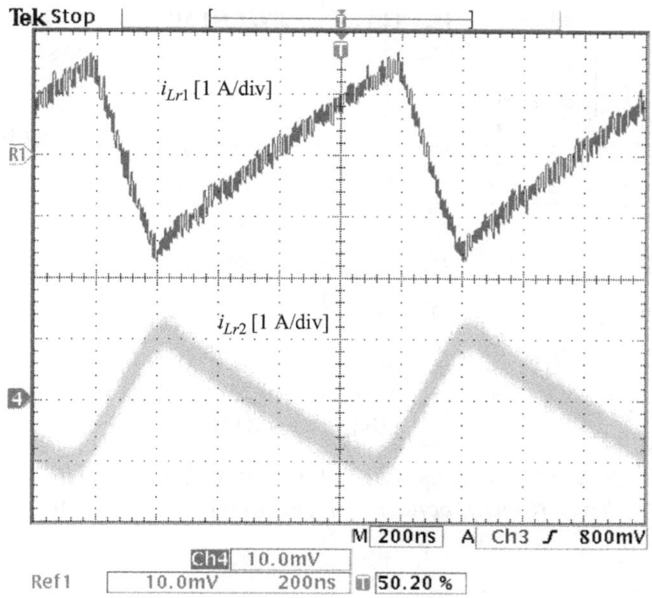

Figure 3.54 Waveforms of resonant inductor currents (with magnetic integration)

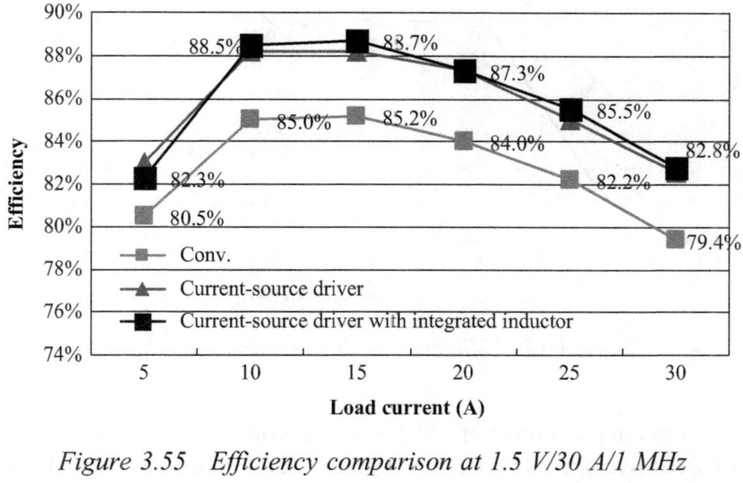

Figure 3.55 Efficiency comparison at 1.5 V/30 A/1 MHz

driver saves approximately 2.2 W (a reduction of 23%) compared to the conventional driver. This loss savings is significant for a multi-phase VRs. For example, a five-phase VR, the total loss savings would be 11 W.

Another interesting observation is that if the power loss per phase is limited to 9.5 W, the buck converter with conventional gate drive can only provide 26 A output current, while the buck converter with the current source gate driver can

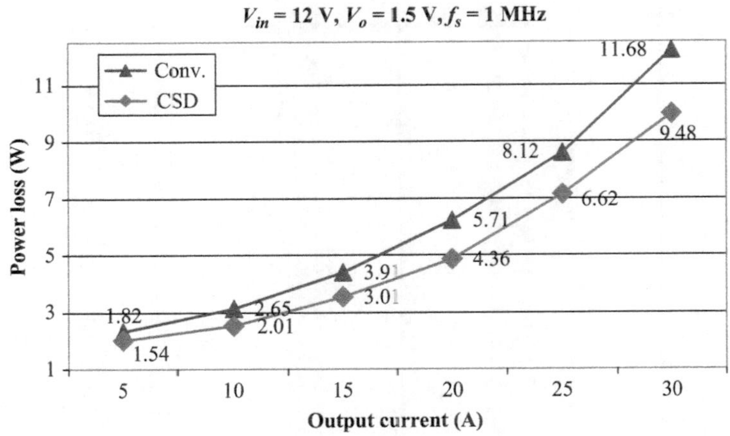

Figure 3.56 Power loss comparison at 1.5 V/30 A condition; top: conventional driver (Conv.), bottom: current-source driver (CSD)

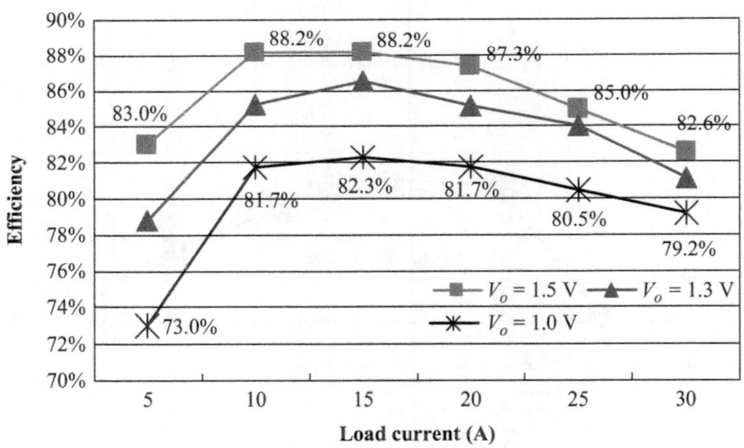

Figure 3.57 Efficiency at different output voltages

provide 30 A (an improvement of 15%). In other words, if the total output current is 120 A, we need five phases (120 A/26 A per phase) for the conventional gate driver and only four phases (120 A/30 A per phase) for the current source driver. This yields a significant potential cost savings and space savings enabling high power density.

Figure 3.57 shows the measured efficiency for the current-source driver at different output voltages as a function of load current. It is observed that at 1.0 V/30 A, the efficiency is 79.2% which is almost same as 79.4% of the conventional driver at 1.5 V/30 A (see Figure 3.55). Therefore, the output voltage can be reduced from 1.5 to 1.0 V using the current-source driver without an efficiency penalty.

This is important since the VR output voltage is reducing to approach sub-1 V in the near future.

3.5 Summary

The continuous CSDs build continuous current of the current source inductors. These drivers feature the following advantages: (a) significant switching loss reduction; (b) gate energy recovery; (c) reduced conduction loss with potential high driver voltage and reverse recovery loss of the body diode; (d) ZVS for the driver switches; and (e) clamp the drive voltage to either source voltage level or zero.

It is important to note that in a synchronous buck converter for VR applications, high-drive current is desired for the control FET to ensure fast switching speed to reduce switching loss, while low drive current is desired for the synchronous FET to achieve gate energy recovery. Therefore, independent CSDs for high-side and low-side are preferred to achieve better performance with different optimal driver current.

References

[1] L. Balogh, "Design and application guide for high speed MOSFET gate drive circuits," Texas Instrument, application notes, slup169.

[2] Y. Ren, M. Xu, F.C. Lee and P. Xu, "The optimal bus voltage study for 12 V two-stage VR based on an accurate analytical loss model," in *Proceedings on IEEE PESC'04*, vol. 6, pp. 4319–4324, 2004.

[3] Y. Xiao, H. Shah, T. P. Chow and R. J. Gutmann, "Analytical modeling and experimental evaluation of interconnect parasitic inductance on MOSFET switching characteristics," in *Proceedings on IEEE APEC'04*, vol. 1, pp. 516–521, 2004.

[4] K. Yao, M. Ye, M. Xu and F. C. Lee, "Tapped-inductor buck converter for high-step-down DC-DC conversion," *IEEE Transactions on Power Electron*, vol. 20, no. 4, pp. 775–780, 2005.

[5] Z. Yang, S. Ye and Y. F. Liu, "A new resonant gate drive circuit for synchronous buck converter," *IEEE Transactions on Power Electron*, vol. 22, no. 4, pp. 1311–1320, 2007.

[6] J. Wei and F. C. Lee, "Two novel soft-switched, high frequency, high-efficiency, non-isolated Voltage Regulators-the phase-shift buck converter and the matrix-transformer phase-buck converter," *IEEE Transactions on Power Electron*, vol. 20, no. 2, pp. 292–299, 2005.

[7] J. Zhou, M. Xu, J. Sun and F. C. Lee, "A self-driven soft-switching voltage regulator for future microprocessors," *IEEE Transactions on Power Electron*, vol. 20, no. 4, pp. 806–814, 2005.

[8] Y. Kawaguchi, T. Kawano, S. Ono, A. Nakagawa and H. Takei, "Design for efficiency improvement and future prediction in Multi Chip Module for DC–DC converter," in *Proceedings on IEEE PESC*, pp. 1–6, 2006.

Chapter 4
Discontinuous current source drivers

4.1 Discontinuous current source driver

This chapter presents different discontinuous current source drivers (CSDs) for the power metal oxide semiconductor field-effect transistors (MOSFETs). These drivers aim to reduce switching loss by providing a near constant gate charging and discharging currents to switch quickly. Furthermore, they can recover a portion of the QV gate energy otherwise lost in conventional voltage source drivers (VSDs).

4.1.1 Proposed low-side discontinuous CSD

The proposed CSD is illustrated in Figure 4.1. It consists of four controlled switches, S_1–S_4 (S_1, S_2 P-channel and S_3, S_4 N-channel) including their body diodes, D_1–D_4, and a small inductance, L. The switches are controlled and diodes are used to allow the inductor current to be discontinuous and allow the power MOSFET to turn on or off beginning from a non-zero pre-charge current. Following charging, or discharging of the power MOSFET, the excess energy stored in the inductor is allowed to return to the supply voltage, V_{cc}, thereby allowing the power MOSFET gate charge energy to be recovered.

The gating waveforms of the four driver switches, S_1–S_4, along with the inductor current, gate current, power MOSFET gate-to-source voltage and the line current are illustrated in Figure 4.2. The key waveforms to note are: (1) the inductor current, labeled i_L, which is discontinuous to minimize conduction loss, (2) the gate-drive current, labeled i_g, which remains approximately constant (in comparison to the conventional VSDs), and (3) the line current, i_{Vcc}, which contains negative intervals, indicating that energy is returned to the line. The operation of the circuit is explained in the following paragraphs.

Initially it is assumed that the power MOSFET is in the off state before time t_0. Initially, before t_0, only switch S_3 and D_4 are on and the gate of M is clamped to zero volts. The current paths during the intervals of the turn-on stage are illustrated in Figure 4.3. It is noted that M has been replaced by its equivalent total gate capacitance, C_g.

1. Mode 1 [t_0, t_1] (Figure 4.1(a)): At t_0, S_2 turns on (with zero-current switching (ZCS)), therefore turning off D_4 by commutation (with ZCS) and allowing the inductor current to ramp up. The current path during this interval is S_2–L–S_3.

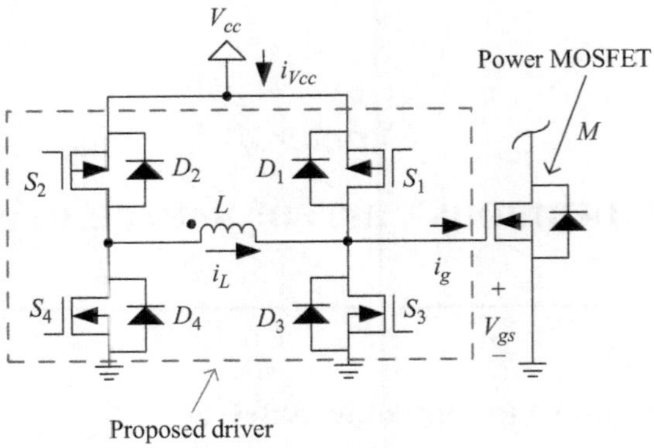

Figure 4.1 Proposed discontinuous CSD

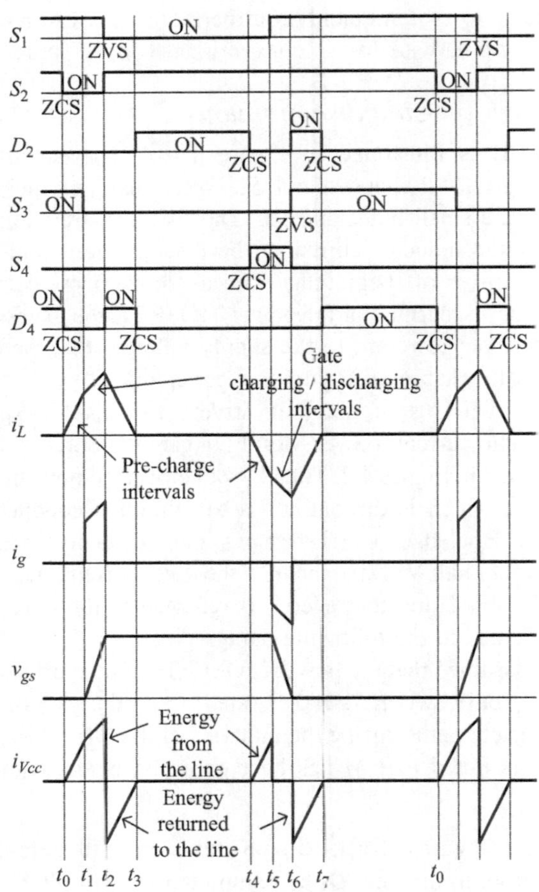

Figure 4.2 Waveforms of the proposed CSD

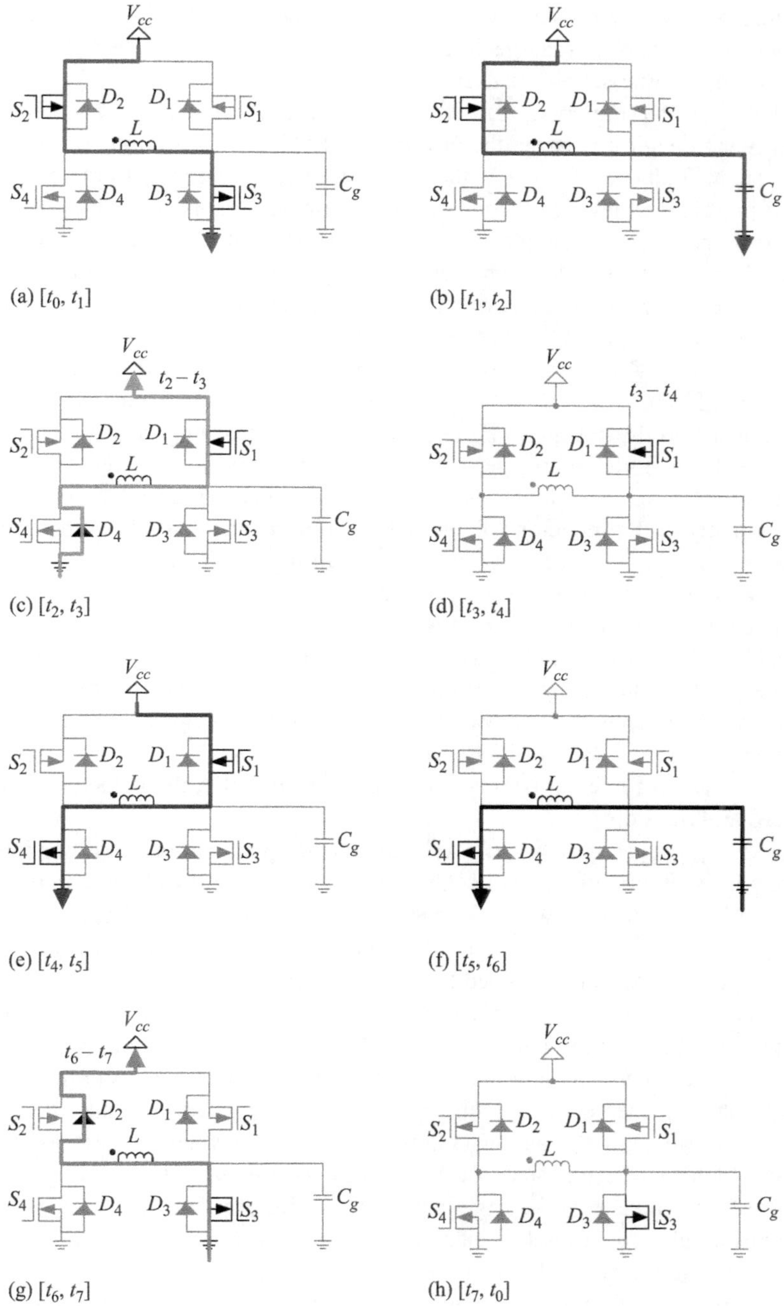

Figure 4.3 Current paths during the intervals of the turn-on stage

Since S_3 is in the on state, the gate of M is clamped low. This interval is the inductor current pre-charge interval and it ends at time t_1, which is a pre-determined time set by the user.

2. Mode 2 [t_1, t_2] (Figure 4.1(b)): At t_1, S_3 is turned off, which allows the inductor current to begin to charge the power MOSFET gate. Assuming a low natural resonant frequency for L–C_g, the inductor current continues to ramp up from the pre-charged level as the voltage across the gate capacitance increases. The current path during this interval is S_2–L–C_g. This interval ends at t_2, when v_{gs} reaches V_{cc}. If the inductor pre-charge current at the end of t_1 is much greater than the ripple amplitude during this interval, then the inductor can assumed to be a current source (CS) charging the power MOSFET gate.
3. Mode 3 [t_2, t_3] (Figure 4.1(c)): At t_2, S_1 is turned on (with ZVS) and S_2 is turned off, therefore driving D_4 on to allow the inductor current to continue to conduct into the dot through the path D_4–L–S_1. Most importantly, during this interval the stored energy in the inductor is returned to the line. This can be observed from the negative portion of the i_{Vcc} curve in Figure 4.2. Also, during this interval, the inductor voltage has become reverse biased, so the inductor current quickly ramps down toward zero. During this interval, the gate voltage of M remains clamped to the line voltage, V_{cc}. The interval ends when the inductor current reaches zero at t_3.
4. Mode 4 [t_3, t_4] (Figure 4.1(d)): At t_3, D_4 turns off (with ZCS). During this interval, the gate voltage of M remains clamped to V_{cc}. The interval ends at t_4 when the pre-charging interval for the turn-off cycle begins as dictated by the pulse width modulation (PWM) signal.

   The turn-off interval begins at the end of t_4. Initially, the inductor current is zero and S_1 is on.
5. Mode 5 [t_4, t_5] (Figure 4.1(e)): At t_4, the turn-off inductor pre-charging interval begins. S_4 is turned on (with ZCS), therefore turning off D_2 by commutation (with ZCS). Since S_1 was previously on, the inductor current begins to ramp negative out of the dot through the path S_1–L–S_4. During this interval, the gate voltage of M remains clamped to V_{cc}. The interval ends at t_5.
6. Mode 6 [t_5, t_6] (Figure 4.1(f)): At t_5, S_1 is turned off, which allows the inductor current to begin to discharge the power MOSFET gate. Assuming a low natural resonant frequency of L–C_g, the inductor current continues to ramp negative from the pre-charged level as the voltage across the gate capacitance decreases. The current path during this interval is C_g–L–S_4, where C_g represents the equivalent gate capacitance of M. This interval ends at t_6, when v_{gs} reaches zero. If the inductor pre-charge current at the end of t_5 is much greater than the ripple amplitude during this interval, then the inductor can assumed to be a CS discharging the power MOSFET gate.
7. Mode 7 [t_6, t_7] (Figure 4.1(g)): At t_6, S_3 is turned on (with ZVS) and S_4 is turned off, therefore driving D_2 on to allow the inductor current to conduct out of the dot through the path D_3–L–D_2. Most importantly, during this interval, the gate discharging energy is returned to V_{cc}. This can be observed from the negative portion of the i_{Vcc} curve. Also, during this interval, the inductor voltage has

become reverse biased, so the inductor current quickly ramps down positive toward zero. During this interval, the gate voltage of M remains clamped to ground. The interval ends when the inductor current reaches zero at t_7.

8. Mode 8 [t_7, t_0] (Figure 4.1(h)): At t_7, D_2 turns off (with ZCS). During this interval, the gate voltage of M remains clamped to ground. The interval ends at t_0 when the pre-charging interval for the turn-on cycle begins and the entire process repeats as dictated by the PWM signal.

4.1.2 Driver loss analysis

The loss components of the proposed CS gate-drive circuit are outlined in the following subsections. The loss components include: conduction loss, gate-drive loss in the driver switches, CV^2 output loss, turn-off loss in the driver switches, core loss in the driver inductor, and loss in the driver logic circuit.

The conduction loss can be determined by analyzing the losses during the three states of the turn-on interval (t_0–t_3) and three states of the turn-off interval (t_4–t_7) when the inductor current is non-zero. The detailed inductor current waveform and power MOSFET gate voltage waveform are shown for the turn-on interval in Figure 4.4. The analysis of the turn-on intervals is explained as follows.

It is assumed that the power MOSFET being driven, M, is represented by an RC network consisting of its parasitic series gate resistance, R_g, and an equivalent gate capacitance, C_g. During the on state, the control switches can be represented by series resistances R_1–R_4. The inductor copper loss can be represented by an equivalent series a.c. resistance, R_L.

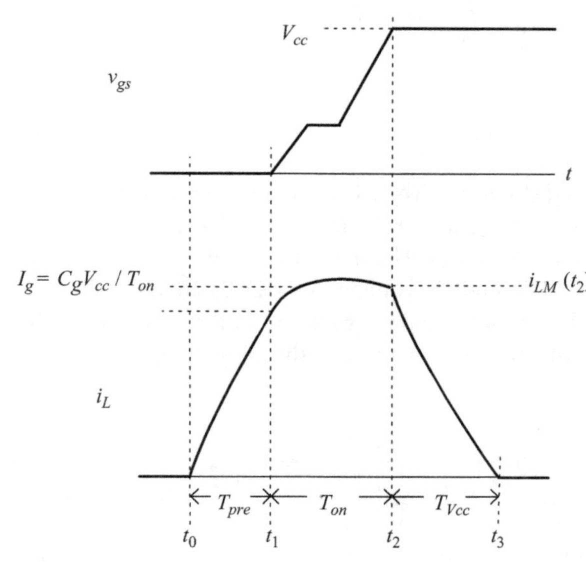

Figure 4.4 Power MOSFET gate-source voltage and CSD inductor current waveforms during the turn-on interval

The driver is designed to charge the gate-source voltage from zero to V_{cc}, during T_{on} by an average CS, I_g. Under this assumption, the turn-on time, T_{on}, is given by (4.1), where Q_g represents the total gate charge of M:

$$T_{on} = \frac{Q_g}{I_g} \tag{4.1}$$

In the following subsections, the equivalent circuits during the pre-charge, turn-on, and energy-return intervals are used and equations are derived to calculate the conduction loss in the proposed driver.

1. Pre-charge $[t_0, t_1]$ (Figure 4.5(a)): During the pre-charge interval, switches S_2 and S_3 are on allowing the inductor current to ramp up into the dot. The equivalent circuit during T_{pre} is given in Figure 4.5, where R_2, R_L, and R_3 have been lumped together as R_{pre} (i.e., $R_{pre} = R_2 + R_L + R_3$). The circuit is a voltage-driven RL circuit with zero initial inductor current. The inductor current is given by (4.2), with the time constant, τ_{pre}, given by (4.3). The root mean square (RMS) current during the pre-charge interval is given by (4.4). Using (4.4), the power consumption, P_{pre}, during the interval is given by (4.27):

$$i_{Lpre} = \frac{V_{cc}}{R_{pre}}\left(1 - e^{-t/\tau_{pre}}\right) \tag{4.2}$$

$$\tau_{pre} = \frac{L}{R_{pre}} \tag{4.3}$$

$$I_{preRMS} = \sqrt{f_s \int_0^{T_{pre}} i_{Lpre}{}^2 dt} \tag{4.4}$$

$$P_{pre} = R_{pre} f_s \int_0^{T_{pre}} i_{Lpre}{}^2 dt \tag{4.5}$$

2. Turn on $[t_1, t_2]$ (Figure 4.5(b)): The circuit is a series RLC circuit consisting of R_2, R_L, R_g, L, and C_g, where R_2, R_L, and R_g have been lumped together as R_{on} (i.e., $R_{on} = R_2 + R_L + R_g$). The inductor contains an initial current $i_L(t_1)$. In the s-domain, the Kirchhoff's voltage law (KVL) equation for the circuit is given by (4.7). Solving for the inductor current in the RLC circuit, the characteristic equation is of the form in (4.8) with poles at p_1 and p_2 as given in (4.9).

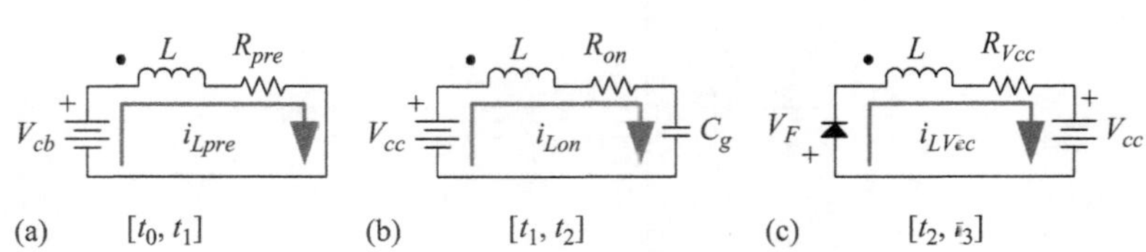

Figure 4.5 Equivalent circuit during the inductor pre-charge interval

The parameters in (4.10) and (4.11) are the neper frequency, α, and the resonant frequency ω_o:

$$i_{Lon}(0) = i_L(t_1) \tag{4.6}$$

$$\frac{V_{cc}}{s} = sLI_L(s) - Li_L(t_1) + I_L(s)R_{on} + \frac{I_L(s)}{sC_g} \tag{4.7}$$

$$s^2 + \left(\frac{R_{on}}{L}\right)s + \left(\frac{1}{LC_g}\right) = (s+p_1)(s+p_2) \tag{4.8}$$

$$\begin{aligned} p_1 &= -\frac{R_{on}}{2L} + \sqrt{\left(\frac{R_{on}}{2L}\right)^2 - \frac{1}{LC_g}} = -\alpha + \sqrt{\alpha^2 - \omega_o{}^2} \\ p_2 &= -\frac{R_{on}}{2L} - \sqrt{\left(\frac{R_{on}}{2L}\right)^2 - \frac{1}{LC_g}} = -\alpha - \sqrt{\alpha^2 - \omega_o{}^2} \end{aligned} \tag{4.9}$$

$$\alpha = \frac{R_{on}}{2L} \tag{4.10}$$

$$\omega_o = \frac{1}{\sqrt{LC_g}} \tag{4.11}$$

There are three possible solutions for the inductor current: (1) overdamped, (2) critically damped, and (3) underdamped. A driver designed with minimal conduction loss will be under damped with $R_{on} \rightarrow 0$ (R_{on} is typically 1 Ω) and $\omega_o > \alpha$. The under damped solution for the inductor current is given by (4.12):

$$\begin{aligned} i_{Lon} = {} & \left[\frac{V_{cc}}{L} - \alpha Li_L(t_1)\right] \frac{e^{-t\alpha}}{\sqrt{\omega_o{}^2 - \alpha^2}} \sin\left(\sqrt{\omega_o{}^2 - \alpha^2 t}\right) \\ & + Li_L(t_1)e^{-t\alpha} \cos\left(\sqrt{\omega_o{}^2 - \alpha^2 t}\right) \end{aligned} \tag{4.12}$$

1. The RMS value of i_{Lon} is given by (4.13). Using (4.13), the conduction loss power, P_{on}, during T_{on} is given by (4.14):

$$I_{onRMS} = \sqrt{f_s \int_0^{T_{on}} i_{Lon}{}^2 dt} \tag{4.13}$$

$$P_{on} = R_{on} f_s \int_0^{T_{on}} i_{Lon}{}^2 dt \tag{4.14}$$

2. Energy return [t_2, t_3] (Figure 4.5(c)): R_L and R_1 have been lumped together as R_{Vcc} (i.e., $R_{Vcc} = R_1 + R_L$). The diode voltage drop is considered constant at V_F.

The inductor current is given by (4.16), with the time constant, τ_{Vcc}, given by (4.17):

$$i_{LVcc}(0) = i_L(t_2) \tag{4.15}$$

$$i_{LVcc} = \frac{-(V_{cc} + V_F)}{R_{Vcc}} + \left(i_L(t_2) + \frac{V_{cc} + V_F}{R_{Vcc}}\right) e^{-t/\tau Vcc} \tag{4.16}$$

$$\tau_{Vcc} = \frac{L}{R_{Vcc}} \tag{4.17}$$

The RMS value of i_{LVcc}, I_{VccRMS} is given by (4.18). The average value of i_{LVcc} during T_{Vcc} over one period is given by (4.19). The conduction loss power, P_{Vcc}, during T_{Vcc} is given by (4.20):

$$I_{VccRMS} = \sqrt{f_s \int_0^{T_{Vcc}} i_{LVcc}{}^2 dt} \tag{4.18}$$

$$I_{VccAVG} = f_s \int_0^{T_{Vcc}} i_{LVcc} dt \tag{4.19}$$

$$P_{Vcc} = R_{Vcc} f_s \int_0^{T_{Vcc}} i_{LVcc}{}^2 dt + V_F f_s \int_0^{T_{Vcc}} i_{LVcc} dt \tag{4.20}$$

To calculate the total conduction loss, it can be assumed that the turn-on and turn-off states of operation are identical. Therefore, under this assumption, the total conduction loss in the proposed CS gate-drive circuit is given by (4.21), which is two times the sum of P_{pre}, given by (4.5), plus P_{on}, given by (4.14), plus P_{Vcc}, given by (4.20):

$$P_{cond} = 2(P_{pre} + P_{on} + P_{Vcc}) \tag{4.21}$$

Equations (4.5), (4.14), and (4.20) can be used in a software package to calculate conduction loss during the pre-charge time. However, under certain conditions, with simple piecewise linear approximations, the conduction loss calculations can be simplified for use in a spreadsheet design file.

In the intended ideal operation of the driver, the inductor current ramps up linearly (during T_{pre}), then charges the power MOSFET gate with a constant current (during T_{on}) and then the excess inductor energy is returned to the driver supply as the inductor current ramps down linearly (during T_{Vcc}) as illustrated in Figure 4.6. To achieve an inductor current close to the desired ideal operation, three conditions are required: (1) the resistance, R_{pre} during the pre-charge interval can be neglected, (2) the current at the end of the pre-charge interval should be much greater than the ripple deviation during T_{on} (i.e., $i_L(t_1) >> |i_L(t_2) - i_L(t_1)|$); therefore, it is assumed that the inductor current remains constant at $i_L(t_1) = I_g$ during T_{on}, and (3) the resistance, R_{Vcc} during the energy return interval can be neglected. The three conditions are explained in the following subsections.

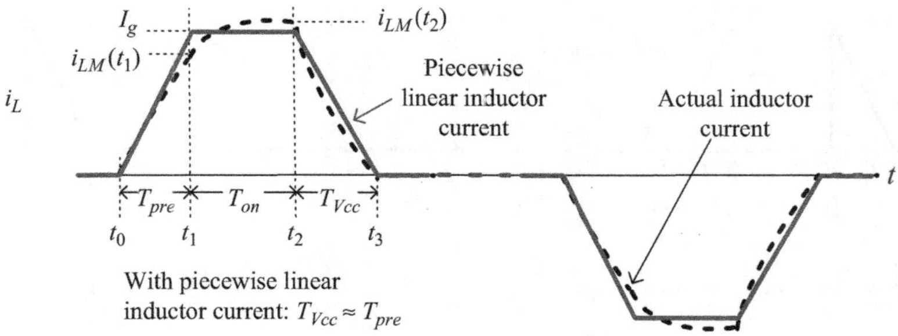

Figure 4.6 Piecewise linear inductor current and actual inductor current

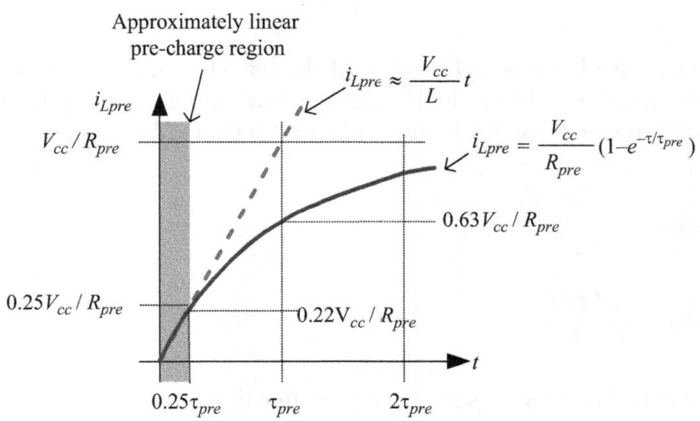

Figure 4.7 Inductor current during the pre-charge time

The inductor current as a function of time is plotted in Figure 4.7 up to two time constants. If it is assumed that $R_{pre} = 0$, then i_{Lpre} increases linearly as given by (4.22) and shown by the dotted line. From Figure 4.7, for driver designs with $T_{pre} < 0.25\tau_{pre}$, it can be assumed that the inductor current is linear as approximated by (4.22), allowing I_g to be expressed by (4.23):

$$i_{Lpre} \approx \frac{V_{cc}}{L}t \tag{4.22}$$

$$I_g \approx \frac{V_{cc}}{L}T_{pre} \tag{4.23}$$

To simplify the design, the inductor current pre-charge ratio a is introduced as given by (4.24):

$$a = \frac{T_{pre}}{T_{on}} \tag{4.24}$$

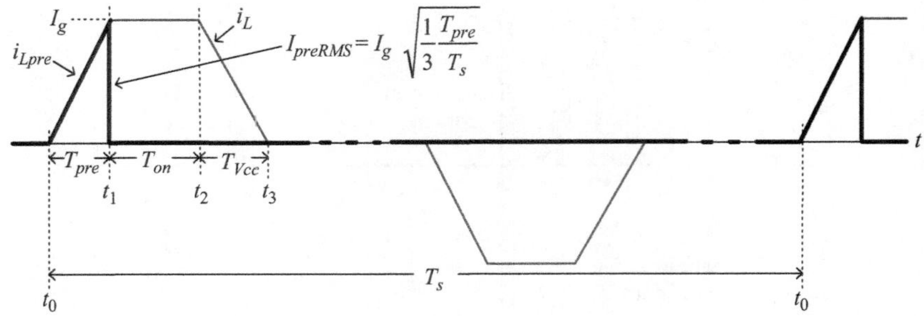

Figure 4.8 Piecewise linear inductor current during the pre-charge interval overlaid on the actual piecewise linear inductor current

The piecewise linear simplification of the inductor current waveform during T_{pre} is highlighted in bold in Figure 4.8. The RMS current during this interval is given by (4.25), which can be expressed by (4.26) using (4.23) and (4.24):

$$I_{preRMS} \approx I_g\sqrt{\frac{T_{pre}f_s}{3}} \tag{4.25}$$

$$I_{preRMS} \approx \sqrt{\frac{I_g^{\ 3}Lf_s}{3V_{cc}}} \tag{4.26}$$

The conduction loss power, P_{pre}, as a function of I_g and T_{pre}, during the interval is approximated by (4.27) using (4.25). P_{pre} as a function of V_{cc}, L, and I_g is approximated by (4.28) using (4.26):

$$P_{pre} \approx R_{pre}I_g^{\ 2}\frac{T_{pre}f_s}{3} \tag{4.27}$$

$$P_{pre} \approx R_{pre}\frac{I_g^{\ 3}Lf_s}{3V_{cc}} \tag{4.28}$$

In the intended operation of the driver, the current at the end of the pre-charge interval should be much greater than the ripple deviation (i.e., $i_L(t_2)–i_L(t_1)$) during T_{on}. Therefore, it is assumed that the inductor current remains constant at $i_{Lon} = i_L(t_1) = i_L(t_2) = I_g$ through T_{on}, following the piecewise linear approximation illustrated in Figure 4.6. The piecewise linear simplification of the inductor current waveform during T_{on} is highlighted in bold in Figure 4.9.

The RMS current during this interval is given by (4.29), which can be expressed by (4.30) using (4.23) and (4.24):

$$I_{onRMS} \approx I_g\sqrt{aT_{pre}f_s} \tag{4.29}$$

$$I_{onRMS} \approx \sqrt{I_gQ_gf_s} \tag{4.30}$$

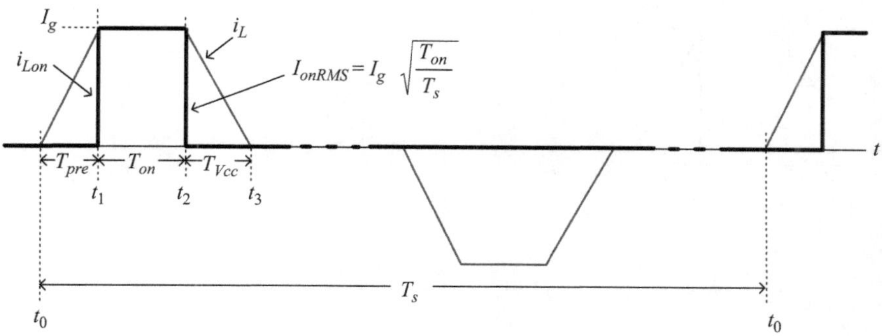

Figure 4.9 Piecewise linear inductor current during the turn-on interval overlaid on the actual piecewise linear inductor current

The conduction loss power, P_{on}, as a function of a, I_g, and T_{pre}, during the interval is approximated by (4.31) using (4.29). P_{on} as a function of V_{cc}, L, and I_g is approximated by (4.32) using (4.30):

$$P_{on} \approx R_{on} I_g^{\ 2} a T_{pre} f_s \tag{4.31}$$

$$P_{on} \approx R_{on} I_g Q_g f_s \tag{4.32}$$

Following the condition presented for the pre-charge interval, for driver designs with $T_{Vcc} < 0.25\tau_{Vcc}$, it can be assumed that the inductor current is linear as approximated by (4.33), where V_F represents the diode forward voltage drop in the body diode of S_4. The energy return interval can be estimated by (4.34):

$$i_{LVcc} \approx I_g - \frac{V_{cc} + V_F}{L} t \tag{4.33}$$

$$T_{Vcc} \approx I_g \frac{L}{V_{cc} + V_F} \tag{4.34}$$

The piecewise linear simplification of the inductor current waveform during T_{Vcc} is highlighted in bold in Figure 4.10. The RMS current during this interval is given by (4.35), which can be expressed by (4.36) using (4.34):

$$I_{VccRMS} \approx I_g \sqrt{\frac{T_{Vcc} f_s}{3}} \tag{4.35}$$

$$I_{VccRMS} \approx \sqrt{\frac{I_g^{\ 3} L f_s}{3(V_{cc} + V_F)}} \tag{4.36}$$

The conduction loss power, P_{Vcc}, as a function of I_g and T_{Vcc}, during the interval is approximated by (4.37) using (4.35). P_{Vcc} as a function of V_{cc}, L, and I_g

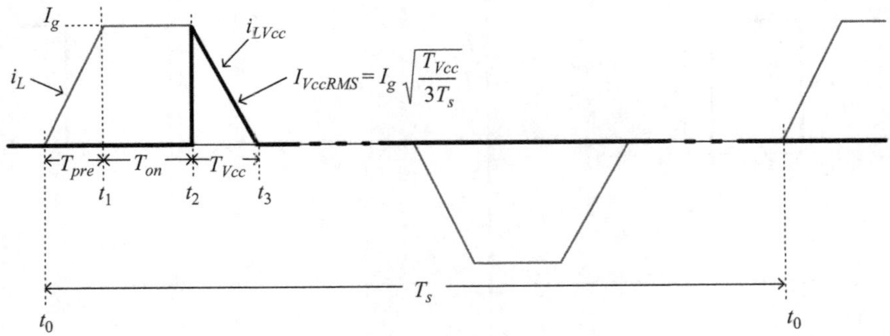

Figure 4.10 Piecewise linear inductor current during the energy return interval overlaid on the actual piecewise linear inductor current

is approximated by (4.38) using (4.36):

$$P_{Vcc} \approx I_g T_{Vcc} f_s \left(\frac{1}{3} R_{Vcc} I_g + \frac{1}{2} V_F \right) \tag{4.37}$$

$$P_{Vcc} \approx \frac{I_g{}^2 L f_s}{V_{cc} + V_F} \left(\frac{R_{Vcc} I_g}{3} + \frac{1}{2} V_F \right) \tag{4.38}$$

To calculate the total estimated conduction loss, it can be assumed that the turn-on and turn-off states of operation are identical. Therefore, under this assumption, the total estimated piecewise linear conduction loss in the proposed CS gate-drive circuit is given by (4.39), which is two times the sum of P_{pre}, given by (4.27), plus P_{on}, given by (4.31), plus P_{Vcc}, given by (4.37). Alternately, P_{cond} can be expressed by (4.40), using (4.28), (4.32), and (4.38):

$$P_{cond} \approx 2 \left[I_g{}^2 \left(R_{pre} \frac{T_{pre}}{3} + R_{on} a T_{pre} + R_{Vcc} \frac{T_{Vcc}}{3} \right) + I_g \frac{V_F T_{Vcc}}{2} \right] f_s \tag{4.39}$$

$$P_{cond} \approx 2 \left[\left(I_g{}^3 \left(\frac{R_{pre}}{3V_{cc}} + \frac{R_{Vcc}}{3(V_{cc} + V_F)} \right) + I_g{}^2 \frac{V_F}{2(V_{cc} + V_F)} \right) L + I_g R_{on} Q_g \right] f_s \tag{4.40}$$

The second significant component of driver circuit loss is driver switch gate loss and is given by (4.41), where Q_{g1}, Q_{g2}, Q_{g3}, and Q_{g4} represent the total gate charge in S_1–S_4 and V_{dd} represents the gate driving voltage for S_1–S_4:

$$P_{S1-S4gate} = (Q_{g1} + Q_{g2} + Q_{g3} + Q_{g4}) V_{dd} f_s \tag{4.41}$$

The CV^2 output loss in S_2 and S_4 at turn on is given by (4.42), where C_{oss2} and C_{oss4} represent the output capacitance values for S_2 and S_4 obtained from the

MOSFET datasheets:

$$P_{out} = \frac{1}{2}(C_{oss2} + C_{oss4})V_{cc}{}^2 f_s \quad (4.42)$$

S_2 and S_4 turn off at the peak inductor current, $i_L(t_2) = I_g$. The turn-off loss in S_2 and S_4 is given by (4.43), where the fall times, T_{f2} and T_{f4}, are obtained from the MOSFET data sheets:

$$P_{off} = \frac{1}{2} V_{cc} I_g (T_{f2} + T_{f4}) f_s \quad (4.43)$$

The core loss can be obtained by standard core loss estimation methods and should be small in comparison to the other loss components. If air core inductors are used, the core loss is zero.

The loss in the logic circuit should be negligible in comparison to the other components, so it can be neglected.

The total loss, P_{tot}, in the proposed driver is the sum of the loss components as given by (4.44):

$$P_{tot} = P_{cond} + P_{S1-S4gate} + P_{out} + P_{off} \quad (4.44)$$

4.1.3 CSD design procedure

The power MOSFET turn-on transition time, T_{on}, (from 0 V to V_{cc}), or average gate current, I_g, must be chosen by the designer for the given application. For the designer, there is a tradeoff between speed (i.e., switching loss reduction, or reduced body diode conduction) and gate energy recovery. Increased gate current (smaller values of T_{on}) results in greater conduction loss in the driver. Typically, T_{on} should be less than 10% of the switching period.

After selecting T_{on} (or I_g), a circuit designer has a choice of the turn-on inductor pre-charge time, T_{pre}, or inductor value, L, using the ratio, a. This is illustrated in Figure 4.4 from t_0–t_1, of the inductor current waveform during the turn-on interval. Typically, T_{pre} should be in the range of 0.25 $T_{on} < T_{pre} <$ 3 T_{on}. Larger values of T_{pre} (and a) yield a larger required inductance and add more delay in the control loop, but provide a more constant gate current and improved gate energy recovery. Smaller values of T_{pre} (and a) yield a smaller required inductance, which is beneficial for component size; however, loss in the drive circuit increases since the inductor pre-charge interval becomes nonlinear and the inductor current ripple during T_{on} becomes large.

Using (4.1), (4.23), and ratio a defined in (4.24), the required inductance can be calculated using (4.45):

$$L = a\frac{V_{cc} T_{on}{}^2}{Q_g} \quad (4.45)$$

Setting $a = 0.5$ is a good starting point assumption for the driver design. With this assumption, using (4.1) and (4.23), the required inductance can be calculated using (4.46):

$$L = 0.5\frac{V_{cc}T_{on}^2}{Q_g} \quad (4.46)$$

4.1.4 Design example

A design and loss analysis of the proposed driver was conducted using the total driver loss in Section 4.1.2 and the design procedure presented in Section 4.1.3. The parameters for the designs are given in Table 4.1. A turn-on time, T_{on}, of 50 ns, was selected along with a pre-charge time, T_{pre}, of 25 ns ($a = 0.5$). Using these values, the calculated driver inductance is 139 nH.

A loss breakdown of the proposed driver is given in Figure 4.11. It is noted that the losses in the conventional driver are 82% greater than those in the proposed CSD.

The largest loss component in the proposed CSD is conduction loss. The largest portion of the conduction loss is dissipated in the power MOSFET parasitic gate resistance, R_g, which is typically about 1 Ω. Curves of CSD loss and conventional driver loss as a function of R_g are given in Figure 4.12 in order to demonstrate the potential benefits of using MOSFETS with lower internal gate resistance.

Table 4.1 CSD design parameters

Parameters		
Switching frequency	1 MHz	
Gate drive voltage, V_{cc}	5 V	
Turn-on time, T_{on}	50 ns	
Pre-charge time, T_{pre}	25 ns	
Energy return time, T_{Vcc}	50 ns	
Driver inductor, L	139 nH	
Driver inductor resistance, R_L	25 mΩ	
MOSFET, M	**IRF6618**	
Total gate charge, Q_g	45 nC	
Internal gate resistance, R_g	1 Ω	
Diodes, D_2–D_4	**MBR0520**	
Diode forward voltage, V_F	0.385 V	
CSD Switches, S_1–S_4	**NDS351AN (S_3–S_4)**	**FDN342P (S_1–S_2)**
Total gate charge, Q_g	1.25 nC	6 nC
On resistance, R_{dson}	90 mΩ	60 mΩ
Output capacitance, C_{oss}	50 pF	200 pF
Fall time, T_f	1 ns	2 ns

4.1.5 Experimental results

The proposed CSD was built using discrete components and compared to the Texas Instruments UCC37322 state-of-the-art driver using the boost converter topology operating at 1 MHz. The boost converter and its input/output specifications were chosen as a simple, ground referenced drive, single switch test application in order to demonstrate the capabilities of the proposed CSD.

The circuit boost converter specifications were as follows: $V_{in} = 5$ V, $V_o = 10$ V, $I_o = 5$ A, $L_{boost} = 1$ mH, $f_s = 1$ MHz. The boost MOSFET was part number IRF6618, rated for 30 V. In addition, an International Rectifier 10TQ035 Schottky

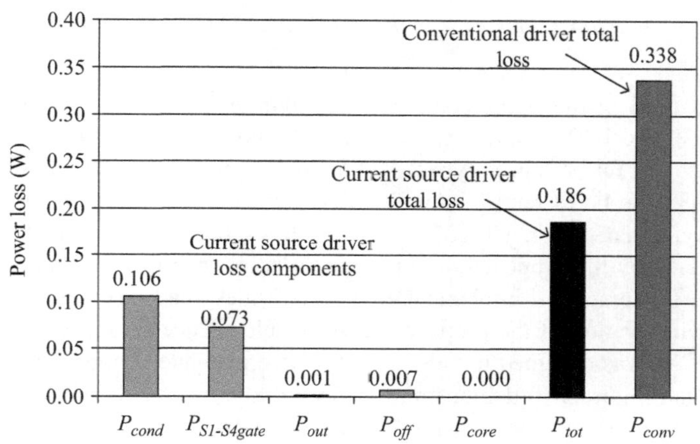

Figure 4.11 Proposed CSD gate-drive loss breakdown with comparison of total losses to total loss in a conventional driver circuit

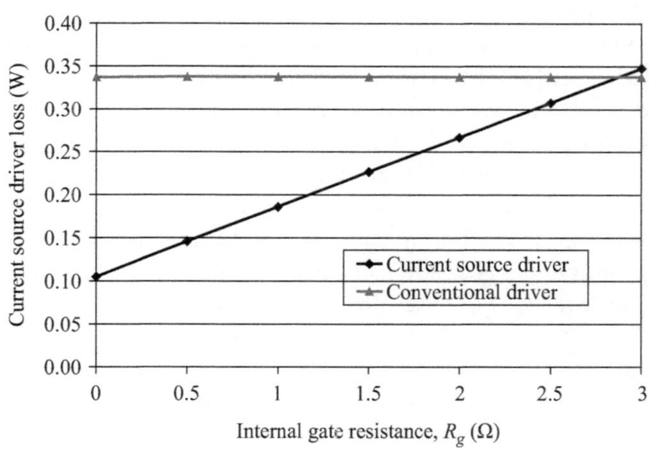

Figure 4.12 CSD loss as a function of MOSFET gate resistance

Figure 4.13 Photo of the proposed CSD in a boost converter topology (left) and the boost converter with UCC37322 driver (right; driver on bottom)

diode was used in the boost converter. The output capacitance consisted of six 10 μF, 16 V, and 1,206 ceramic capacitors. A six-layer, 1.5 oz printed circuit board (PCB) was used for both prototypes. A photo displaying both prototypes is given in Figure 4.13. The boost converter with CSD is on the left and the boost converter with conventional driver, UCC37322, is on the right. It is noted that the circuit area occupied by the proposed driver is much larger than the conventional driver. However, for practical implementation and widespread use, complete semiconductor integration of the proposed driver would reduce its size significantly.

The CSD was designed to deliver 1.25 A average gate charging current to the IRF6618 gate in order to demonstrate the switching loss reduction. In order to do so, a turn-on time of approximately 40 ns ($T_{on}=Q_g/I_g=45$ nC/1.25 A$\,=36$ ns) was selected with a pre-charge time, T_{pre}, of 20 ns. Dallas Semiconductor DS1100 tapped delay lines were used to implement the delay times. Fairchild UHS series discrete logic components were used to implement the logic. Fairchild FMBT3646 BJT pair pre-drivers were used to drive the CSD driver switches, S_1–S_4. A 100 nH Coilcraft 1812SMS-R10L air core inductor was used for the driver inductance, L. Fairchild NDS351AN N-channel MOSFETs were used for S_3 and S_4. Fairchild FDN342P P-channel MOSFETs were used for S_1 and S_2.

Detailed waveforms of the turn-on transition are provided in Figure 4.14(a). The three control signals for S_1, S_2, and S_3 are at the top (S_4 is not shown and remains off for the entire turn-on transition). The middle waveform is the IRF6618 gate voltage and the bottom waveform is the driver inductor current. The turn on of S_2 (active low) initiates the turn-on inductor pre-charge interval where the inductor current ramps up during T_{pre}. After the designed 20 ns, S_3 turns off allowing the inductor to charge the gate of the IRF6618 power MOSFET during T_{on}. During this interval, the gate current remains at approximately 1.25 A and the power MOSFET voltage charges from 0 V to $V_{cc}=5$ V. After 40 ns, the gate is clamped high when S_1 (active low) turns on. After S_1 turns on, the inductor current ramps back down to zero while the inductor energy is returned to V_{cc}.

Detailed waveforms of the turn-off transition are provided in Figure 4.14(b). The three control signals for S_1, S_3, and S_4 are at the top (S_2 is not shown and

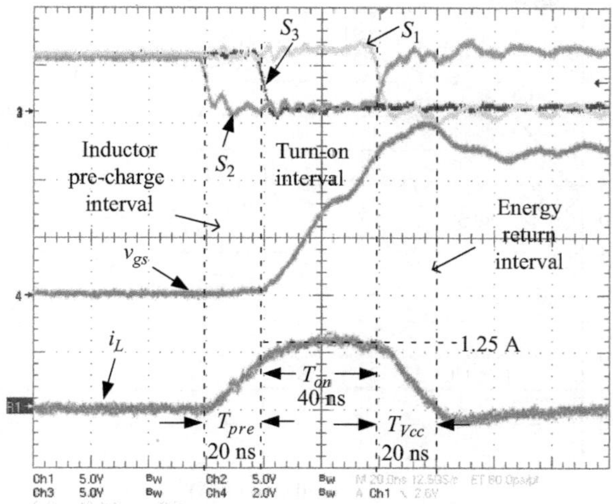

(a)

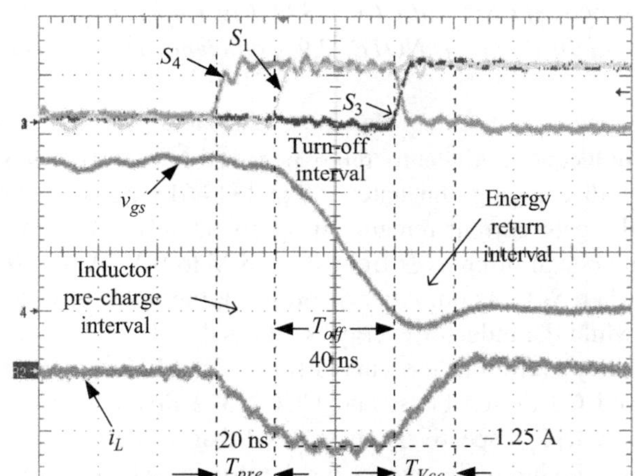

(b)

Figure 4.14 Detailed turn-on and turn-off waveforms; top: gate signals for S_1, S_2 and S_3 (5 V/div, 20 ns/div), second: IRF6618 gate-source voltage (2 V/div), bottom: CSD inductor current (1 A/div); (a) turn-on interval and (b) turn-off interval

remains off for the entire turn-on transition). The middle waveform is the IRF6618 gate voltage and the bottom waveform is the driver inductor current. The turn on of S_4 initiates the turn-off inductor pre-charge interval where the inductor current ramps negative during T_{pre}. After the designed 20 ns, S_1 turns off (active low)

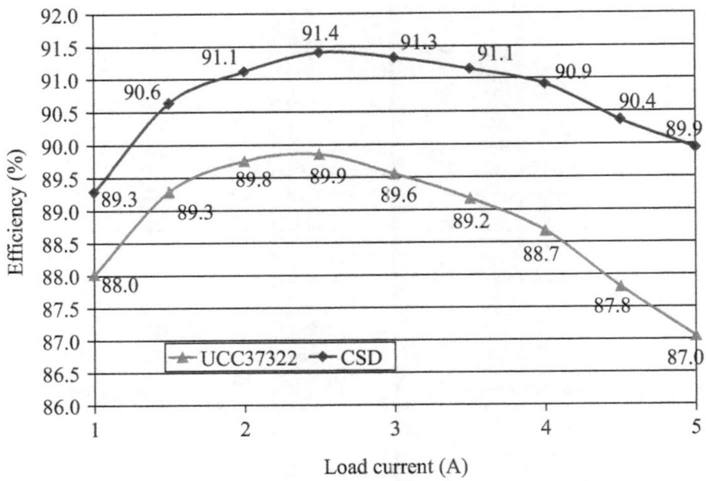

Figure 4.15 Efficiency curves as a function of load for the boost converter with proposed CSD and UCC37322 driver operating at 1 MHz, 5 V input and 10 V output; NOTE: 2.9% efficiency improvement at full load

allowing the inductor continue to ramp negative at a decreasing rate while at the same time discharging the gate of the IRF6618 power MOSFET. During this interval, the gate current remains at approximately −1.25 A and the power MOSFET gate voltage discharges from $V_{cc} = 5$ V to 0 V. After 40 ns, the gate is clamped low when S_3 turns on. After S_3 turns on, the inductor current ramps back down to zero while the inductor energy is returned to V_{cc}.

The efficiency and total loss curves (driver and boost converter powertrain) for the proposed CS driven boost and UCC37322 driven boost are provided in Figures 4.15 and 4.16, respectively. It is noted that in the testing, all components were the same in both circuits except for the drivers. The CSD maintains greater efficiency and lower power loss across the entire load range. Furthermore, the efficiency improvement at full load is 2.9%, representing a total power loss reduction of 1.85 W. It is noted that the loss reduction of 1.85 W is very close to the 1.8 W predicted and simulated. The loss reduction with the proposed driver is achieved by reduced gate-drive loss and reduced switching loss. In addition, at 1 A load current in Figure 4.16, the proposed CSD achieves a loss reduction of 0.16 W (1.36–1.20 W). At 1 A load current, switching loss and conduction loss are very small, so this loss reduction can be attributed to gate energy recovery. The loss reduction of 0.16 W is very close to the 0.15 W predicted in the analysis results reported in Figure 4.11.

The proposed driver was tested at other turn-on times of 20, 50, 80, 110, and 140 ns. There is a design tradeoff for the proposed driver between speed and driver loss. Greater switching speed (i.e., smaller T_{on}) requires greater driver current, which increases conduction loss in the driver. On the other hand, an added benefit

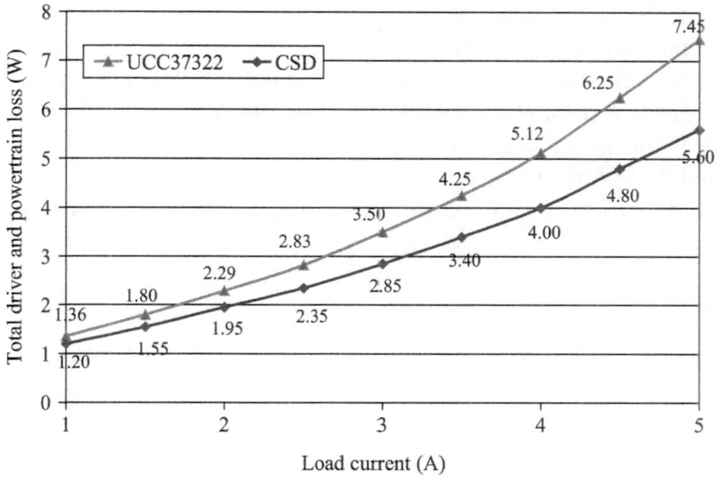

Figure 4.16 Total loss curves as a function of load for the boost converter with proposed CSD and UCC37322 driver operating at 1 MHz, 5 V input and 10 V output; NOTE: 1.8 W loss reduction achieved at full load

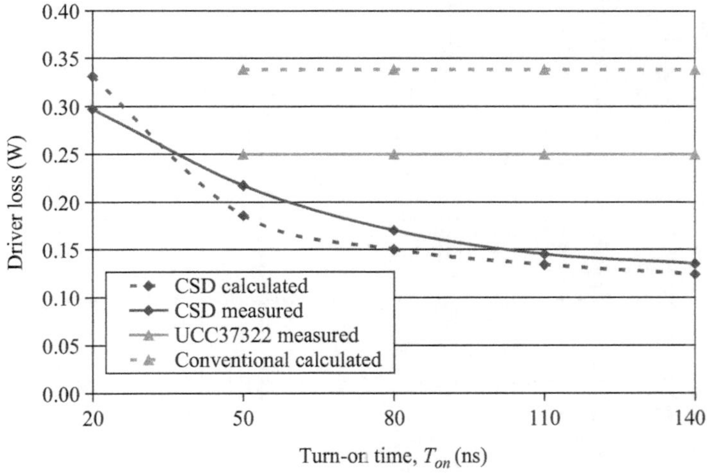

Figure 4.17 Curves of driver loss as a function of turn-on time

is that as turn-on time decreases, the required driver inductance also decreases. Curves of the proposed driver loss are given as a function of turn-on time, T_{on}, in Figure 4.17. Curves of the actual measured driver loss are also included. It is noted that the proposed driver operates with lower driving loss than the UCC37322 for all turn-on times achievable with the UCC37322 driver. While the driving loss reduction is small (50–100 mW for the example given), in other applications at higher switching frequencies, or with multiple switches, the potential for gate energy savings can become a significant.

The proposed CSD operates correctly for duty cycles ranging from 0% to 100%. Driver waveforms at 3% duty cycle are given in Figure 4.18(a). The top waveform is the PWM signal. The second waveform is the IRF6618 gate-source voltage and the bottom waveform is the driver inductor current. The peak inductor current during turn on is 1.25 A as expected. However, at turn off the peak inductor

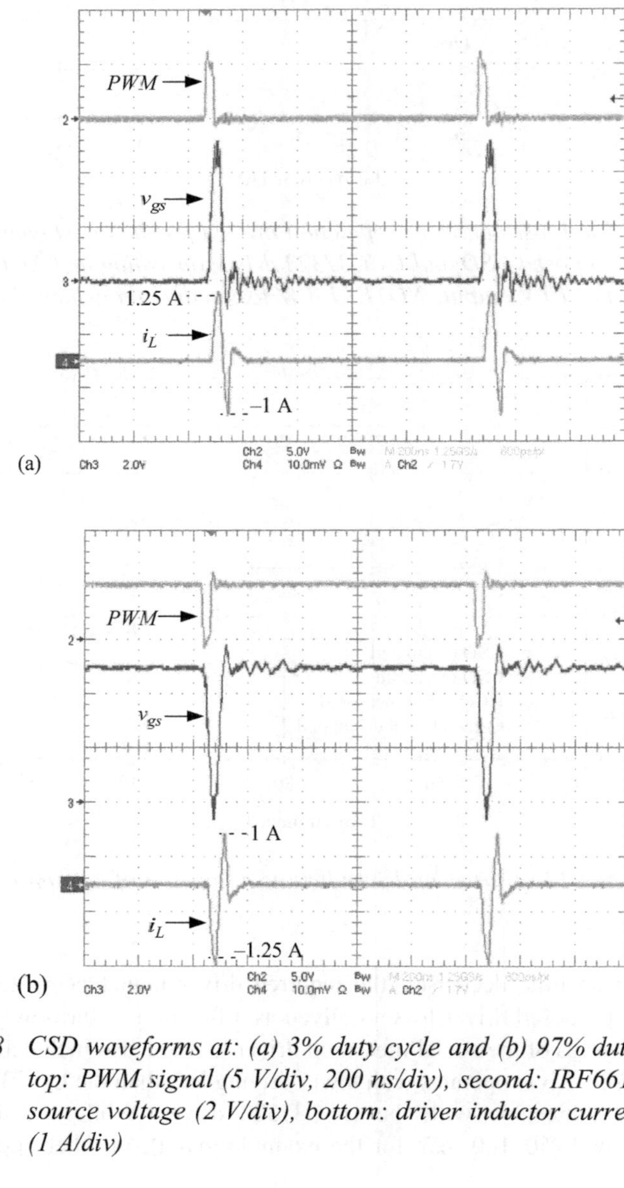

Figure 4.18 CSD waveforms at: (a) 3% duty cycle and (b) 97% duty cycle; top: PWM signal (5 V/div, 200 ns/div), second: IRF6618 gate-source voltage (2 V/div), bottom: driver inductor current (1 A/div)

current is −1 A. The lower peak inductor current at narrow duty cycles is expected since the turn-on energy recovery time, T_{Vcc} and the turn-off pre-charge time, T_{pre} overlap. This overlap causes the turn-off time to be slightly slower than expected. However, if the power MOSFET gate is not fully discharged by the inductor current after T_{off}, it quickly discharges when S_3 turns on. It is also noted that when the PWM input duty cycle is 0%, the power MOSFET gate-source voltage will remain clamped at 0 V.

Driver waveforms at 97% duty cycle are given in Figure 4.18(b). The top waveform is the PWM signal. The second waveform is the IRF6618 gate-source voltage and the bottom waveform is the driver inductor current. In contrast to the 3% duty cycle example, the peak inductor current during turn off is −1.25 A as expected and the peak inductor current at turn on is 1 A. At wide duty cycles, the turn-off energy recovery time, T_{Vcc} and the turn-on pre-charge time, T_{pre} overlap, resulting in the turn-on time to be slightly slower than expected. However, if the power MOSFET gate is not fully charged by the inductor current after T_{on}, it quickly charges to V_{cc} when S_1 turns on.

4.2 High-side discontinuous CSD

4.2.1 Proposed CSD for synchronous buck converter and operation

The proposed synchronous buck CS gate-drive circuit is illustrated in Figure 4.19. The circuit consists of the synchronous buck and two CSDs similar to the ground referenced driver proposed in Section 4.1. Each driver has four small MOSFETs and two very small inductors that carry a discontinuous current, which minimizes

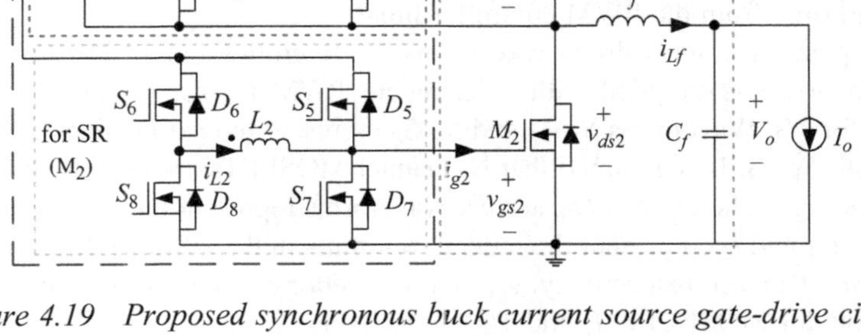

Figure 4.19 Proposed synchronous buck current source gate-drive circuit

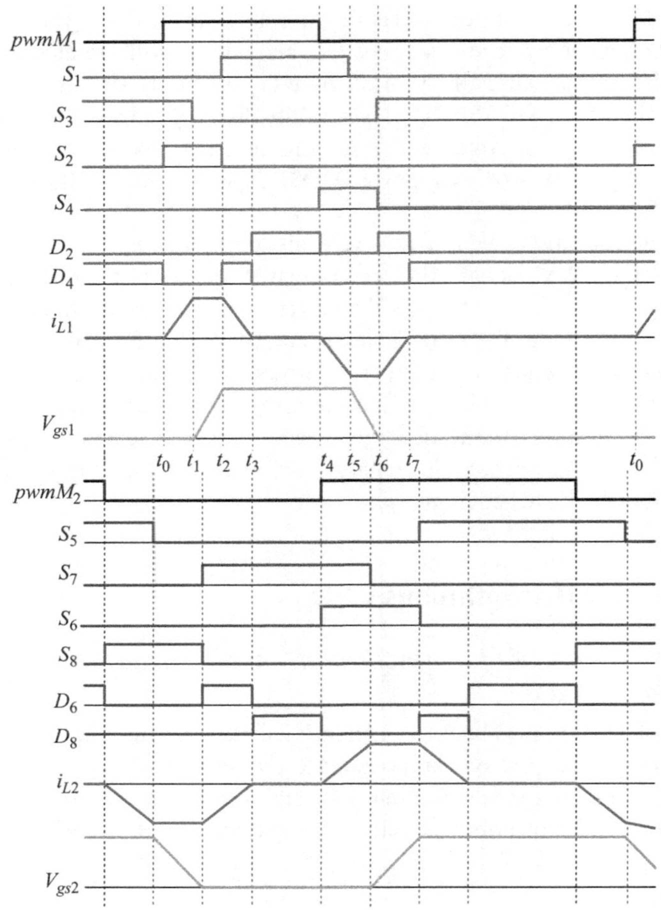

Figure 4.20 Proposed driver waveforms

conduction loss and allows the driver to work effectively over a wide duty cycle range. The driver for the high-side (HS) MOSFET M_1 contains a bootstrap diode D_b and capacitor C_b. The operation of the driver is similar for M_1 and M_2 since each driver operates using independent control signals ($pwmM_1$ and $pwmM_2$) derived with dead time from the PWM controller input.

The piecewise linear driver waveforms are given in Figure 4.20. It is assumed that each driver is supplied with independent PWM inputs, $pwmM_1$ for M_1 and $pwmM_2$ for M_2. Waveforms S_1–S_4 and S_5–S_8 represent the gating signals for driver MOSFETs S_1–S_8. It is assumed that N-channel MOSFETs are used, so all devices are active high. The D_2, D_4, D_6, and D_8 waveforms represent the diode conduction intervals. i_{L1} and i_{L2} represent the inductor currents in the HS MOSFET driver and SR MOSFET driver, respectively. v_{gs1} and v_{gs2} represent the gate-to-source voltage waveforms for MOSFETs M_1 and M_2, respectively.

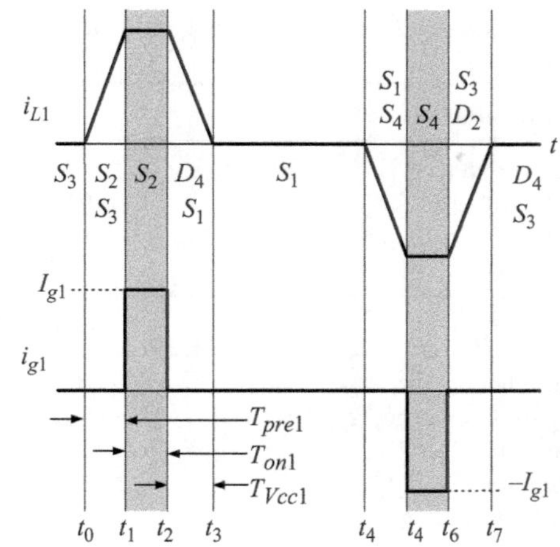

Figure 4.21 CSD inductor (top) and gate (bottom) piecewise linear current waveforms for the HS MOSFET (M_1)

The key to the driver operation is the controlling of the driver switches and use of the body diodes to generate the discontinuous inductor current waveforms, i_{L1} and i_{L2}. A portion of the inductor current waveform at its peak is then used to charge the power MOSFET gate as a nearly constant CS. This concept is illustrated in Figure 4.21, for the HS driver of M_1.

4.2.2 Design example

This section includes design examples for the HS MOSFET driver for M_1 and SR MOSFET driver, for M_2.

The parameters for the design of HS MOSFET CSD are given in Table 4.2. An inductor value of 68 nH was selected for L_1.

Using the values in Table 4.2, the loss curves for the driver, P_{dr1}, switching loss, P_{sw1}, and total gate current dependent loss, P_{tot1}, are given in Figure 4.22. The minimum power loss point, $P_{tot1min}$ corresponds to a gate current of $I_{g1opt} = 3.25$ A, enabling the pre-charge inductor current to be calculated as 20 ns.

The parameters for the design of the SR MOSFET CSD are given in Table 4.3. An inductor value of 307 nH was selected for L_2.

Using the values in Table 4.3, the loss curves for the driver, P_{dr2}, body diode conduction loss, P_{BD2}, and total gate current dependent loss, P_{tot2}, are given in Figure 4.23. The minimum power loss point, $P_{tot2min}$, corresponds to a gate current of $I_{g1opt} = 1.3$ A, enabling the pre-charge inductor current to be calculated as 40 ns.

4.2.3 Experimental results

In this section, experimental results are presented to compare the proposed CSD to a benchmark synchronous buck converter with conventional driver.

Table 4.2 HS MOSFET CSD design parameters

Circuit parameters	
Switching frequency, f_s	1 MHz
Gate drive voltage, V_{cc}	10 V
Load current, I_o	30 A
Input voltage, V_{in}	12 V
MOSFET, M_1	**IRF6617**
Total gate charge, Q_{g1}	20 nC
Internal gate resistance, R_{g1}	1 Ω
Gate charge at V_{pl1}, Q_{pl1}	4 nC
Gate charge at V_{th1}, Q_{th1}	2 nC
Gate-drain charge, Q_{gd1}	4 nC
Driver Switches, S_1–S_4	**NDS351AN**
Diode forward voltage, V_F	0.7 V
R_{dsS1}–R_{dsS4}	140 mΩ
Fall time, T_{f2}, T_{f4}	1 ns
Driver inductor, L_1	**1812SMS-68N**
Inductor value, L_1	68 nH
Driver inductor resistance, R_{L1}	20 mΩ

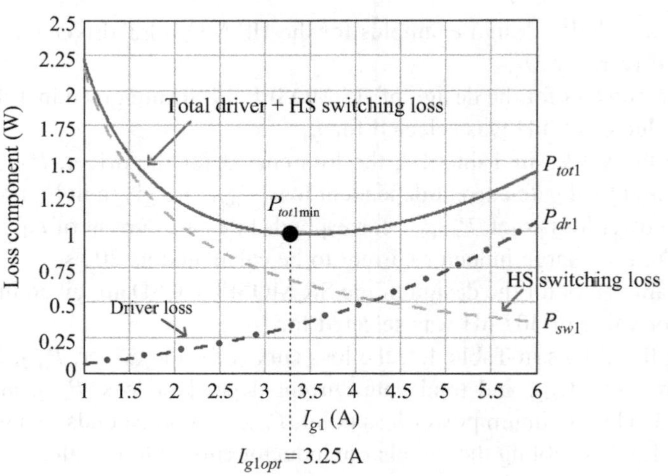

Figure 4.22 Optimization curves for the control MOSFET M_1; power loss vs. gate current, I_{g1}

Table 4.3 SR MOSFET CSD design parameters

Circuit parameters	
Switching frequency, f_s	1 MHz
Gate drive voltage, V_{cc}	10 V
Load current, I_o	30 A
Input voltage, V_{in}	12 V
MOSFET, M_2	**IRF6691**
Total gate charge, Q_{g2}	60 nC
Internal gate resistance, R_{g2}	1 Ω
Gate charge at V_{pl2}, Q_{pl2}	18 nC
Diode forward voltage, V_{F_2}	0.5 V
Driver Switches, S_1–S_4	**NDS351AN**
Diode forward voltage, V_F	0.7 V
R_{dsS5}–R_{dsS8}	140 mΩ
Fall time, T_{f6}, T_{f8}	1 ns
Driver inductor, L_2	**132-16SM**
Inductor value, L_2	307 nH
Driver inductor resistance, R_{L2}	70 mΩ

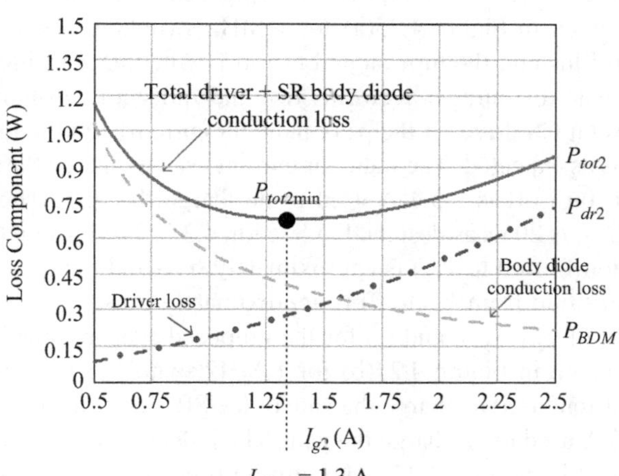

Figure 4.23 Optimization curves for the SR MOSFET M_2; power loss vs. gate current, I_{g2}

A prototype of the synchronous buck converter with CSD was built on a six-layer PCB as shown in the photo in Figure 4.24. The driver was built using discrete components with an Altera MaxII EPM240 CPLD used to generate the control signals. Circuit, driver and MOSFET parameters from Tables 4.2 and 4.3 in

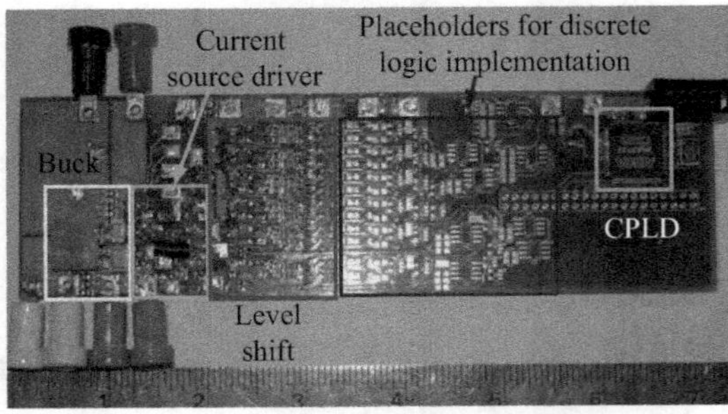

Figure 4.24 Photo of the proposed synchronous buck converter with CSD prototype

Section 4.2.2 were used. The buck inductor was IHLP5050FD, 330 nH from Vishay. Load voltages were tested at 1.5, 1.3, 1.2, and 1.0 V up to a full-load current of $I_o = 30$ A. The output voltage was regulated within ± 30 mV of the nominal value using a function generator with 1 ns resolution control of the duty cycle. The proposed synchronous buck with CSD was compared to an identical synchronous buck with the conventional UCC27222 driver from Texas Instruments as the benchmark. The powertrain components and layout were the same for the two circuits.

Waveforms of v_{gs1}, v_{gs2}, and i_{L1} for the proposed synchronous buck converter with CSD are given in Figure 4.25(a) for 1 MHz switching frequency operation. The waveforms illustrate the turn-on and turn-off switching transitions of M_1. It is noted that the gate-to-source waveforms rise and fall with constant linear slopes due to the constant CS drive (at the peak inductor current) and the inductor current is discontinuous as expected. The inductor current used to charge the gate at turn on is 2.9 A and at turn off is -3.2 A used to discharge the gate. The inductor pre-charge time, T_{pre1}, is 20 ns as designed in Section 4.2.2. The turn-on time, T_{on1}, (for the v_{gs1} transition from 0 to V_{cc}) is approximately 6 ns and the turn-off time, T_{off1}, (for the v_{gs1} transition from V_{cc} to 0) is approximately 5 ns.

Waveforms of v_{gs1}, v_{gs2}, and i_{L2} for the proposed synchronous buck converter with CSD are given in Figure 4.25(b) for 1 MHz switching frequency operation. The inductor current used to charge the gate of the SR, M_2, at turn on is 1.3 A and at turn off is -1.3 A used to discharge the gate. The inductor pre-charge time, T_{pre2}, is 40 ns as designed in Section 4.2.2. The turn-on time, T_{on2}, (for the v_{gs1} transition from 0 to V_{cc}) is approximately 50 ns and the turn-off time, T_{off2}, (for the v_{gs2} transition from V_{cc} to 0) is also approximately 50 ns.

A waveform of v_{gsM1} for the synchronous buck converter with the proposed CSD is given in Figure 4.26(a) illustrating the turn-on and turn-off switching transitions of M_1. The measurement was taken at $V_o = 1.3$ V, $V_{in} = 12$ V and $I_o = 20$ A. The measured fall time is approximately 4 ns, which is half of the fall time for the UCC27222 driver implying that the turn-off switching loss is reduced by half with the CSD.

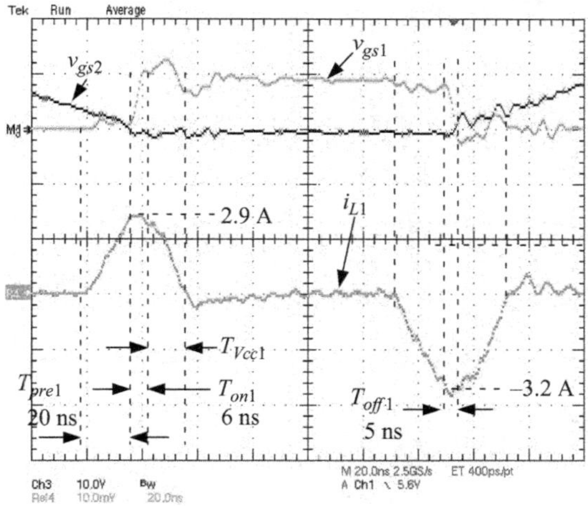

(a)

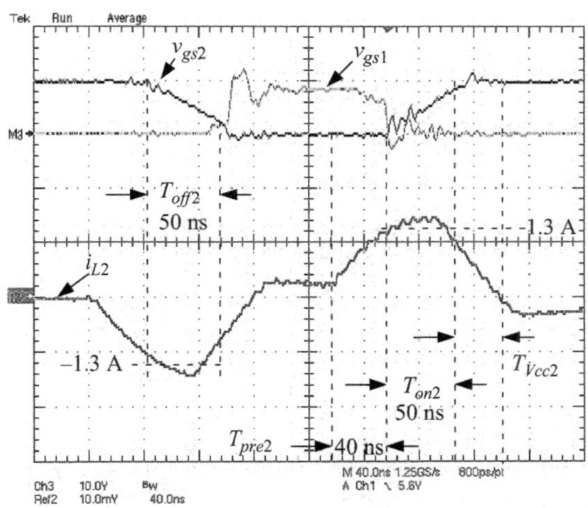

(b)

Figure 4.25 Top: CSD HS MOSFET (M_1) and SR (M_2) gate-to-source voltages at 1 MHz (y-axis: 10 V/div and x-axis 20 ns/div); bottom: M_1 CSD inductor current (y-axis: 2 A/div); with (a) proposed CSD and (b) conventional driver

A waveform of v_{gs1} for the synchronous buck converter with conventional UCC27222 driver is given in Figure 4.26(b) illustrating the turn-on and turn-off switching transitions of M_1. The measurement was taken at $V_o = 1.3$ V, $V_{in} = 12$ V and $I_o = 20$ A. The rise and fall times cannot be accurately measured since the common source inductance voltage is included in the probed measurement; however, the turn-off plateau is clearly evident. The measured fall time is approximately 8 ns.

The efficiency as a function of load is given in Figure 4.27 for the synchronous buck converter with CSD and synchronous buck converter with UCC27222 driver at 1 MHz switching frequency and 1.5 V output. The converter with CSD achieved

(a)

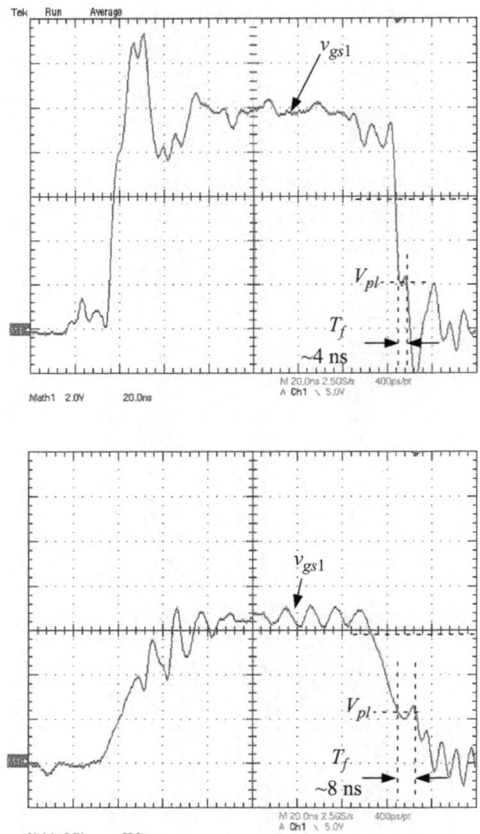

(b)

Figure 4.26 Synchronous buck converter with proposed CSD and UCC27222 driver HS MOSFET (M_1) gate-to-source voltage at 1 MHz (y-axis: 2 V/div and x-axis 20 ns/div); with (a) proposed CSD and (b) conventional driver

an efficiency of 87.5% at 15 A load and 83.3% at 30 A load, compared to 85.2% and 79.4%, respectively, for the UCC27222 driver.

The total power loss including powertrain and gate-drive loss for both converters is given in Figure 4.28. It is noted that at 30 A load, the proposed CSD saves 2.6 W, or 21.9% compared to the UCC27222 conventional driver. This loss reduction is significant for a multi-phase converter. For example, in five-phase buck converters, the total loss reduction would be 13 W.

The efficiency as a function of load is given in Figure 4.29 for the synchronous buck converter with CSD and the synchronous buck converter with UCC27222 driver at 1 MHz switching frequency and 1.2 V output. The converter with CSD achieved an efficiency of 85.5% at 15 A load and 80.5% at 30 A load, compared to 82.5% and 73.1%, respectively, for the UCC27222 driver.

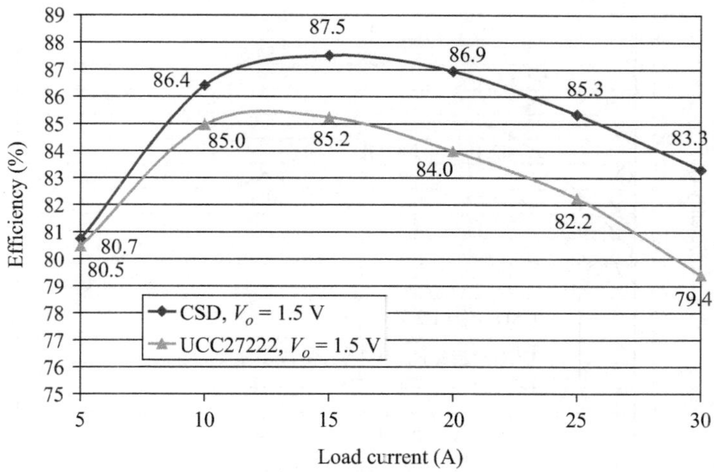

Figure 4.27 Efficiency as a function of load for the CSD and UCC27222 gate driver at 1.5 V output and 1 MHz switching frequency

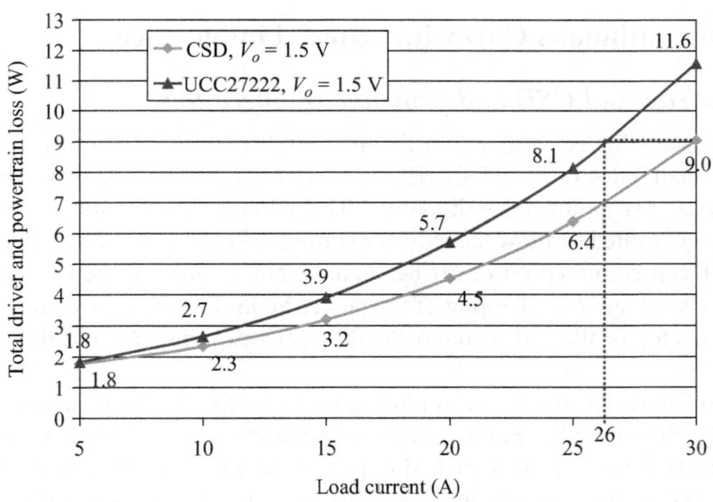

Figure 4.28 Total measured loss as a function of load for the CSD and UCC27222 gate driver at 1.5 V output and 1 MHz switching frequency

The semiconductor integration approach has a significant advantage since the parasitic inductance of the package leads between the buck HS and SR MOSFETs is negligible. It is believe that if the proposed synchronous buck converter with CSD were similarly integrated, the efficiency of the proposed solution would be improved further.

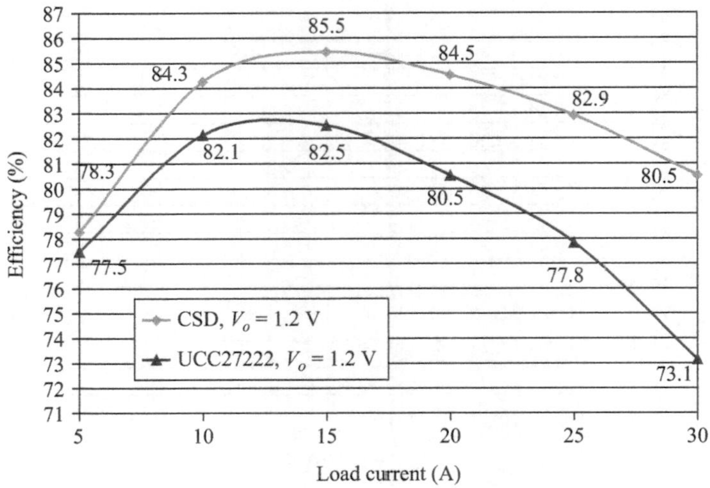

Figure 4.29 Efficiency as a function of load for the CSD and UCC27222 gate driver at 1.2 V output and 1 MHz switching frequency

4.3 Discontinuous CSD with reduced inductance

4.3.1 Proposed CSD and principle of operation

The proposed CSD is illustrated in Figure 4.30. It consists of four drive switches, S_1–S_4, a small inductor L_r and a series capacitor C_s. V_D is the gate-drive voltage. In the analysis, it is assumed that the same MOSFETs (N-channel) are used for S_1–S_4. S_1–S_4 are controlled to allow the inductor current to be discontinuous and the power MOSFET can be turned on or off beginning from a non-zero pre-charge current. Flowing charging, or discharging of the power MOSFET, the excess stored energy in the inductor is allowed to return to the series capacitor C_s and drive-voltage source.

Figure 4.31 illustrates the control gating signals, inductor current i_{Lr}, gate current i_G and power MOSFET gate-to-source voltage v_{GS}. The key waveforms to note are: (1) S_1 and S_2 are switched out of phase with complimentary control to drive Q; (2) the inductor current i_{Lr} is discontinuous to minimize conduction loss compared to the continuous circulating current as shown in dotted line; and (3) the gate-drive current i_G is relative constant during turn-on and turn-off transitions, which achieves fast switching speed of the power MOSFET.

There are eight switching modes in one switching period. The operation of the circuit is explained in the following paragraphs. The equivalent circuits of turn-on transition are illustrated in Figure 4.32(a)–(d). D_1–D_4 are the body diodes of S_1–S_4. C_1–C_2 are the intrinsic drain-to-source capacitors of S_1 and S_2. C_{gs} is the intrinsic gate-to-source capacitor of the main power MOSFET Q. The switching transitions of charging and discharging C_{gs} are during the intervals of [t_1, t_2] and [t_5, t_6], respectively. The peak current i_G during [t_1, t_2] and [t_5, t_6] are constant during the

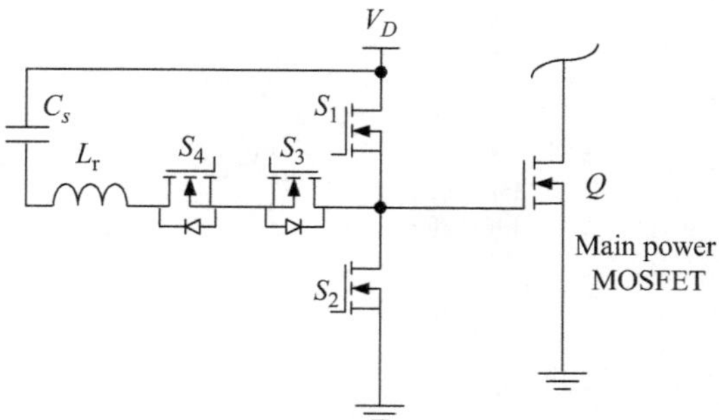

Figure 4.30 The proposed discontinuous CSD

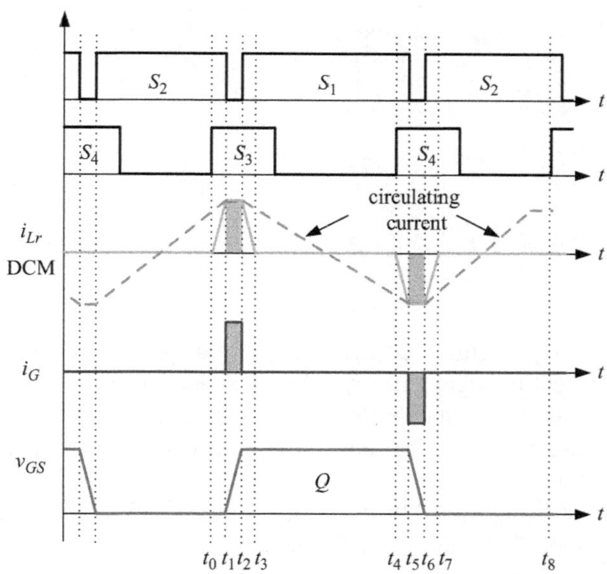

Figure 4.31 Key waveforms of the proposed CSD

switching transition, which ensures fast charging and discharging of Q gate capacitor including the Miller capacitor. Initially, it is assumed that the power MOSFETs is in the off state before time t_0.

1. Mode 1 [t_0, t_1] (Figure 4.32(a)): Prior to t_0, S_2 is on and the gate of Q is clamped to ground. At t_0, S_3 turns on (with ZCS) allowing the inductor current i_{Lr} to ramp up through D_4. The current path during this interval is C_s–L_r–S_3–D_4–S_2. This interval is the inductor current pre-charge interval and it ends at time t_1, which is a predetermined time set by the user. Since S_2 is in the on state, the gate of Q is always clamped low.

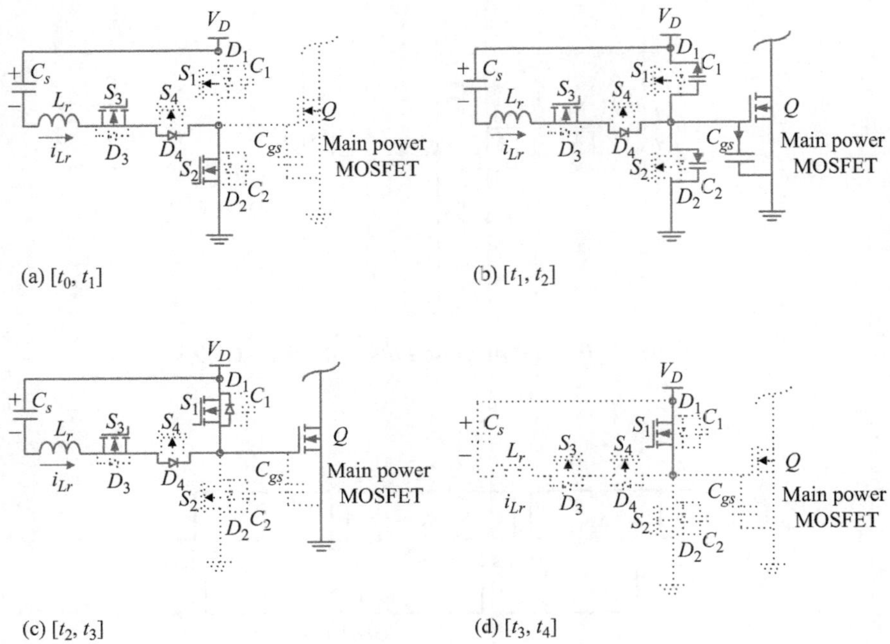

Figure 4.32 Equivalent circuits: turn-on intervals

2. Mode 2 $[t_1, t_2]$ (Figure 4.32(b)): At t_1, S_2 is turned off, which allows the inductor current to begin to charge the gate capacitor C_{gs}. i_{Lr} charges C_2 plus the input capacitor C_{gs} and discharges C_1 simultaneously. Due to C_1 and C_2, S_2 is zero-voltage turn off. The inductor current continues to ramp up from the pre-charged level.
3. Mode 3 $[t_2, t_3]$ (Figure 4.32(c)): At t_2, v_{c2} rises to V_D and v_{c1} decays to zero. The body diode D_1 conducts and S_1 turns on under zero-voltage condition. The inductor current continues to conduct through the path C_s–L_r–S_3–D_4–S_1. This interval continues for a short duration until t_3. During this interval, the gate of Q is clamped to the drive voltage V_D. This interval ends when the inductor current reaches zero at t_3. It is noted that it is during this interval when the stored energy in the inductor is returned to C_s. Also during this interval, the inductor voltage has become reverse biased, so the inductor current quickly ramps down toward zero.
4. Mode 4 $[t_3, t_4]$ (Figure 4.32(d)): At t_3, D_4 turns off (with ZCS) and the inductor current is zero. During this interval, the gate of Q remains clamped high. This interval ends at t_4 when the pre-charged interval for the turn-off cycle begins as dictated by the PWM control signals.

The equivalent circuits of turn-off intervals are illustrated in Figure 4.33(a)–(d).

5. Mode 5 $[t_4, t_5]$ (Figure 4.33(a)): At t_4, S_4 turns on (with ZCS). Since S_1 was previously on, the inductor current i_{Lr} begins to ramp negative through the path

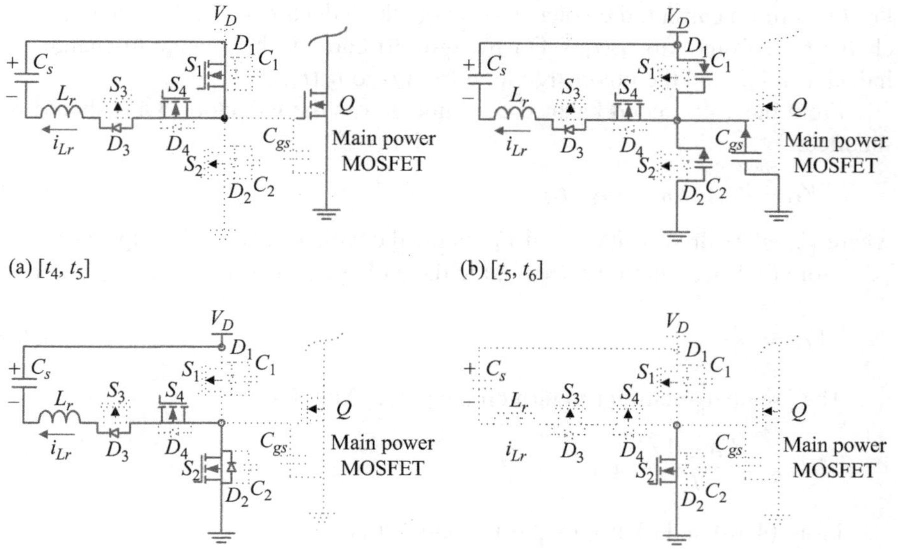

(a) $[t_4, t_5]$ (b) $[t_5, t_6]$

(c) $[t_6, t_7]$ (d) $[t_7, t_8]$

Figure 4.33 Equivalent circuits: turn-off intervals

C_s–S_1–S_4–D_3–L_r. The energy to charge the inductor is provided by C_s. During this interval, the gate of Q remains clamped to V_D. This interval ends at t_5.

6. Mode 6 $[t_5, t_6]$ (Figure 4.33(b)): At t_5, S_1 is turned off, which allows the inductor current to begin to discharge the gate capacitor C_{gs}. i_{Lr} discharges C_2 plus the input capacitor C_{gs} and charges C_1 simultaneously. Due to C_1 and C_2, S_1 is zero-voltage turn off. The inductor current continues to ramp negative from the pre-charged level.
7. Mode 7 $[t_6, t_7]$ (Figure 4.33(c)): At t_6, v_{c1} rises to V_D and v_{c2} decays to zero. The body diode D_2 conducts and S_2 turns on under zero-voltage condition. The inductor current continues to conduct through the path C_s–L_r–D_3–S_4–S_2. This interval continues for a short duration until t_7. During this interval, the inductor voltage has become reverse biased, so the inductor current quickly ramps down toward zero. It is noted that it is during this interval when the stored energy in the inductor is returned to the drive-voltage source. During this interval, the gate of Q is clamped low. This interval ends when the inductor current reaches zero at t_7.
8. Mode 8 $[t_7, t_8]$ (Figure 4.33(d)): At t_7, D_3 turns off (with ZCS) and the inductor current is zero. During this interval, the gate of Q_1 remains clamped low. This interval ends at t_8 when the pre-charged interval for the turn-on cycle begins and the entire process repeats as dictated by the PWM control signal.

The pre-charge current to turn on and turn off the power MOSFET is decided by the voltage to charge the current-source inductor and the pre-charge time.

For the turn-on current, the voltage to charge the inductor is (V_D–V_{Cs}) and the pre-charge time from t_0 to t_1 (t_{10}). For the turn-off current, the voltage to charge the inductor is V_{Cs} and the pre-charge time from t_4 to t_5 (t_{54}).

From the volt-second balance condition across the inductor, (4.47) should be satisfied:

$$(V_D - V_{Cs}) \cdot t_{10} = V_{Cs} \cdot t_{32} \tag{4.47}$$

where V_D is the drive voltage and V_{Cs} is the d.c. voltage across the capacitor.

From (4.47), assuming $t_{10} = t_{32}$, the d.c. voltage across the series capacitor is

$$V_{Cs} = \frac{V_D}{2} \tag{4.48}$$

The pre-charge current to turn on the power MOSFT is

$$I_{G\_on} = \frac{V_D - V_{Cs}}{L_r} \cdot t_{10} \tag{4.49}$$

From (4.48) and (4.49), the turn on current is

$$I_{G\_on} = \frac{V_D}{2L_r} t_{10} \tag{4.50}$$

From (4.48), the pre-charge current to turn off the power MOSFET is

$$i_{G\_off} = \frac{V_{Cs}}{L_r} \cdot t_{54} = \frac{V_D}{2L_r} \cdot t_{54} \tag{4.51}$$

From (4.49) and (4.50), by changing pre-charge time, the turn-on gate current and turn-off gate current can be decided. It is also noted that compared to the discontinuous CSD in Section 4.1, since the actual voltage over the inductor is reduced by half as $V_D/2$, for the same pre-charge time and gate-drive current, the proposed driver can reduce the inductance value by half.

The advantages of the new CSD are highlighted as follows:

1. Significant reduction of the switching transition time and switching loss
   The key idea of the proposed CSD is to control the four drive switches to create a constant CS to drive the main power MOSFETs. During the switching transition [t_1, t_2] and [t_5, t_6] (see Figure 4.31), the proposed CSD uses the pre-charge inductor current to drive the control MOSFET and absorbs the parasitic inductance. This reduces the propagation impact of the parasitics during the switching transition, which leads to a reduction of the switching transition time and switching loss. At the same time, the discontinuous current does not increase the circulating loss compared to other continuous CSDs.
2. Gate energy recovery
   The stored energy in the inductor is returned to the series capacitor C_s during [t_2, t_3] and is returned to the drive-voltage source during [t_6, t_7] (see Figure 4.31). One benefit of the gate energy recovery capability is that high gate-drive voltage can be used to further reduce $R_{DS(on)}$ conduction loss.

3. Small current-source inductance
   One of the most important advantages of the proposed CSD is the small inductance. Compared to the continuous CSDs in Section 4.1, which have the inductance value around 1 μH at the switching frequency of 1 MHz, the inductance of the proposed circuit is only about 10–20 nH. This is a significant reduction of the inductance value.
4. Wide range of duty cycle and switching frequency
   In a high-frequency buck converter, the duty cycle is required to response fast during a transient event. At the same time, in order to improve the efficiency in a wide load range, the switching frequency of a buck converter may need to vary according to the load condition. The proposed CSD operates correctly for duty cycles ranging from 0% to 100%. The gate-drive current (current-source inductor current) only depends on the pre-charge time and is independent of duty cycle and switching frequency, so it is suitable for different types of control and wide operating conditions.
5. High noise immunity
   With the new CSD, the gate terminal of the power MOSFETs are clamped to either the drive-voltage source via a low impedance path (S_1 fairly small $R_{DS(on)}$) or the source terminal (S_2). This offers high noise immunity and leads to the alleviation of dv/dt effect.

4.3.2 Proposed high-side CSD and hybrid gate-drive scheme

Figure 4.34 illustrates the proposed HS CSD for non-ground referenced power MOSFET. It uses a bootstrap circuit consisting of a diode D_f and a bootstrap capacitor C_f. This CSD can be used to the control MOSFET in a buck converter to achieve fast switching and reduced switching loss.

Figure 4.35 illustrates another version of the HS CSD using C_{s1} and C_{s2} in series as the bootstrap capacitor, where C_{s1} and C_{s2} also serve as the bootstrap capacitors.

Figure 4.36 shows the loss breakdown of buck converter with the conventional voltage-gate driver. Carefully investigating the switching behavior of the

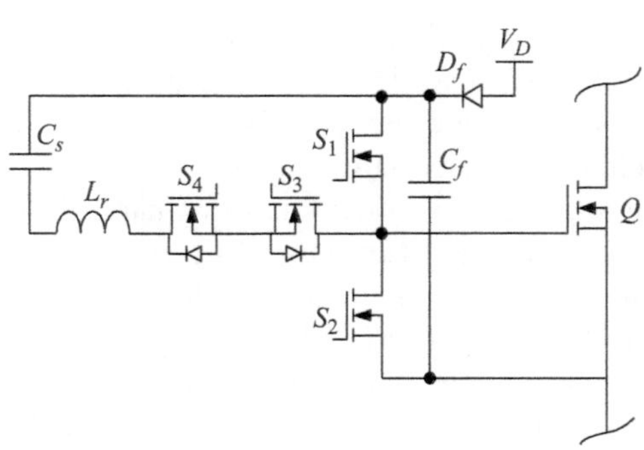

Figure 4.34 Proposed high-side CSD

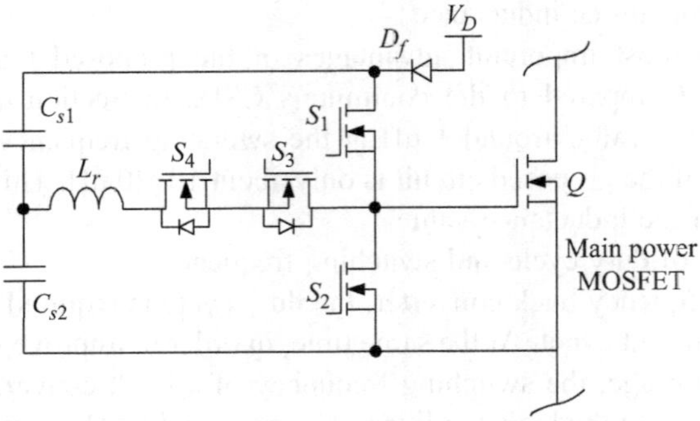

Figure 4.35 Proposed high-side CSD using series capacitors

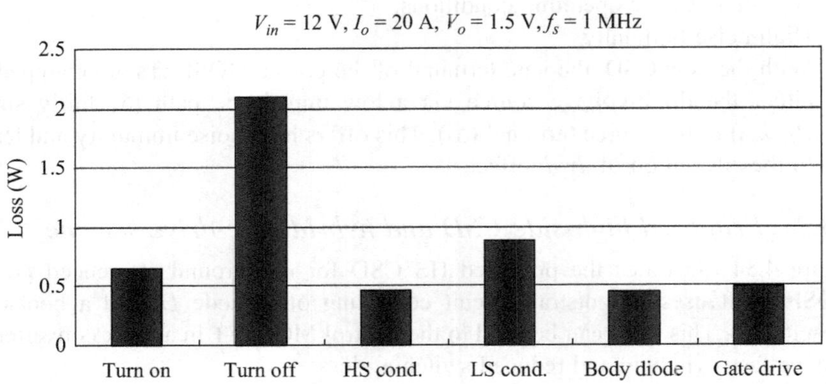

Figure 4.36 Loss breakdown of the buck converter with the conventional voltage-gate driver ($L_s = 1$ nH, $L_D = 2$ nH, control MOSFET: Si7860DP and SR: Si7336ADP)

MOSFETs in a synchronous buck converter, it is interesting to observe that the switching loss of the control MOSFET is dominant among the total loss breakdown as shown in Figure 4.36. For the control MOSFET, the common source inductance results in high switching loss, especially, turn-off loss, which makes the fast turn-off transition more desirable. However, on the other hand, for the SR, it almost has no switching loss since its output capacitor has been discharged to zero voltage before it turns on, which can be regarded to achieve ZVS. Moreover, due to the ZVS condition, no Miller charge is present and the gate charge is saved by around 30%. Also, the common-source inductance could help to improve the *dv*/*dt* immunity of the SR MOSFET [1]. So the conclusion is that control MOSFET and the SR MOSFET have different switching behavior as far as the parasitics are concerned.

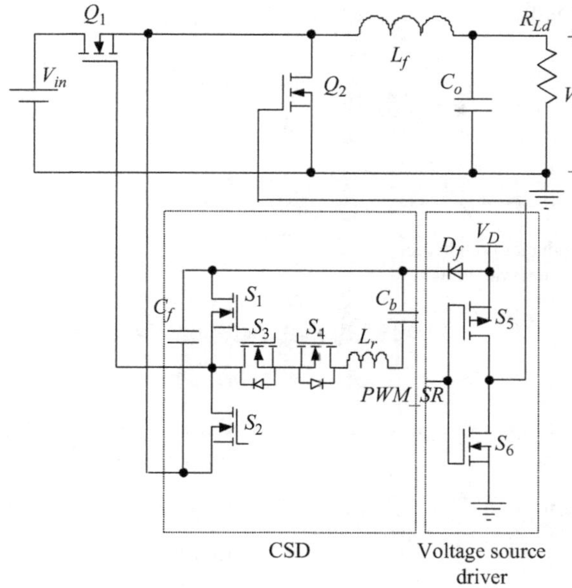

Figure 4.37 The proposed discontinuous CSD diagram in the experiment

4.3.3 Experimental results

The first experiment is to verify the switching loss reduction with the HS CSD for a 12 V input buck converter. The circuit diagram of the experimental prototype is shown in Figure 4.37.

The idea to use the proposed CSD is to achieve fast switching speed and reduce the switching loss of the control MOSFET in a buck converter. Conventional voltage-source driver is used for the SR MOSFET for its simplicity, good immunity and alleviation of *dv/dt* effect as the switching loss for SR is minimum.

The specifications of the buck converter prototype are as follows: input voltage $V_{in} = 12$ V; output voltage $V_o = 1.0$ V–1.5 V; output current $I_o = 25$ A; switching frequency $f_s = 1$ MHz; gate-driver voltage $V_D = 5$ V. The PCB is six-layer with 4 oz copper. The components used in the circuit are listed as follows: Q_1: Si7860DP; Q_2: irf6691; output filter inductance: $L_f = 300$ nH (IHLP-5050CE-01, Vishay); current-source inductor: $L_r = 18$ nH (SMT 1812SMS-22N, Coilcraft); drive switches S_1–S_4: FDN335.

Figure 4.38(a) shows the gate-drive signal for the four drive MOSFETs S_1–S_4 from the CPLD. All waveforms agree with the theory in Figure 4.31. Most notably, the pre-charge intervals are indicated as 20 ns.

Figure 4.38(b) shows the inductor current i_{Lr} and gate-drive signals v_{gs_Q1} (control MOSFET). Its peak current value is 2.0 A, which is the optimized value of the CSD drive current. The inductor current is discontinuous as expected. During the pre-charge time, the current ramps up linearly. After the pre-charge time, the

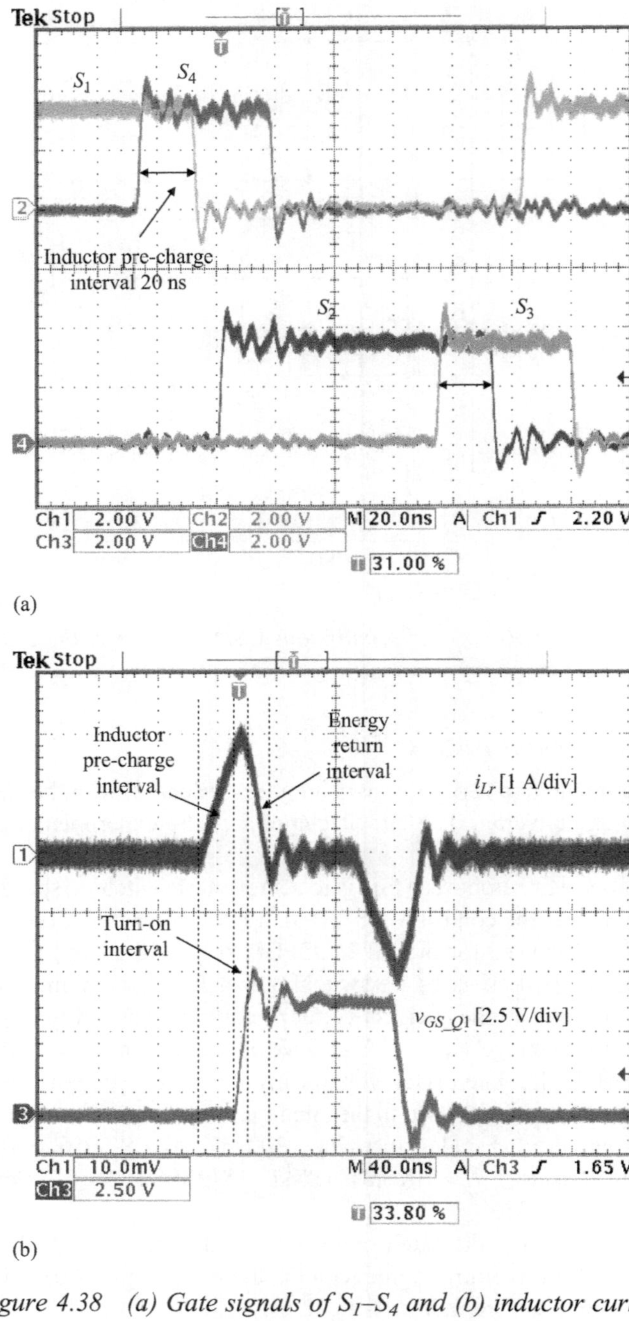

Figure 4.38 (a) Gate signals of S_1–S_4 and (b) inductor current

inductor current continues to ramp up while charging the gate capacitance of the Si7860DP power MOSFET during the turn-on intervals. During this interval, the average drive current is approximately 2.0 A and the power MOSFET voltage charges from 0 V to $V_D = 5$ V. After the power MOSFET turns on, the inductor current ramps back down to zero while the inductor energy is returned to *drive-voltage source*.

Figure 4.39(a) shows the gate-drive signals v_{gs_Q1} (control MOSFET) and v_{gs_Q2} (SR). It is observed that v_{gs_Q1} is smooth since the Miller charge is removed fast by the constant inductor drive current. Moreover, the total rise time and fall time of v_{gs_Q1} is less than 15 ns, which means fast switching speed. The dead time between two drive voltages is fixed to avoid shoot-through and is minimized to reduce the SR body diode conduction loss.

Figure 4.39(b) shows the drain-to-source voltage v_{ds_Q2} of the SR at the load current of 25 A. It can be seen from v_{ds_Q2} that the body diode conduction time is small, which reduces the conduction loss and the reverse recovery loss of the body diode. It should be also noted that adaptive control or predictive of the voltage-source driver can also be applied to the hybrid gate driver.

A benchmark of a synchronous buck converter with the conventional gate driver was built. The predictive gate drive UCC 27222 from Texas Instruments was used as the conventional voltage driver. Figure 4.40 shows the measured efficiency comparison for the hybrid gate driver and the conventional gate driver at 1.3 V output. It is observed that at 25 A, the efficiency is improved from 80.7% to 85.4% (an improvement of 4.7%) and at 30 A, the efficiency is improved from 77.9% to 83.9% (an improvement of 6%).

Figure 4.41 gives the measured efficiencies for the CSD with different output voltages and load currents at 1 MHz respectively.

The second experiment is to verify the gate energy recovery of the proposed CSD. Typically, high gate-drive voltage helps to reduce the MOSFET $R_{DS(on)}$ and conduction loss. The SR MOSFET usually has high total gate charge. However, it also increases the gate-drive loss. Typically, the gate-drive voltage of 6 V–7 V gives a good tradeoff between the conduction loss and the gate-drive loss. In addition, in high-current application, more SR MOSFETs are often paralleled to reduce $R_{DS(on)}$ and thus the conduction loss. In this experimental test, the proposed low-side CSD is used to drive two paralleled IRF6691 as SRs. The CSD with the same components and parameters is tested. The conventional gate-driver chip ISL6208 [2] from Intersil is used for the same SR MOSFETs for comparison.

Figure 4.42(a) and (b) illustrates the gate-drive voltages of the SR with the proposed CSD at 1 and 2 MHz, respectively. In this figure, it is observed that the turn-on and turn-off transition time is less than 20 ns, which means that the proposed CSD achieves fast turn-on and turn-off speed of the SR MOSFET.

The measured losses are listed in Table 4.4. Three different drive voltages of 5, 6, and 7 V are tested under different switching frequency condition. It is observed that when the gate-driver voltage increases, the loss difference (Δ loss) increases, which means the CSD is more effective with high-drive gate-drive voltages to reduce $R_{DS(on)}$ conduction loss. For example, $R_{DS(on)}$ of IRF6691 [3] is reduced

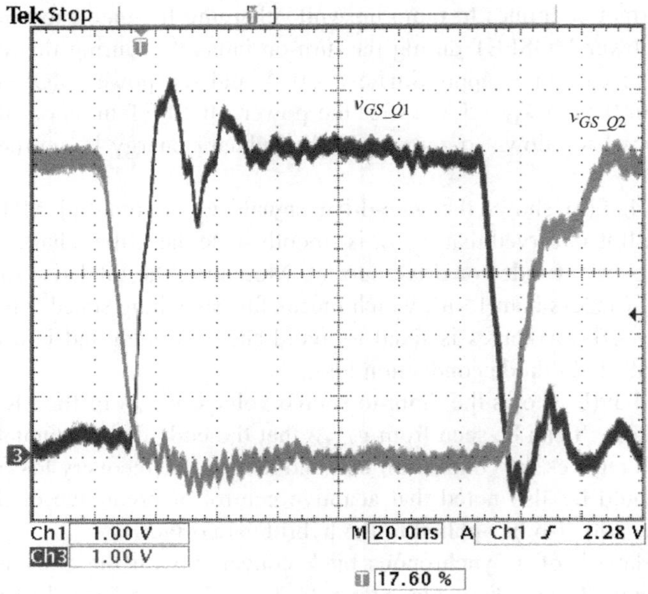

(a)

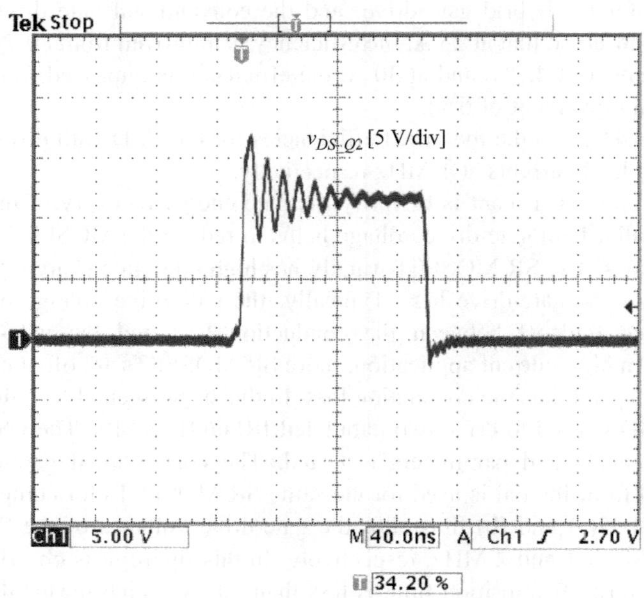

(b)

Figure 4.39 *(a) Gate signals v_{GS_Q1} (control MOSFET), v_{GS_Q2} (SR) and (b) drain-to-source voltage v_{DS_Q2} (SR) at load current of 25 A*

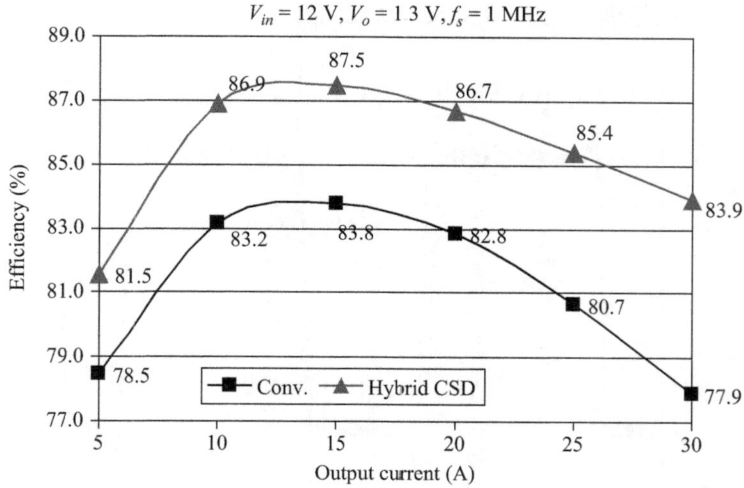

Figure 4.40 Efficiency comparison: top: hybrid CSD; bottom: conventional voltage driver (Conv.)

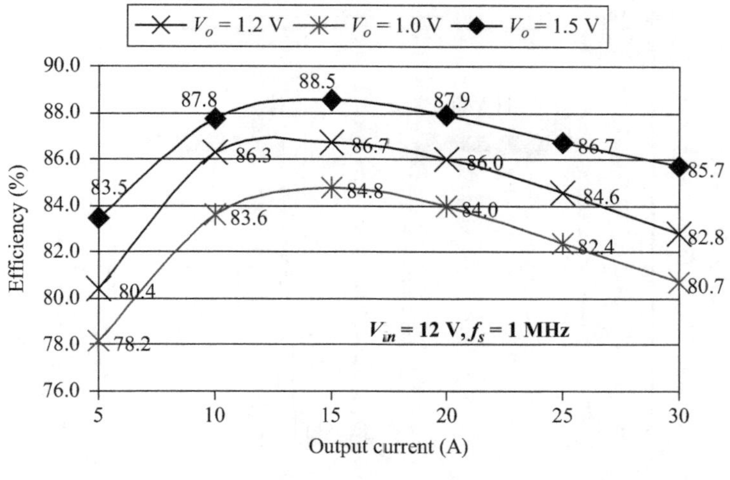

Figure 4.41 Efficiency with different output voltages and currents at 1 MHz

from 2.5 mΩ@V_{GS} = 5 V drive voltage to 1.9 mΩ@ V_{GS} = 7 V (a reduction of 24%). In Table 4.4, it is noted that with V_D = 7 V and the switching frequency of 2 MHz, the gate loss reduction with the CSD is as much as 2.24 W, a reduction of 63% with the voltage-source driver.

For the step change of the duty cycle, Figure 4.43(a) and (b) illustrates the duty cycle changes from D = 0.1 to D = 0.8 and from D = 0.8 to D = 0.1, respectively. It is observed that the proposed CSD can responds fast when duty cycle has a step

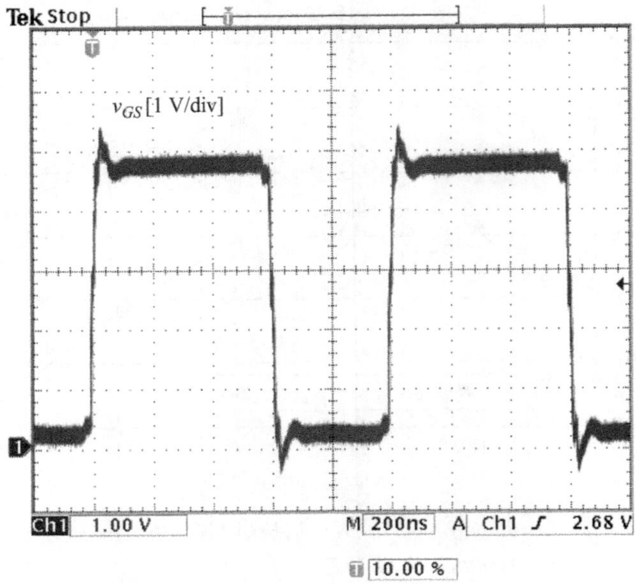

(a)

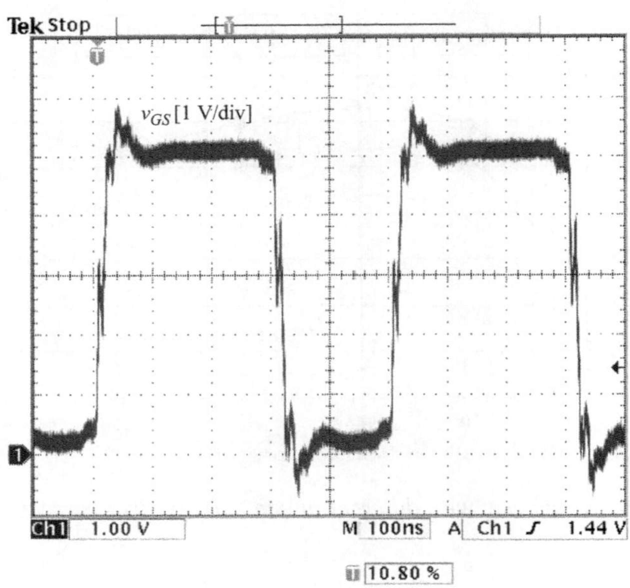

(b)

Figure 4.42 Gate signals v_{GS}: (a) 1 MHz and (b) 2 MHz

Table 4.4 Measured loss comparison of proposed CSD and conventional (Conv.) voltage driver

Switching frequency	$V_D = 5$ V			$V_D = 6$ V			$V_D = 7$ V		
	CSD (W)	Conv. (W)	Δ *Loss* (W)	CSD (W)	Conv. (W)	Δ *Loss* (W)	CSD (W)	Conv. (W)	Δ *Loss* (W)
800 kHz	0.28	1.07	0.79	0.40	1.76	1.36	0.56	2.40	1.84
1.0 MHz	0.35	1.10	0.75	0.49	1.98	1.49	0.69	2.70	2.01
1.2 MHz	0.41	1.30	0.89	0.61	2.06	1.45	0.83	2.80	1.97
1.5 MHz	0.52	1.40	0.88	0.75	2.30	1.55	1.02	3.13	2.11
1.8 MHz	0.60	1.50	0.90	0.87	2.46	1.59	1.19	3.35	2.16
2.0 MHz	0.65	1.62	0.97	0.95	2.60	1.65	1.30	3.54	2.24

change. This is of great benefit for voltage regulator (VR) applications with fast dynamic response requirement.

4.4 Current diversion

4.4.1 Introduction

The CSDs can reduce the switching loss by charging and discharging the MOSFET with a nearly constant current. Another advantage of the CSDs is that the gate-drive waveforms of CSDs are bipolar [4], which means that CSDs can turn off the power MOSFET with a negative voltage. The typical CSDs presented previously can turn off the power MOSFET with −0.7 V, while the CSD presented in [5] can turn off the power MOSFET with −3.5 V. In comparison, conventional VSDs can only turn off the power MOSFET with +0.5 V. Therefore, compared to VSDs, CSDs can significantly reduce the turn-off loss, which is the dominant part of the whole switching loss.

However, during the switching transitions, the current in the CS inductor is diverted, which reduces the effective drive current to the MOSFET. This is known as the current diversion problem, which commonly exists in the CSDs. In order to maintain the performance of CSDs compared with VSDs, the current diversion problem needs to be analyzed mathematically to predict and optimize the performance of the CSDs more accurately.

An analytical loss model, which thoroughly analyzes the impact of the parasitic inductance in CSDs, is presented in [6] to evaluate the performance of the CSDs. In addition, a generalized method to optimize the overall performance of the buck converter with a CSD is analyzed in the proposed model. A piecewise model that enables easy calculation and estimation of the switching loss is also proposed in [7]. However, the current diversion problem, which reduces the effective gate current and the switching speed, has not been investigated yet.

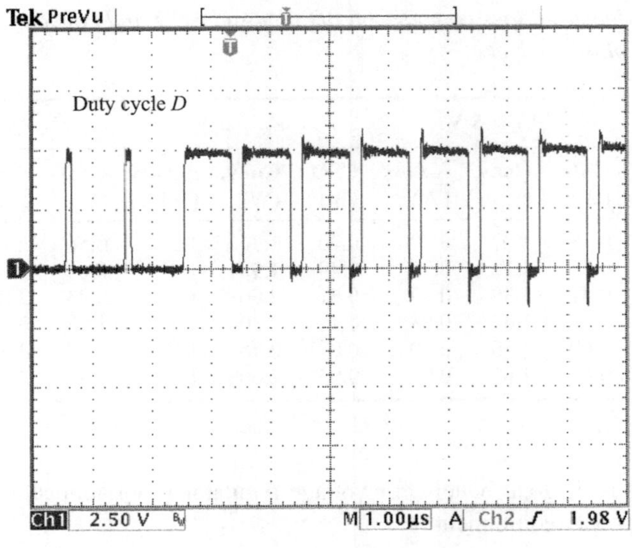

(a)

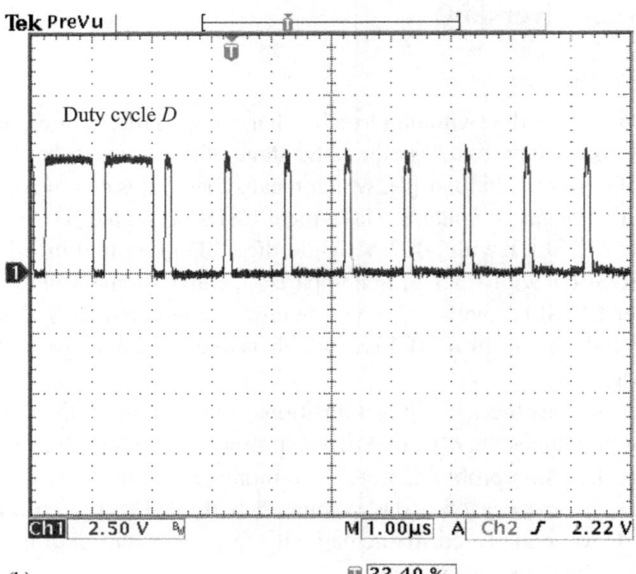

(b)

Figure 4.43 Gate signals v_{gs_Q1} (control MOSFET): (a) from $D = 0.1$ to $D = 0.8$ and (b) $D = 0.8$ to $D = 0.1$

4.4.2 Proposed switching loss model considering current diversion

The equivalent circuit of the MOSFET driven by proposed CSD is shown in Figure 4.44, where the power MOSFET Q is represented by a typical capacitance model, L_S is the common source inductance including the PCB track and the bonding wire inside the MOSFET package and L_D is the switching loop inductance. For the purpose of the transient analysis, the following assumptions are made [8]:

1. $i_{DS} = g_{fs}(v_{CGS} - V_{th})$ and MOSFET is ACTIVE, provided $v_{CGS} > V_{th}$ and $v_{DS} > i_{DS}R_{DS(on)}$
2. For $v_{CGS} < V_{th}$, $I_{DS} = 0$, and MOSET is OFF
3. When $g_{fs}(v_{CGS} - V_{th}) > v_{DS}/R_{DS(on)}$, the MOSFET is fully ON.

where i_{DS} is the drain current of the Q, g_{fs} is the transconductance, v_{DS} is the voltage across the drain-source capacitance of the Q, v_{CGS} is the voltage across the gate-source capacitance of the Q, V_{th} is the threshold voltage of Q, $R_{DS(on)}$ is the drain-source on-state resistance. During the Active State when switching loss happens:

$$i_{DS} = g_{fs}(v_{CGS} - V_{th}) \tag{4.52}$$

According to Figure 4.44, i_G is the effective current to charge or discharge Q as shown below,

$$i_G = (C_{GS} + C_{GD})\frac{dv_{GS}}{dt} - C_{GD}\frac{dv_{DS}}{dt} \tag{4.53}$$

v_{DS} is given as,

$$v_{DS} = V_{in} - L_D\frac{di_{DS}}{dt} - L_S\frac{d(i_{DS} + i_G)}{dt} \tag{4.54}$$

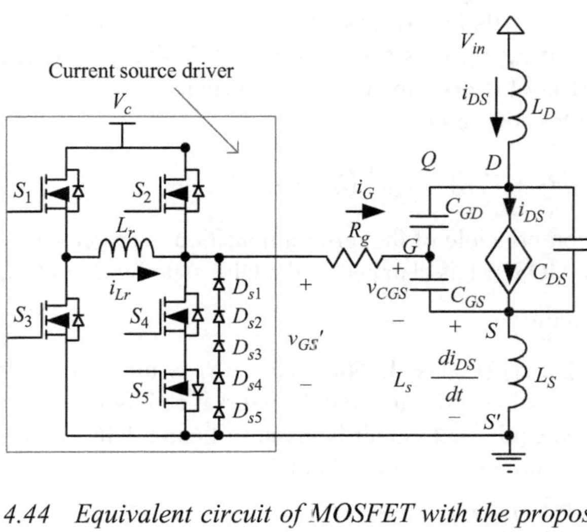

Figure 4.44 Equivalent circuit of MOSFET with the proposed CSD

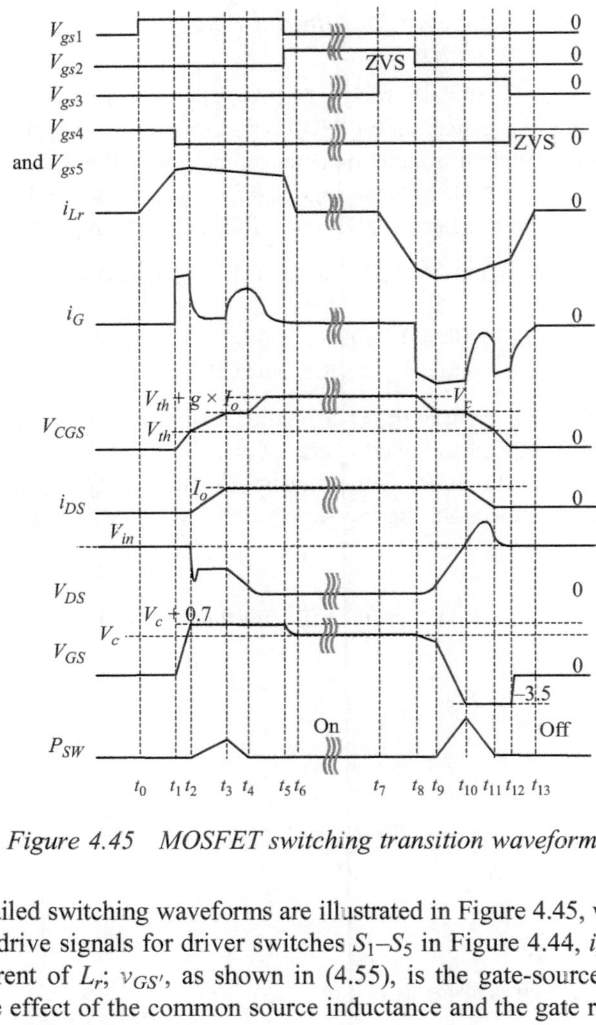

Figure 4.45 MOSFET switching transition waveforms

The detailed switching waveforms are illustrated in Figure 4.45, where v_{gs1}–v_{gs5} are the gate-drive signals for driver switches S_1–S_5 in Figure 4.44, i_{Lr} is the driver inductor current of L_r; $v_{GS'}$, as shown in (4.55), is the gate-source voltage of Q including the effect of the common source inductance and the gate resistance, p_{SW}, is the switching loss of the Q:

$$v_{GS'} = -i_G R_g + V_{CGS} - L_S \frac{d}{dt}(i_{DS} + i_G) \tag{4.55}$$

The operation principle of the turn-on transition is presented as follows. Prior to t_0, the power MOSFET is clamped in the OFF state by S_4 and S_5.

Turn-ON Transition:

1. Pre-charge $[t_0, t_1]$ (Figure 4.46(a)): At t_0, S_1 is turned on, and the inductor current i_{Lr} rises almost linearly and the interval ends at t_1 which is preset by the designer. The equivalent circuit is given in Figure 4.46 turn-on transition (a). The inductor current i_{Lr} is given in (4.56):

$$i_{Lr} \approx \frac{V_c \cdot (t - t_0)}{L_r} \tag{4.56}$$

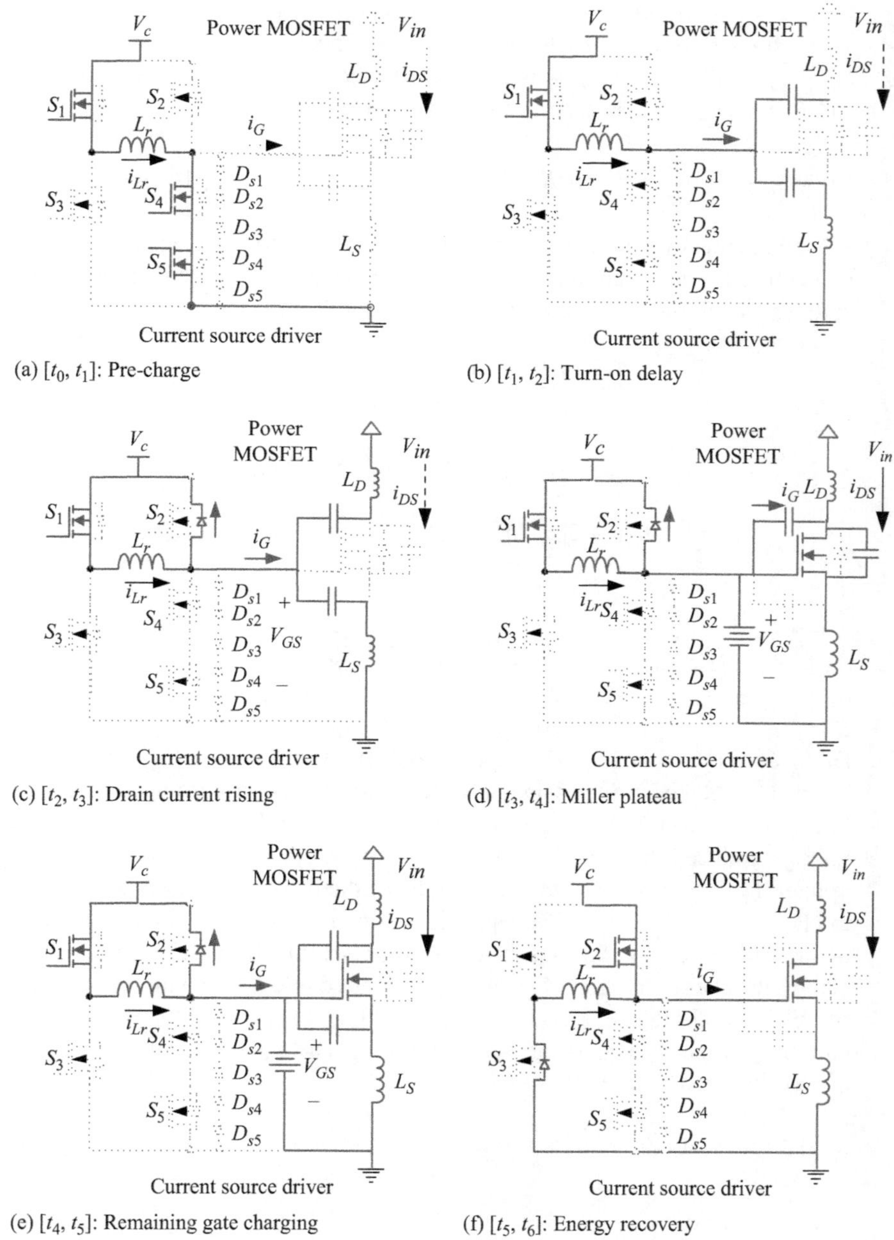

(a) $[t_0, t_1]$: Pre-charge

(b) $[t_1, t_2]$: Turn-on delay

(c) $[t_2, t_3]$: Drain current rising

(d) $[t_3, t_4]$: Miller plateau

(e) $[t_4, t_5]$: Remaining gate charging

(f) $[t_5, t_6]$: Energy recovery

Figure 4.46 Turn-on transition

2. Turn-on delay $[t_1, t_2]$ (Figure 4.46(b)): At t_1, S_4 and S_5 are turned off, the inductor current i_{Lr} starts to charge the gate capacitance of Q – the equivalent circuit is given in Figure 4.46 turn-on transition (b). At this interval, the effective charge current i_G equals i_{Lr}. The initial condition of this interval is

$i_{Lr\_t1} = i_{Lr}(t_1 - t_0)$, *and* $v_{CGS\_t1} = 0$. This interval ends when v_{CGS} reaches V_{th}. The differential equations for the circuit are given in (4.57), where $R_{on} = R_{dson\_S2} + R_{Lr} + R_g$. R_{dson_S2} is the on-resistance of S_2, R_{Lr} is the DCR of L_r and R_g is the gate-resistance of the power MOSFET:

$$V_c = (L_r + L_s)\frac{d}{dt}(i_{Lr}) + i_{Lr}R_{on} + v_{CGS}, i_{Lr} = (C_{GS} + C_{GD})\frac{d}{dt}(v_{CGS}) \qquad (4.57)$$

Mathematically, there are three possible forms for the equation of the inductor current: overdamped, critically damped and underdamped – for practical situations, $\omega_0 > \alpha_0$, the equations for i_G, i_{Lr}, and v_{CGS} are given in (4.58)–(4.59):

$$v_{CGS} = \left[A_0 \cos\left(\sqrt{\omega_0{}^2 - \alpha_0{}^2 t}\right) + B_0 \sin\left(\sqrt{\omega_0{}^2 - \alpha_0{}^2 t}\right)\right]e^{-\alpha_0 t} + C_0 \qquad (4.58)$$

$$i_G = i_{Lr} = (C_{GS} + C_{GD})\left[\left(-A_{01}\sqrt{\omega_0^2 - \alpha_0^2} + B_0\alpha_0\right)\sin\left(\sqrt{\omega_0^2 - \alpha_0^2 t}\right) + \left(-A_0\alpha - B_0\sqrt{\omega_0^2 - \alpha_0^2}\right)\cos\left(\sqrt{\omega_0^2 - \alpha_0^2 t}\right)\right]e^{-\alpha_0 t} \qquad (4.59)$$

where
$\alpha_0 = R_{on}/2(L_r + L_S)$, $\omega_0 = 1/\sqrt{(L_r + L_S)(C_{GS} + C_{GD})}$,
$A_0 = -[2\alpha_0 i_{Lr\_t1} + (V_C - 2i_{Lr\_t1}R_{on})/(L_r + L_S)]/[(C_{GS} + C_{GD})\omega_0{}^2]$,
$B_0 = (i_{Lr\_t1} + (C_{GS} + C_{GD})\alpha_0 A_0/[(C_{GS} + C_{GD})\omega_0{}^2])$, $C_0 = -A_0$.

3. Drain current rising [t_2, t_3] (Figure 4.46(c)): At t_2, $v_{CGS} = V_{th}$. During this interval, v_{CGS} keeps increasing, and i_{DS} starts to rise according to the relationship in (4.52). Since i_{DS} flows through L_S, according to (4.55), the large voltage induced across L_S makes $v_{GS'}$ far larger than the driver supply voltage V_c. Therefore, D_2, the body diode of the driver switch S_2, is driven on to clamp $v_{GS'}$ at $V_c + 0.7$. The equivalent circuit is shown in Figure 4.46 turn-on transition (c). At this interval, i_G drops sharply because of the voltage clamping. The subtraction of i_{Lr} and i_G is diverted into D_2. The initial condition of this interval is $i_{G\_t2} = i_G(t_2 - t_1)$, $i_{DS\_t2} = 0$, and $v_{CGS\_t2} = V_{th}$. The interval ends at t_3 when i_{DS} equals the load current, I_o. The equations for i_G, i_{Lr}, v_{CGS}, and v_{DS} are given in (4.60)–(4.64):

$$v_{CGS} = A_1 e^{\left(-\alpha_1 - \sqrt{\alpha_1{}^2 - \omega_1{}^2}\right)t} + B_1 e^{\left(-\alpha_1 + \sqrt{\alpha_1{}^2 - \omega_1{}^2}\right)t} + C_1 \qquad (4.60)$$

$$i_{DS} = g_{fs}(v_{CGS} - V_{th}) = g_{fs}\left(A_1 e^{\left(-\alpha_1 - \sqrt{\alpha_1{}^2 - \omega_1{}^2}\right)t} + B_1 e^{\left(-\alpha_1 + \sqrt{\alpha_1{}^2 - \omega_1{}^2}\right)t} + C_1 - V_{th}\right) \qquad (4.61)$$

$$i_G = A_1\left(-\alpha_1 - \sqrt{\alpha_1{}^2 - \omega_1{}^2}\right)e^{\left(-\alpha_1 - \sqrt{\alpha_1{}^2 - \omega_1{}^2}\right)t} + B_1\left(-\alpha_1 + \sqrt{\alpha_1{}^2 - \omega_1{}^2}\right)e^{\left(-\alpha_1 + \sqrt{\alpha_1{}^2 - \omega_1{}^2}\right)t} \qquad (4.62)$$

$$i_{Lr} = (I_{G\_t2} + 0.7/R_g)e^{(-R_g/L_r)t} - 0.7/R_g \tag{4.63}$$

$$v_{DS} = V_{in} - L_D\frac{d}{dt}(i_G) - (L_S + L_D)\frac{d}{dt}(i_{DS}) \tag{4.64}$$

where $\omega_1 = 1/\sqrt{L_S(C_{GS} + C_{GD})}$, $\alpha_1 = [R_g(C_{GS} + C_{GD}) + L_S g_{fs}]/[L_S(C_{GS} + C_{GD})]$, $C_1 = 0.7 + V_c$, $A_1 = [(V_{th} - 0.7 - V_c)(-\alpha_1 + \sqrt{\alpha_1^2 - \omega_1^2}) - I_{G\_t2}]/[2\sqrt{\alpha_1^2 - \omega_1^2}]$, $B_1 = V_{th} - A_1 - C_1$.

4. Miller plateau $[t_3, t_4]$ (Figure 4.46(d)): At t_3, $i_{DS} = I_o$. During this interval, v_{CGS} is held at the Miller plateau voltage. i_G mainly flows through the gate-to-drain capacitance of Q, and v_{DS} decreases accordingly. It is noted that i_G starts to increase rapidly since the EMF across L_s falls sharply due to the unchanged i_{DS}; however, part of the inductor current is still diverted through D_2. The equivalent circuit is given in Figure 4.46 turn-on transition (d). The initial values of the interval are $i_{G\_t3} = i_G(t_3 – t_2)$, $v_{CGS\_t3} = v_{CGS}(t_3 – t_2)$, and $v_{DS\_t3} = v_{DS}(t_3 – t_2)$. The interval ends when v_{DS} equals zero at t_4. The equations for i_G, v_{CGS} and v_{DS} are given in (4.65)–(4.66). i_{Lr} remains the same as the previous interval:

$$v_{CGS} = V_{CGS\_t3} \tag{4.65}$$

$$i_G = [I_{G\_t3} - (V_c + 0.7 - V_{CGS\_t3})/R_g]e^{(-R_g/L_s)t} + (V_c + 0.7 - V_{CGS\_t3})/R_g \tag{4.66}$$

$$\begin{aligned} v_{DS} = &\left(\frac{I_{G\_t3} - (V_c + 0.7 - V_{CGS\_t3})/R_g}{C_{GD}R_g/L_s}\right)e^{(-R_g/L_s)t} \\ &- \frac{(I_{G\_t3} - (V_c + 0.7 - V_{CGS\_t3})/R_g)t}{C_{GD}} \\ &+ \left(V_{DS\_t3} - \frac{I_{G\_t3} - (V_c + 0.7 - V_{CGS\_t3})/R_g}{C_{GD}R_g/L_s}\right) \end{aligned} \tag{4.67}$$

5. Remaining gate charging $[t_4, t_5]$ (Figure 4.46(e)): At t_4, $v_{DS} = 0$ and v_{CGS} starts to rise again until it reaches V_c. $v_{GS'}$ remains at $V_c + 0.7$, and due to the rising of the v_{CGS}, i_G decreases gradually. The equivalent circuit is given in Figure 4.46 turn-on transition (e). The initial values of this interval are: $i_{G\_t4} = i_G(t_4 – t_3)$, $v_{CGS\_t4} = v_{CGS\_t3}$, and $v_{DS\_t4} = v_{DS}(t_4 – t_3)$. This interval ends at t_5 when $v_{CGS} = V_c$. The equations for i_G, v_{CGS}, and v_{DS} are given in (4.68)–(4.70) and i_{Lr} is the same as the previous interval:

$$v_{CGS} = \left[A_2\cos\left(\sqrt{\omega_1^2 - \alpha_1^2 t}\right) + B_2\sin\left(\sqrt{\omega_1^2 - \alpha_1^2 t}\right)\right]e^{-\alpha_1 t} + C_2 \tag{4.68}$$

$$\begin{aligned} i_G = &(C_{GS} + C_{GD})\left[(-A_1\sqrt{\omega_1^2 - \alpha_1^2} + B_1\alpha_1)\sin(\sqrt{\omega_1^2 - \alpha_1^2 t}) \right. \\ &\left. + (-A_1\alpha_1 - B_1\sqrt{\omega_1^2 - \alpha_1^2})\cos(\sqrt{\omega_1^2 - \alpha_1^2 t})\right]e^{-\alpha_1 t} \end{aligned} \tag{4.69}$$

$$v_{DS} = V_{DS\_t4} - \frac{(V_{CGS} - V_{DS\_t4})(V_{DS\_t4} - I_o R_{on@Vc})}{V_c - V_{CGS\_t4}} \tag{4.70}$$

where $A_2 = V_{CGS\_t3} - C_2$, $B_2 = [I_{G\_t4}/(C_{GS}+C_{GD}) - A_2\alpha_1]/\sqrt{\alpha_1{}^2 - \omega_1{}^2}$, $C_2 = V_C + 0.7$, and $R_{on@Vc}$ is the on-resistance of the MOSFET when $V_{CGS} = V_c$.

6. Energy recovery [t_5, t_6] (Figure 4.46(f)): At t_5, S_2 is turned on to recover the energy stored in the inductor to the source as well as actively clamping Q to V_c. The initial value of this interval is $i_{Lr\_t5} = i_{Lr}(t_5\_t_2)$, and this interval ends when i_{Lr} becomes zero. The equivalent circuit is illustrated in Figure 4.46 turn-on transition (f). The equation for i_{Lr} is in (4.71):

$$i_{L_r} = [I_{G\_t5} + (V_c + 0.7)/(R_{on\_S2} - R_{lr})]e^{[-(R_{on\_S2}+R_{lr})/Lr]t} - (V_c + 0.7)/(R_{on\_S2} + R_{lr}) \tag{4.71}$$

Prior to t_7, the power MOSFET is clamped in the ON state by S_2.

Turn-OFF Transition:

1. Predischarge [t_7, t_8] (Figure 4.47(a)): At t_7, S_3 is turned on, and the inductor current i_{Lr} rises almost linearly and the interval ends at t_8 which is preset by the designer. The equivalent circuit is shown in Figure 4.47(a). The equation for i_{Lr} is given in (4.72):

$$i_{Lr} \approx -\frac{V_c \cdot (t - t_7)}{L_r} \tag{4.72}$$

2. Turn-off delay [t_8, t_9] (Figure 4.47(b)): At t_8, S_2 is turned off. In this interval, v_{CGS} decreases until $V_{th} + I_o{}^*g_{fs}$ which ends the interval. The equivalent circuit is given in Figure 4.47(b). The initial condition of this interval is $i_{G\_t8} = i_{Lr\_t8} = i_{Lr}(t_8)$, $v_{CGS\_t8} = V_c$. The way to solve the equations for i_G, i_{Lr}, and v_{CGS} are the same as *turn-on delay* interval.
3. Miller plateau [t_9, t_{10}] (Figure 4.47(c)): At t_9, $v_{CGS} = V_{th} + I_o{}^*g_{fs}$. In this interval, v_{CGS} holds at the Miller plateau voltage, $V_{th} + I_o{}^*g_{fs}$. i_G (equal to i_{Lr}) strictly discharges the gate-to-drain capacitance C_{gd} of Q, and v_{DS} rises until it reaches V_{in} at t_{10}. The equivalent circuit is illustrated in Figure 4.47(c). The equations of this interval can be obtained in the same way as the *Millerplateau* in turn-on interval.
4. Drain current drop [t_{10}, t_{11}] (Figure 4.47(d)): At t_{10}, $v_{DS} = V_{in}$ and v_{CGS} continues to decrease from $V_{th} + I_o{}^*g_{fs}$ to V_{th}. i_{DS} falls from I_o to zero according to relationship in (4.52). According to (4.55), due to the induction EMF across L_s, the series connected diodes D_{s1}–D_{s5} are driven on to clamp v_{GS}' at around −3.5 V. The voltage across the CS inductor becomes −3.5 V, so i_{Lr} decreases at a higher rate than in the turn-on transition. The equivalent circuit of this interval is given in Figure 4.47(d).

   By comparison, the CSD previously proposed only can clamp $v_{GS'}$ to −0.7 V during this interval. This means that the turn-off speed of the CSD proposed in this section (Figure 4.44) is more than three times that of the CSD proposed previously. It is worth mentioning that v_{DS} in this interval will keep rising due

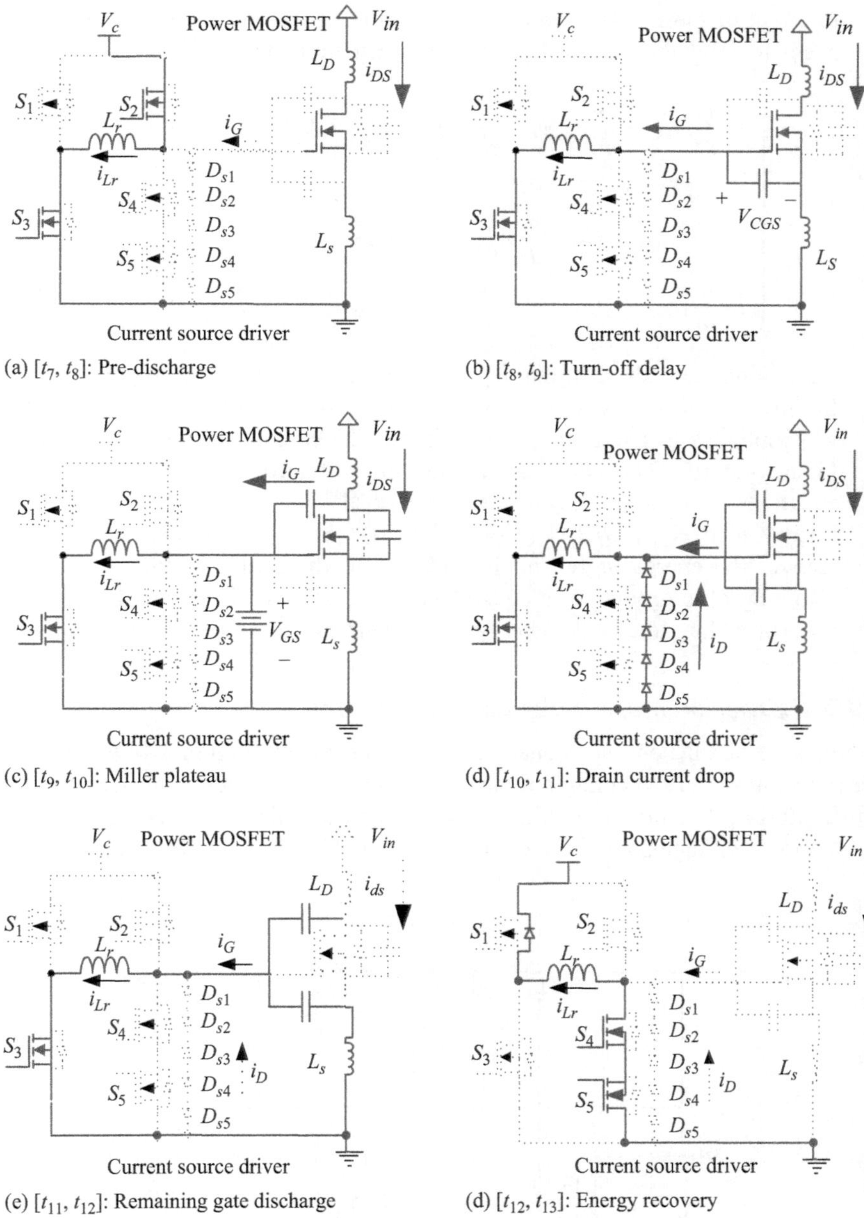

(a) $[t_7, t_8]$: Pre-discharge

(b) $[t_8, t_9]$: Turn-off delay

(c) $[t_9, t_{10}]$: Miller plateau

(d) $[t_{10}, t_{11}]$: Drain current drop

(e) $[t_{11}, t_{12}]$: Remaining gate discharge

(d) $[t_{12}, t_{13}]$: Energy recovery

Figure 4.47 Turn-off transition

to effect of the *Ls*. Therefore, the derivation of the equations in this interval needs to solve the third-order differential equations in (4.73):

$$\begin{cases} L_S\dfrac{d}{dt}(i_{DS}+i_G)+v_{CGS}+i_GR_g+V_f=0 \\ i_{DS}=g_{fs}(v_{CGS}-V_{th}) \\ i_G=(C_{GS}+C_{GD})\dfrac{d}{dt}(v_{CGS})-C_{GD}\dfrac{d}{dt}(v_{DS}) \\ L_S\dfrac{d}{dt}(i_G)+(L_D+L_D)\dfrac{d}{dt}(i_{DS})+v_{DS}-V_{in}-0.7=0 \end{cases} \tag{4.73}$$

5. Remaining gate discharging [t_{11}, t_{12}] (Figure 4.47(e)): At t_{10}, $v_{CGS}=V_{th}$. In this interval, v_{CGS} continues to decrease until it equals zero; it is noted that v_{DS} continues to rise during this interval. The equivalent circuit is shown in Figure 4.47(e). The equations in this interval have the same form as the equations in *remaining gate charging* interval.
6. Energy recovery [t_{12}, t_{13}] (Figure 4.47(f)): At t_{11}, S_4–S_5 are turned on to recover the energy stored in the inductor to the source as well as actively clamping Q to ground. The equivalent circuit is given in Figure 4.47(f). This interval is the same as the *energy recovery* at this turn-on transition.

4.4.3 Experimental results and discussions

A prototype of a synchronous buck converter as shown in Figure 4.48 was built to verify the proposed switching loss model and the optimal design of the CS inductor. The control FET is driven with the proposed CSD, while the SR is driven with a conventional voltage-source driver since the switching loss for SR is very small.

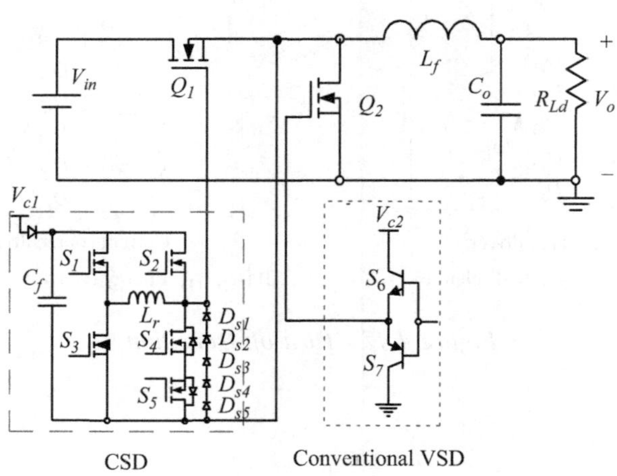

Figure 4.48 Synchronous buck converter with proposed CSD

The PCB consists of six-layer 4-oz copper. The components used are: Q_1: Si7386DP; Q_2: IRF6691; output filter inductance: $L_f = 330$ nH (IHLP-5050CE-01); current-source inductor: $L_r = 23$ nH (Coilcraft 2508-23N_L); drive switches S_1–S_4: FDN335; Anti-diodes D_{s1}–D_{s5}: MBR0520. Altera Max II EPM240 CPLD is used to generate the PWM signals with accurate delays since the CPLD can achieve time resolution as high as 1/3 ns per gate. For common practice, the driver voltages for the control FET and SR are both set to be 5 V. The operating conditions are: input voltage V_{in}: 12 V; output voltage V_o: 1.2–1.5 V; switching frequency f_s: 500 kHz–1 MHz.

The gate-driver signals for V_{gs_Q1} and V_{gs_Q2} are shown in Figure 4.49. It is observed that during turn-off transition, V_{GS_Q1} is clamped to about −3.5 V. The current waveform of the CSD inductor is impossible to obtain without breaking the setup of the prototype.

Figure 4.50 summarizes the efficiencies of the CSD with the optimal inductor for 1.2, 1.3, and 1.5 V output respectively at 1 MHz switching frequency. It can be observed that at 30 A load current, when output voltage increases from 1.2 to 1.5 V, the efficiency increases from 82.8% to 85.1%; and the peak efficiency at 1.5 V output is 89.8% for a 15 A load.

In order to validate the accuracy of the proposed switching loss model, the calculated total loss of the synchronous buck converter as a function of load current is compared to the experimentally measured loss in Figure 4.51. It is observed that the calculated loss by the loss model proposed in this paper matches the actual loss

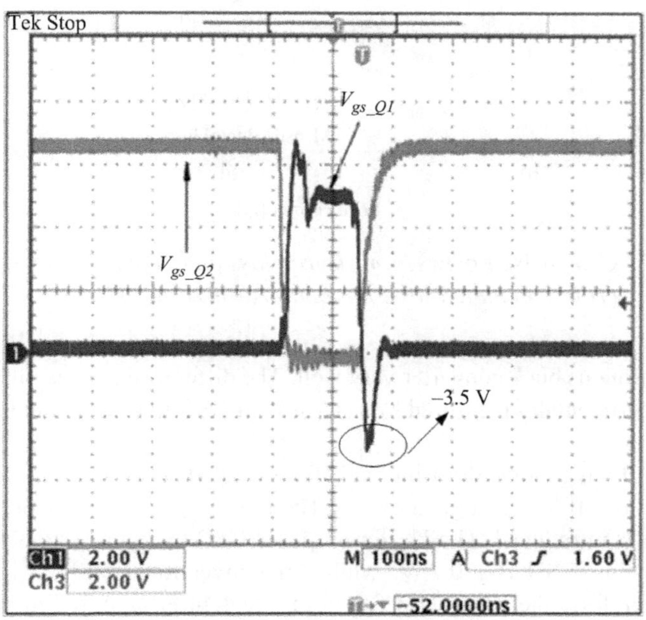

Figure 4.49 The waveforms of driver signals V_{gs_Q1} and V_{gs_Q2}

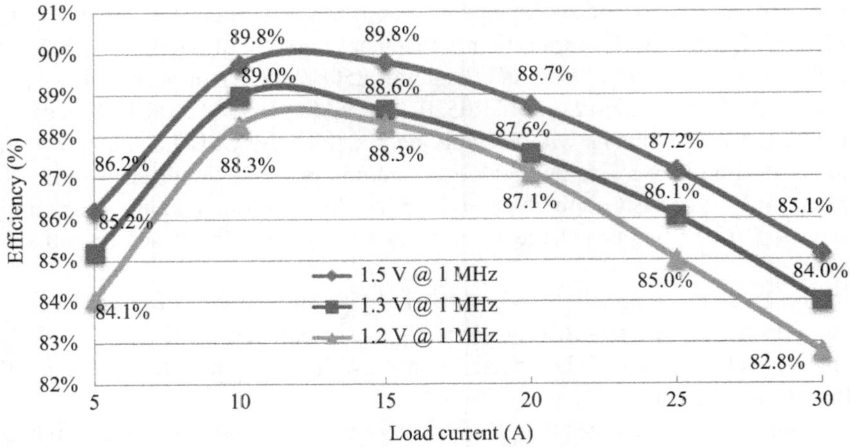

Figure 4.50 Efficiencies of 1.2 V, 1.3 V, 1.5 V output for 12 V input, 1 MHz frequency

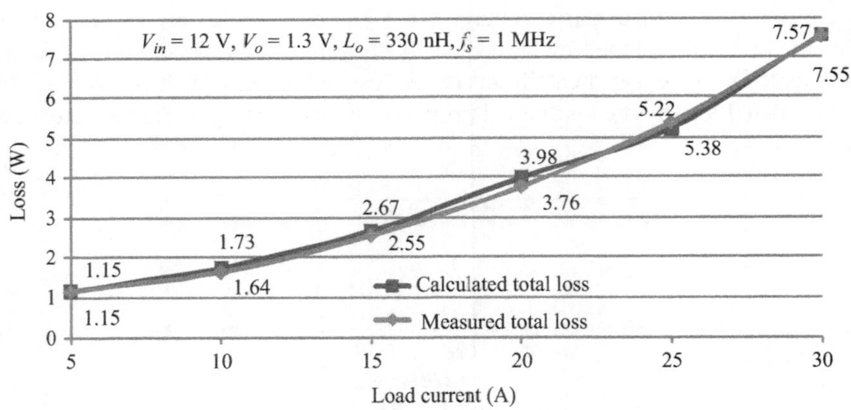

Figure 4.51 Comparison between measured loss and calculated loss (Top: calculated loss; bottom: measured loss)

of the synchronous buck converter very well. The difference between the calculated loss and experimentally measured loss is less than 10% across all load levels from 5 to 30 A.

During drain current drop interval, the clamped gate-to-source voltage, V_{off} affects the turn-off loss to a great extent. The more negative V_{off} is, the smaller the turn-off loss is introduced. The V_{off} for proposed CSD in Figure 4.44 is −3.5 V, and V_{off} for existing CSD is −0.7 V, while for conventional VSD, V_{off} is roughly +0.5 V. Therefore, the proposed CSD in Figure 4.44 is supposed to have smaller switching loss than existing CSD; both CSDs introduce smaller switching loss than conventional VSD. The calculated loss comparison among the proposed CSD in

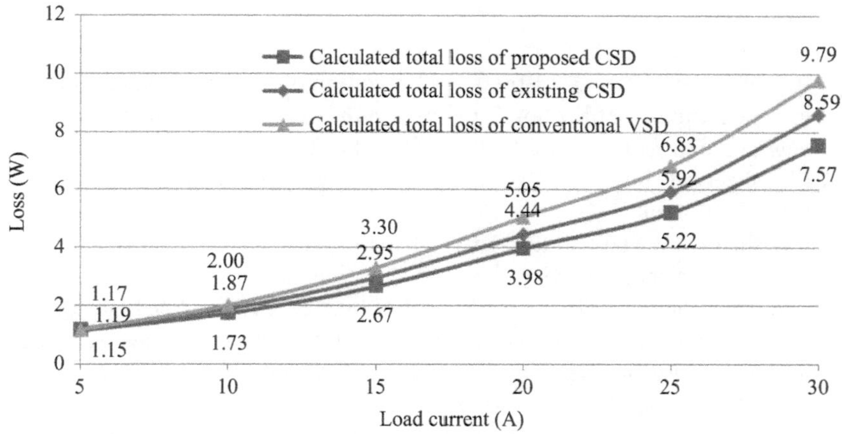

Figure 4.52 Calculated loss comparison between proposed CSD (Figure 4.44), existing CSD and VSD

Figure 4.44, existing CSD and conventional VSD is shown in Figure 4.52. It is verified that the proposed CSD in Figure 4.44 has the smallest total loss due to the negative gate-to-source voltage during turn-off transition.

4.5 Summary

Discontinuous CSDs are proposed for the power MOSFETs. They can achieve quick turn-on and turn-off transition times to reduce high-frequency switching loss and conduction loss in power MOSFETs. In addition, it can recover a portion of the *QV* gate energy normally dissipated in a conventional driver. The circuit consists of four controlled switches and a small inductor (100 nH or less, typical). The current through the inductor is discontinuous in order to minimize circulating current conduction loss in the driver. This also allows the driver to operate effectively over a wide range of duty cycles with constant peak current – a significant advantage for many applications since turn-on and turn-off times do not vary with the operating point. The driver provides superior performance in comparison to conventional drivers and solves the problems of existing resonant gate drivers. These drivers can drive both the HS MOSFET and synchronous MOSFET in a synchronous buck converter. A significant loss reduction and efficiency improvement are achieved.

The current diversion phenomenon happens in the CSDs is analyzed in detail. An accurate switching loss model considering the current diversion is presented for the power MOSFET driven by a CSD, and the analytical equations for each interval to predict the switching loss are derived. Based on the proposed model, the optimal CS inductor is obtained to maximize overall efficiency of the synchronous buck converter.

References

[1] B. Yang and J. Zhang, "Effect and utilization of common source inductance in synchronous rectification," in *Proceedings on IEEE Applied Power Electronics Conference (APEC)*, pp. 1407–1411, 2005.

[2] Datasheet of Intersil 6208, http://www.intersil.com/cda/deviceinfo/0,1477, ISL6208,0.html

[3] Datasheet of IRF6691, http://www.irf.com/productinfo/datasheets/data/irf6691.pdf, International Rectifier.

[4] J. Fu, Z. Zhang, A. Dickson, Y. F. Liu, and P. C. Sen, "A high efficiency current source driver with negative gate voltage for buck voltage regulators," in *Proceedings on IEEE Applied Power Electronics Conference and Exposition (APEC)*, pp. 702–709, 2010.

[5] J. Fu, Z. Zhang, W. Eberle, Y. F. Liu, and P. C. Sen, "A high efficiency current source driver with negative gate voltage for buck voltage regulators," in *Proceedings on IEEE Energy Conversion Congress and Exposition (ECCE)*, pp. 1663–1670, 2009.

[6] Z. Zhang, W. Eberle, Z. Yang, Y. F. Liu, and P. C. Sen, "Optimal design of current source gate driver for a buck voltage regulator based on a new analytical loss model," *IEEE Transactions Power Electron*, vol. 23, no. 2, pp. 653–666, 2008.

[7] W. Eberle, Z. Zhang, Y. F. Liu, and P. C. Sen, "A practical switching loss model for buck voltage regulators," *IEEE Transactions Power Electron*, vol. 24, no. 3, pp. 700–713, 2009.

[8] D. A. Grant and J. Gowar, *Power MOSFET Theory and Applications*, New York: Wiley, 1989.

Chapter 5
Adaptive current source drivers

5.1 Adaptive current source driver for buck converters

5.1.1 Introduction

For the current source driver (CSD) technology, high-drive current normally leads to lower switching loss. A stronger drive current is desired to reduce the switching loss further when the power metal oxide semiconductor field-effect transistor (MOSFET) carries higher current. Nevertheless, higher drive current also results in higher circulating loss. However, the present CSD circuits normally use constant-drive current and voltage. For the control field-effect transistors (FETs), stronger drive current means faster switching speed and lower switching loss, but results in higher drive circuit loss. This leads to a trade-off between the switching loss reduction and gate-drive loss. For the sync FETs, stronger drive current means faster switching speed and then lower body diode loss, but results in higher gate-drive loss. This leads to a trade-off between the body diode loss and gate-drive loss. However, constant-drive current and voltage limit the optimal operating conditions of the loss reduction. To achieve high-efficiency curves through wide load range becomes an important topic for high-frequency voltage regulators (VRs). High gate-drive current and voltage are beneficial to the nominal load conditions, but hurt light-load efficiency, which is of great importance for buck VRs. Therefore, the adaptive-drive current would be beneficial to improve the performance of the CSDs and help to achieve optimal design of the switching loss reduction and gate-drive loss reduction.

5.1.2 Adaptive continuous current source driver

5.1.2.1 Synchronous buck converter with the continuous current source driver

Figure 5.1 shows the continuous CSD applied to the synchronous buck converter, where Q_1 is the control FET and Q_2 is the sync FET. V_{c1} and V_{c2} are the drive supply voltage. In Figure 5.1, there are two sets of the drive circuits (CSD #1 and CSD #2) and each of them has the structure of the half-bridge (HB) topology, consisting of drive MOSFETs S_1–S_2 and S_3–S_4, respectively. CSD #1 can also be regarded as a level-shift version of CSD #2.

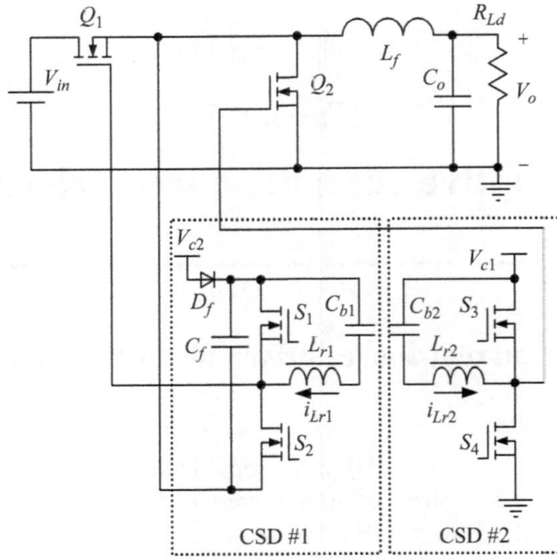

Figure 5.1 Buck converter with the continuous CSD

5.1.2.2 Optimal design and adaptive control with the continuous current source driver

For a given application, in order to achieve fast switching speed, the gate-drive current should be chosen properly. The design trade-off is between switching speed, which translates into reduced switching loss and gate-drive loss. In order to find the optimal design point, the object function adding the switching loss and the CSD circuit loss together is used. The object function is a U-shaped curve as function of the drive current I_G, and the optimization solution is simply located at the lowest point of the curve. Based on the above idea, Figure 5.2(a) illustrates the optimal curve for the continuous CSDs with different load current, which includes the switching loss $P_{switching\ loss}$, the CSD circuit loss $P_{CSD\ circuit}$ and the objective function $F(I_G) = P_{switching\ loss} + P_{CSD\ circuit}$ as function of the gate-drive current I_G. The specifications of the buck converter are: $V_{in} = 12$ V; $V_o = 1.3$ V; $I_o = 30$ A; $V_c = 5$ V; $f_s = 1$ MHz; control FET Q_1: Si7386DP; Q_2: IRF6691 and $L_f = 330$ nH.

From Figure 5.2(a), at 30 A and with the increase of I_G, the switching loss $P_{switching\ loss}$ keeps reducing due to stronger drive current while the CSD circuit loss $P_{CSD\ circuit}$ keeps increasing due to high circulating current. The summation of these two, i.e., $F(I_G)$, is a U-shaped curve, and therefore, the optimization solution can be found at the lowest point of the curve. At the load current of 30 A, the optimal-drive current I_G is chosen as 1.8 A for the continuous CSD. However, when the load changes, the optimal point will also change depending on the switching loss associated with the drain current in the MOSFET. It is interesting to note that when the load current increase from 10 to 30 A, the optimal-drive current increases accordingly from 1.2 to 1.8 A. This means that higher drive current leads to lower switching losses. So the desired CSD current should be able to adjust

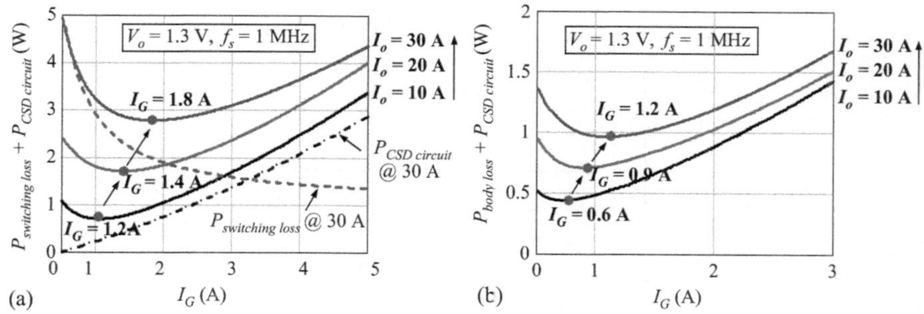

Figure 5.2 Optimization curves: power loss vs. gate current; (a) control FET Q_1 and (b) Sync FET Q_2

adaptively as the load current increases. Although the optimal design curves are calculated from the above mentioned specifications, it should be pointed that the trend of the optimal design curves are independent with the specification used, and the physical meaning of adaptive concept and design method is general.

Similarly, Figure 5.2(b) gives the optimal curves for the Sync FET according to the load conditions, where $P_{body\ loss}$ is the body diode loss, $P_{CSD\ circuit}$ is the CSD circuit loss and $F(I_G) = P_{body\ loss} + P_{CSD\ circuit}$ is the objective function. The sync FET, Q_2, operates with zero-voltage switching (ZVS) since its output capacitance is discharged to zero voltage before it turns on. Therefore, for the sync FET, the optimal design involves a trade-off between body diode conduction loss and gate-drive loss. When the load current increases from 10 to 30 A, the optimal-drive current increases from 0.6 to 1.2 A, accordingly. High-drive current leads to lower body diode conduction and body diode loss.

5.1.3 Adaptive discontinuous current source drivers

5.1.3.1 Synchronous buck converter with the discontinuous current source driver

Figure 5.3 shows the proposed hybrid gate-drive scheme for a buck converter. As we know, for a buck converter, the dominant loss is the switching loss. Therefore, for the control FET Q_1, the proposed high-side CSD is used to achieve the switching loss reduction. For the synchronous rectifier (SR) Q_2, the conventional voltage source driver (VSD) is used for low cost and simplicity. PWM_SR is the signal fed into the VSD. The details of the operation and optimal design of the buck converter with the discontinuous CSD have been addressed in Chapter 4.

5.1.3.2 Optimal design and adaptive control with the discontinuous current source driver

Figure 5.4(a) shows the power loss as the function of the adaptive-drive current and Figure 5.4(b) shows the adaptive-gate current with different load current. As the load current increases from 10 to 30 A, the optimal-drive current increases from 1.8

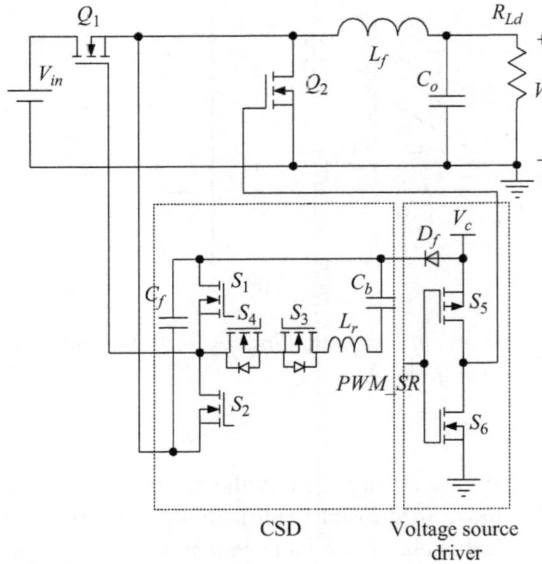

Figure 5.3 Buck converter with the discontinuous CSD

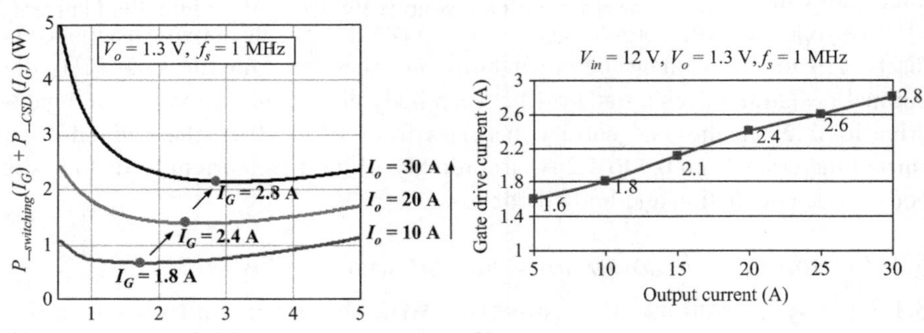

Figure 5.4 Optimization curves for the control FET Q_1: discontinuous; (a) power loss vs. gate current and (b) gate-drive current vs. load current

to 2.8 A accordingly. The switching current in the control FET increases as the load current increases, which leads the switching loss to increase accordingly. Therefore, the switching time needs to be minimized with stronger gate-drive current to reduce the switching loss. This results in the relationship between the gate-drive current and output current as shown in Figure 5.4(a). It should be also noted that at the same load current of 30 A, the optimal-drive current I_G is chosen as 2.8 A (see Figure 5.4) for the discontinuous CSD while the optimal-drive current is 1.2 A (see Figure 5.2(b)) for the continuous one. This is because the discontinuous CSD has lower drive circulating loss over the continuous one, which helps to achieve high-drive current.

For the discontinuous CSD, the pre-charge current to turn on the power MOSFET is

$$I_{Gon} = \frac{V_C}{2L_r} \cdot \Delta t_{10} \tag{5.1}$$

where V_C is the drive voltage and Δt_{10} is the pre-charge interval from t_0 to t_1.

Similarly, the pre-charge current to turn off the power MOSFET is

$$I_{Goff} = \frac{V_{Cb}}{L_r} \cdot \Delta t_{54} = \frac{V_C}{2L_r} \cdot \Delta t_{54} \tag{5.2}$$

where V_{Cb} is the d.c. voltage across the capacitor and Δt_{54} is the pre-charge interval from t_4 to t_5.

5.1.4 Implementation of the adaptive drive voltage for the current source drivers

As discussed in Section 4, for both of the continuous and discontinuous CSDs, the drive current is proportional to the drive voltage. One method to build the adaptive-drive current is to adjust the drive voltage adaptively according to the load current, or input voltage, etc.

Normally, to achieve adaptive voltage function, a lower power switching converter or a linear regulator can be used. The switching converters have higher efficiency over the linear regulators. However, it generally needs output filter inductors, such as buck converters, which are difficult to be integrated into a drive chip. The advantage of the linear regulator is that no inductors are required. This makes the integration of the linear regulator become achievable. As an example, UCC27222 from Texas Instrument (TI) has an integrated linear regulator to achieve desired voltage for its drive circuitry.

Figure 5.5 gives schematic of the adaptive control circuit using the basic linear regulator. The advantage of using the linear regulator is compact structure and fast voltage change rate. In Figure 5.5(a), R_{sens} is the sensor resistor and the voltage across R_{sens} is proportional to the load current. This voltage is sent into the amplifying stage. Three operational amplifiers are used to amplify the input signal linearly. The first and third amplifiers are used as inverting amplifiers and the second amplifier is used to realize the addition of the amplified input signal and the bias voltage V_a. The relationship between the reference voltage and the output current is

$$V_{REF} = \frac{R_{sens} R_2 R_6 R_9}{R_1 R_5 R_8} I_o + \frac{R_6 R_9}{R_5 R_8} V_a \tag{5.3}$$

Figure 5.5(b) shows the basic linear regulator with the reference voltage V_{REF}. The output voltage V_d of the linear regulator is used to supply the CSD drive

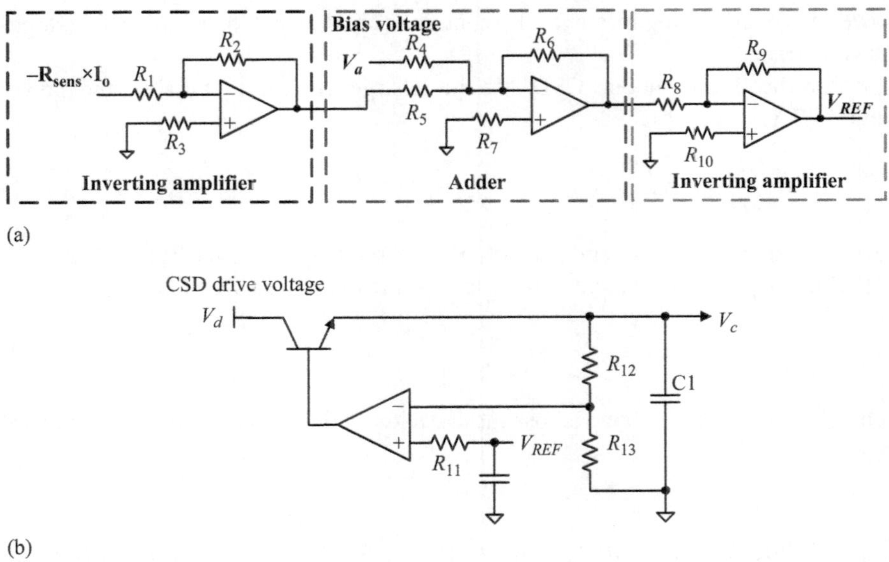

Figure 5.5 Adaptive drive voltage circuit using the basic linear regulator; (a) circuitry of the adaptive control voltage V_{REF} and (b) linear regulator with adaptive control voltage

circuitry. When V_{REF} varies according to the load current I_o from (5.3), the output voltage V_d will change adaptively.

Using linear regulator to achieve adaptive voltage control results in additional loss and the efficiency of the linear regulator is $\eta = V_d/V_c$. The loss of the linear regulator is $P_{Vc} \cdot (1-\eta)$, where P_{Vc} is the drive supply power. However, the total loss reduction of the switching loss, the conduction loss and the gate-drive loss from the main power stage weigh much higher than the introduced loss with linear regulator. Therefore, the overall loss reduction improves the converter efficiency effectively in a wide load range.

5.1.5 Experimental results and discussion

In order to verify the proposed adaptive concept, the synchronous buck converters with the continuous and discontinuous CSDs were implemented. The specifications of the buck converter prototype are as follows: input voltage $V_{in} = 12$ V; output voltage $V_o = 1.3$ V; output current $I_o = 30$ A; switching frequency $f_s = 1$ MHz. The printed circuit board (PCB) is six-layered with 4 oz copper. The components used in the circuit are listed as follows: Q_1: Si7386DP; Q_2: IRF6691; output filter inductance: $L_f = 330$ nH (IHLP-5050CE-01, Vishay); for the continuous CSD, $L_r = 1$ μH; for the discontinuous CSD, $L_r = 22$ nH; drive switches S_1–S_4: FDN335.

A benchmark of a synchronous buck converter with the conventional gate driver was also built. The Predictive Gate Drive UCC 27222 from TI was used as the conventional voltage driver. Figure 5.6(a) shows the measured efficiency

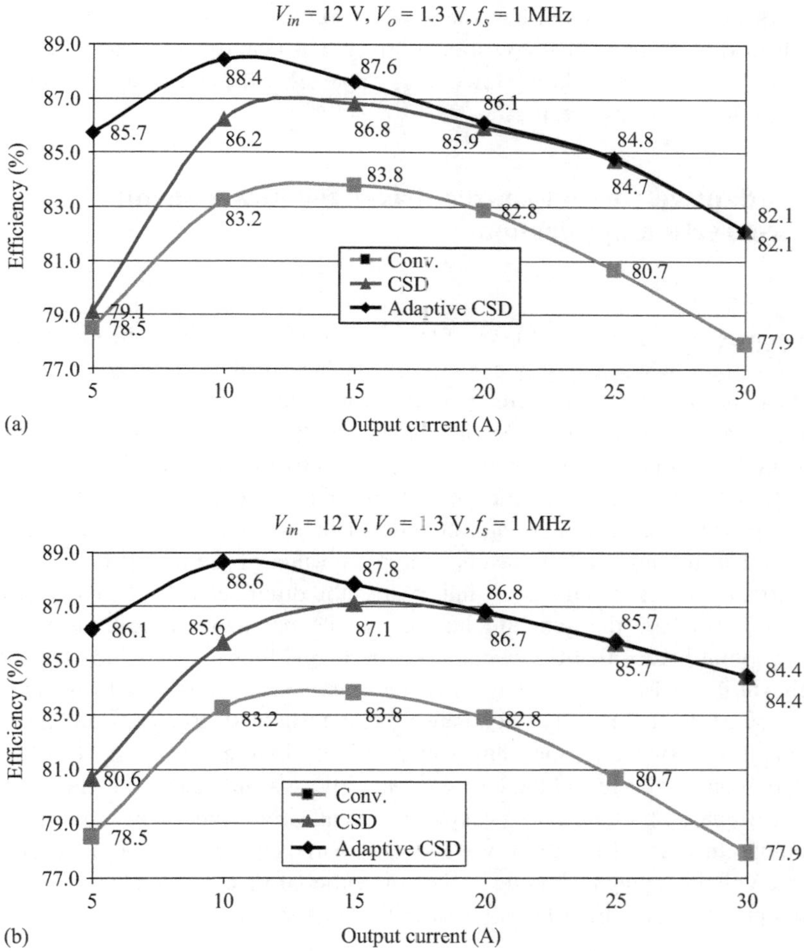

Figure 5.6 Efficiency comparison: (a) continuous and (b) discontinuous

comparison between the continuous CSDs with and without the adaptive voltage at 1.3 V output. It is observed that at 5 A, the efficiency is improved from 79.1% to 85.7% (an improvement of 6.6%) with the drive voltage of 4 V. The loss of the linear regulator is 0.13 W. At 10 A, the efficiency is improved from 86.2% to 88.4% (an improvement of 2.2%) with the drive voltage of 5 V. The loss of the linear regulator is 0.22 W. It should be also noted that the CSDs improve the efficiency effectively in full-load range over the conventional driver.

Figure 5.6(b) shows the measured efficiency comparison between the discontinuous CSDs with and without the adaptive voltage at 1.3 V output. It is observed that at 5 A, the efficiency is improved from 80.6% to 86.1% (an improvement of 5.5%) with the drive voltage of 4 V. At 10 A, the efficiency is improved from 85.6% to 88.6% (an improvement of 3%) with the drive voltage of

5 V. It should be also noted that the CSDs improve the efficiency effectively in the full-load range over the conventional driver. The adaptive CSD achieves higher efficiency improvements during lower load range since the switching losses happen with light load reduces greatly.

5.2 Adaptive current source driver for power factor correction application

5.2.1 *Introduction*

Most of present work related to the CSDs is to investigate their applications in d.c.–d.c. converters, where the duty cycle normally has steady state value. In a.c.–d.c. applications, power factor correction (PFC) technique is widely used. Different from d.c. to d.c. converters, the duty cycle of the PFC converters needs to be modulated fast and has wide operation range. Normally, the switching loss is the proportional to the switching current and the drain-to-source voltage. For a boost PFC converter, the input line current follows the input line voltage in the same phase. When the line voltage reaches the peak value of the power MOSFET, the line current also reaches the peak and so does the drain current (i.e., the switching current). This means the switching loss reaches its maximum value at this moment. The key point here is that the switching current, and thus the switching loss varies widely during a half-line period when the duty cycle modulates to realize unit power factor (PF). Normally, high gate-drive current leads to fast switching speed and thus lower switching loss, but results in high drive-circuit loss. Therefore, in order to minimize the switching loss and gate-drive loss in wide range, the adaptive-drive current is preferred to provide different drive capability according the switching current and the drain voltage. Compared to the adaptive-drive current, the gate-voltage-swing technique of the conventional voltage drivers was proposed to minimize the gate-drive loss dynamically depending on load conditions [1].

5.2.2 *Analysis of current source driver circuits and proposed adaptive current source driver for boost power factor correction converters*

5.2.2.1 The proposed current source driver for boost power factor correction converter

Figure 5.7 shows the proposed CSD circuit for the boost PFC converter. S_2 and S_4 are used to remove the blocking capacitor C_b, which forms a full-bridge (FB) CSD structure. Since there is no blocking capacitor C_b anymore, the proposed CSD can be suitable for the boost PFC converters with the modulated duty cycle. Figure 5.8 shows the key waveforms. As seen from Figure 5.8, S_1–S_3 and S_2–S_4 are controlled complementarily with dead time. This is similar to the control of the synchronous buck converters, so the commercial-off-the-shelf buck drivers can be directly used instead of the discrete components in other CSD circuits. This much reduces the complexity of the control circuit and level shift circuits.

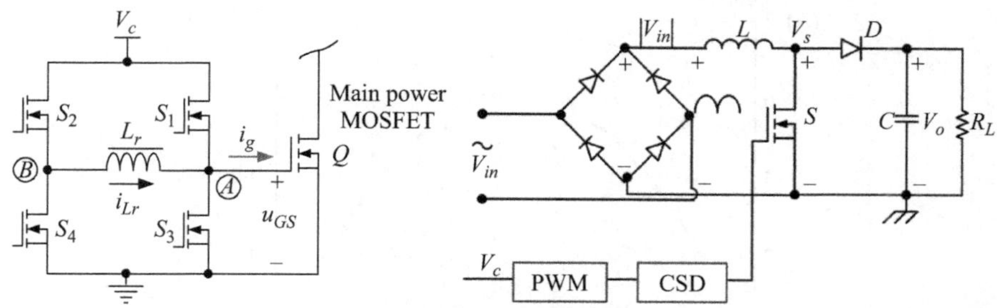

Figure 5.7 Proposed CSD solution for PFC applications

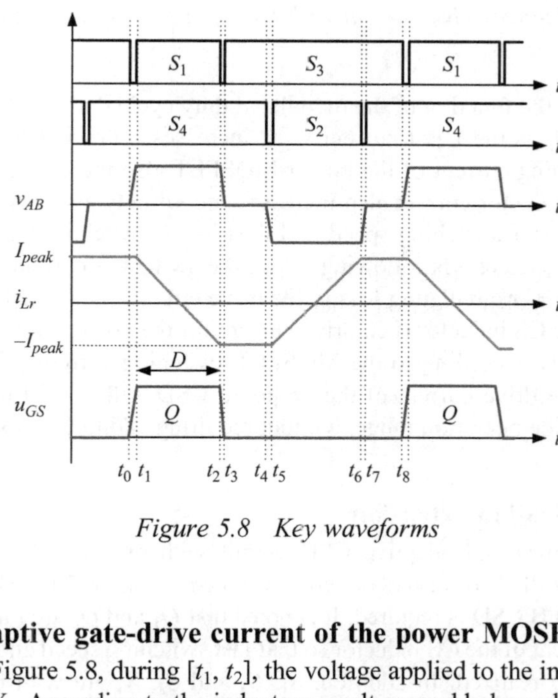

Figure 5.8 Key waveforms

5.2.2.2 Adaptive gate-drive current of the power MOSFET

As seen from Figure 5.8, during $[t_1, t_2]$, the voltage applied to the inductor v_{AB} is the drive voltage V_c. According to the inductance volt-second balance law, the relationship between the inductor value L_r and the peak current I_{peak} of the CS inductor is

$$I_{peak} = \frac{V_c \cdot D}{2 \cdot L_r \cdot f_s} \qquad D < 0.5 \tag{5.4}$$

$$I_{peak} = \frac{V_c \cdot (1 - D)}{2 \cdot L_r \cdot f_s} \qquad D > 0.5 \tag{5.5}$$

For a boost PFC converter with 110 Vac input and 380 V output, the minimum duty cycle is $D_{min} = 1 - 120 \cdot \sqrt{2}/380 = 0.55$. When $D > 0.5$, the peak value I_{peak} of the inductor current is governed by (5.5) and is a monotone decreasing function of the duty cycle. Figure 5.9 shows the peak current value I_{peak} of the current source

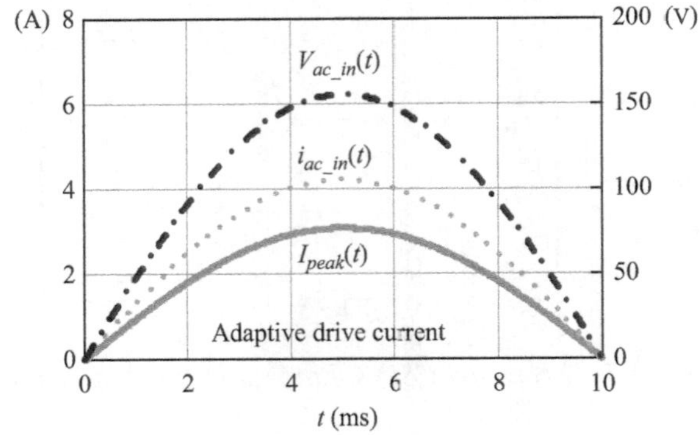

Figure 5.9 Adaptive-drive current with the input line voltage and input current

(CS) inductor as the function of the modulated duty cycle during a half-line period. From Figure 5.9, as the input current i_{ac_in} increases following the input voltage v_{ac_in}, the switching current of the power MOSFET also increases, the peak value I_{peak} of the CS inductor current also increases accordingly and this leads to higher driver current, faster switching speed and thus lower switching loss. On the other hand, as the i_{ac_in} decreases following v_{ac_in}, the switching current also decreases, which leads to lower circulating loss in the drive circuit. As a conclusion, the peak value I_{peak} of the CS inductor (i.e., drive current for the power MOSFET) is able to behave adaptively according to the MOSFET switching current. This property of inherent adaptive-drive current of the proposed CSD will benefit the overall efficiency during wide operation range, without requiring additional auxiliary circuitry and control.

5.2.2.3 Application extension

Moreover, the presented adaptive CSD circuit with one inductor can drive two interleaved boost PFC converters directly as shown in Figure 5.10. The advantage is that only single FB CSD is required. It is noted that Q_1 and Q_2 are turned on and off by the peak current of the CS inductor so that fast switching speed and switching loss reduction can be realized. In addition, S_1, S_3 and S_2, S_4 are with complementary control, respectively, and therefore, the commercial buck drivers with two complementary drive signals can be directly used to the CSD circuit. This much reduces the complexity of the CSD circuit implementation with the discrete components. Other benefits of this proposed topology include low current stress of the power MOSFETs, ripple cancellation of input current and reduced boost inductances.

5.2.3 Loss analysis of proposed adaptive current source driver for boost power factor correction converter

5.2.3.1 Loss analysis of power stage

The efficiency of the PFC stage is determined by the losses of the input diode bridge, the boost inductor, the main power MOSFET, and the rectifier diode.

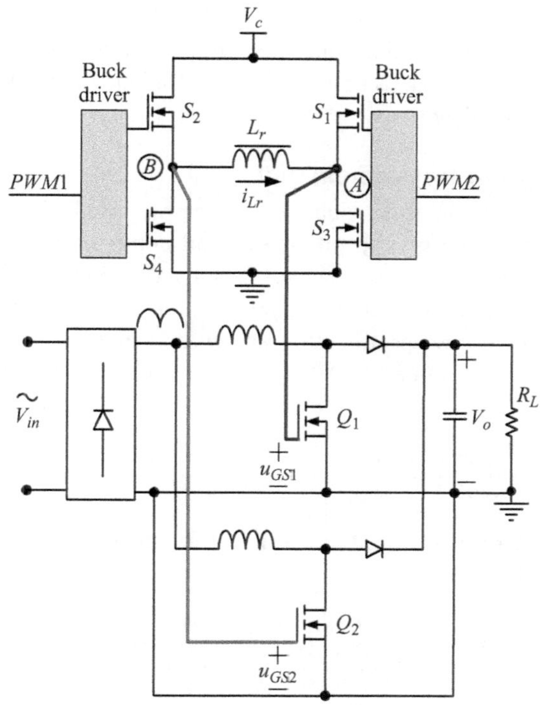

Figure 5.10 Interleaving boost PFC converters with the proposed adaptive CSD

Table 5.1 Loss distribution comparison with 100 kHz and 1 MHz: $V_{in} = 110$ Vac, $V_o = 380$ V and $P_o = 300$ W

	Driving loss	**FET switching loss**	**FET conduction loss**	**Rectifier loss**	**Bridge loss**	**Inductor loss**	**Total loss**
100 k	0.12 W	2.4 W	4.4 W	1.5 W	4.8 W	2.1 W	15.3 W
1 M	1.2 W	24 W	4.4 W	4.0 W	4.8 W	2.1 W	40.5 W

Because the power MOSFET and boost diode are with hard switching condition, the switching loss is the dominant at high frequency.

Table 5.1 provides the calculated loss breakdown comparison of the 110 Vac input, 380 V output, and 300 W boost PFC converter with 100 kHz and 1 MHz, respectively. The 600 V/11 A CoolMOS™ SPA11N60 from Infineon is used for the power MOSFET and SiC CSD06060 from CREE is used for the diode. The benefits of the MHz PFC converters include high power density and fast dynamic performance owing to significant reduction of the passive components and EMI filter size [2–4].

From Table 5.1, it is noted that when the switching frequency increases from 100 kHz to 1 MHz, the switching loss becomes the dominant loss and increases as much as 25.2 W. This results in a significant efficiency reduction of 6.9% from 95% to 88.1% as calculated.

5.2.3.2 Example of loss comparison

In order to demonstrate the loss reduction with the proposed adaptive CSD, a boost PFC converter with the same specifications and components is presented as an example. For the CSD, the drive voltage V_c is 12 V and the CS inductor is 1 μH (DS3316P, Coilcraft). It should be pointed that the experimental prototype in Section 5 is with the same specifications and components.

With $V_{in} = 110$ Vac and $V_o = 380$ V, the duty cycle of the boost PFC converter is:

$$D(t) = 1 - \frac{V_{in}|\sin \omega_L t|}{V_o} \tag{5.6}$$

Substituting (5.6) into (5.5), the CS inductor peak current I_{peak} (i.e., the gate-drive current I_g) is

$$I_g(t) = I_{peak}(t) = \frac{V_C[1 - D(t)]}{2f_s L_r} = \frac{V_C V_{in}|\sin \omega_L t|}{2f_s L_r V_o} \tag{5.7}$$

where V_c is the gate-drive voltage.

With the proposed adaptive CSD, the switching loss of the power MOSFET is

$$P_{sw\_csd}(t) = \frac{1}{2} f_s V_o I_L(t) \left[T_{rcsd}(t) + T_{fcsd}(t) \right] \tag{5.8}$$

$$T_{rcsd}(t) = \frac{Q_{pl} - Q_{th} + Q_{gd}}{I_g(t)} \tag{5.9}$$

$$T_{fcsd}(t) = \frac{Q_{pl} - Q_{th} + Q_{gd}}{|I_g(t)|} \tag{5.10}$$

where Q_{pl} is the MOSFET total gate charge at the beginning of the plateau; Q_{th} is the total gate charge at the threshold, and Q_{gd} is the gate-to-drain charge. T_{rcsd} is the switching rising time and T_{fcsd} is the switching falling time. For the power MOSFET of SPA11N60C3 from Infineon, $Q_{th} = 3.2$ nC, $Q_{pl} = 6$ nC, and $Q_{gd} = 22$ nC.

Figure 5.11 illustrates the switching transition time comparison between the CSD and conventional VSD. T_{sw_CSD} and T_{sw_conv} represent the switching transition time with the CSD and the conventional VSD respectively. During $[t_1, t_2]$ during half of line period, it is noted that T_{sw_CSD} is always less than T_{sw_conv}, and moreover, the reduction of the switching transition time between T_{sw_CSD} and T_{sw_conv} increases when I_g increases. Based on the adaptive-drive current concept described in Section 2, as I_g modulates with the input line voltage and current (i.e., the switching current), the switching transition time and the switching loss with the CSD are also reduced adaptively.

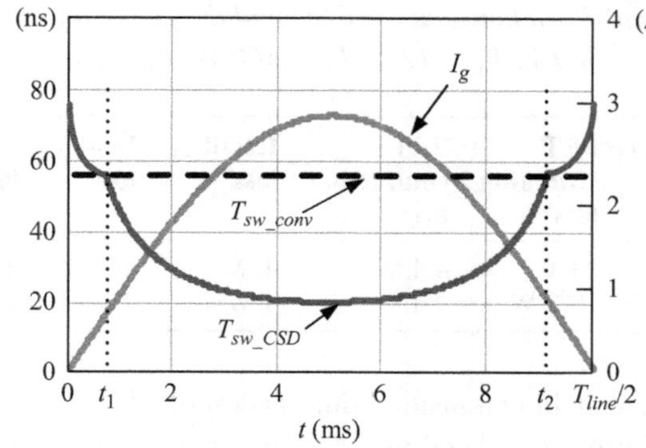

Figure 5.11 The switching transient time comparison in half of line period (V_{in} = 110 Vac, V_o = 380 V, V_c = 12 V, P_o = 300 W, and L_r = 1 μH)

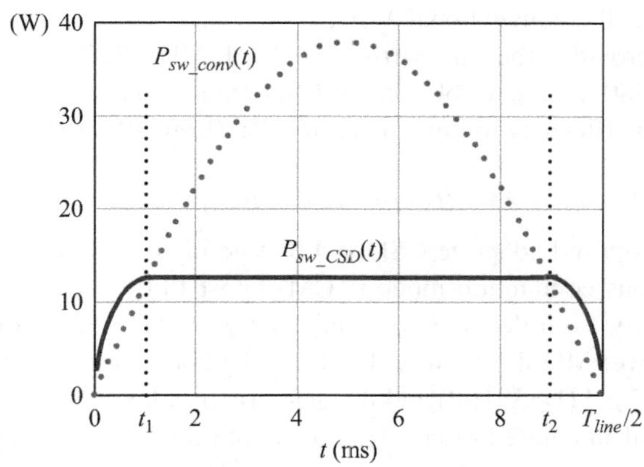

Figure 5.12 The calculated switching loss comparison with the CSD and the conventional VSD in half of line period (V_{in} = 110 Vac, V_o = 380 V, V_c = 12 V, P_o = 300 W, and L_r = 1 μH)

Another thing should be noted that during [0, t_1] and [t_2, $T_{line}/2$], T_{sw_CSD} is higher than T_{sw_conv}, which means the switching transition time with the CSD is longer than that with the voltage driver. This is because during these intervals, the gate-drive current I_g is not large enough to switch the power MOSFET completely and the CSD behaves as a VSD in part with an effective drive current of about 0.8 A in this case. However, during these intervals, because the switching current is quite low as the input line voltage and current are low, the switching loss is also limited under this hybrid drive condition.

Based on the loss analysis, the calculated switching loss comparison using the Mathcad software is shown in Figure 5.12. P_{sw_CSD} and P_{sw_conv} represent the switching loss with the CSD and the conventional VSD, respectively. Similar to

Table 5.2 Loss breakdown between the CSD and the conventional VSD: $V_{in} = 110$ Vac, $V_o = 380$ V, $V_c = 12$ V, $P_o = 300$ W, $L_r = 1$ μH, and $f_s = 1$ MHz

	Driving loss	**FET switching loss**	**FET conduction loss**	**Rectifier loss**	**Bridge loss**	**Inductor loss**	**Total loss**
Con. driver	1.2 W	24 W	4.4 W	4 W	4.8 W	2.1 W	40.5 W
CSD	2.8 W	10.5 W	4.4 W	4 W	4.8 W	2.1 W	28.6 W

the reduction of the switching transition time as described above, P_{sw_CSD} is always less than P_{sw_conv} during $[t_1, t_2]$ and furthermore, the loss reduction becomes larger when I_g modulates with the input line voltage and current. During $[0, t_1]$ and $[t_2, T_{line}/2]$, it should be noted that P_{sw_CSD} is a slightly higher than P_{sw_conv}. This is because that the input current and voltage are both quite low during these intervals, then the switching loss is limited. Overall, the CSD can reduce the switching loss significantly over the conventional VSD.

Table 5.2 provides the comparison of the 1 MHz PFC loss distribution. At 110 Vac input, 380 V output voltage, the CSD reduces the total loss of 12 W. This translates into an efficiency improvement of 3.2% (from 89% to 92.2%) at full load.

5.2.4 Experimental results and discussion

To verify the proposed adaptive CSD, a 110 Vac input, 380 V/300 W output and 1 MHz continuous conduction mode (CCM) boost PFC converter was built. The specifications are as follows: boost inductor $L = 100$ μH; output capacitance $C = 220$ μF; power MOSFET: SPA11N60C3, the boost diode: CSD06060, the CS inductor $L_r = 1.5$ μH (DS3316P) and the gate driver voltage $V_c = 12$ V. Since there is no commercial integrated circuit (IC) available for MHz PFC applications, the discrete components were used to build the controller, saw tooth generator and CSD circuit. Two synchronous buck driver ISL6209 from Intersil are used to drive CSD switches. The discrete multiplies AD633 and comparators are used to implement PFC current loop and generate saw tooth. In order to minimize the negative impact of the parasitics of the discrete components and layout at the switching frequency of 1 MHz, the commercial drive IC LTC4442 from Linear Technology was chosen as the conventional voltage driver to the main power MOSFET of the boost converter. This drive IC is a fully integrated conventional voltage driver in SOIC-8 package and is able to provide the peak-drive current of around 2 A.

The measured PF values are given in Table 5.3 according to different loads. The measured PF values are all above 0.99, which are compatible to the industrial standards. The functionality of the PFC is realized with the proposed adaptive CSD.

Figure 5.13 illustrates the CS inductor current i_{Lr} and the gate-to-source voltage u_{GS} with different duty cycle. It is noted that with different duty cycle, the peak current I_{peak} of the CS inductor also changes adaptively and at the same time, the peak current is used to switch the main power MOSFET to achieve fast switching speed as labeled in Figure 5.13(a).

Table 5.3 Measured PF values at different loads under 100 Vac

Load conditions (W)	25%(75 W)	50%(150 W)	75%(225 W)	100%(300 W)
PF	0.992	0.994	0.998	0.999

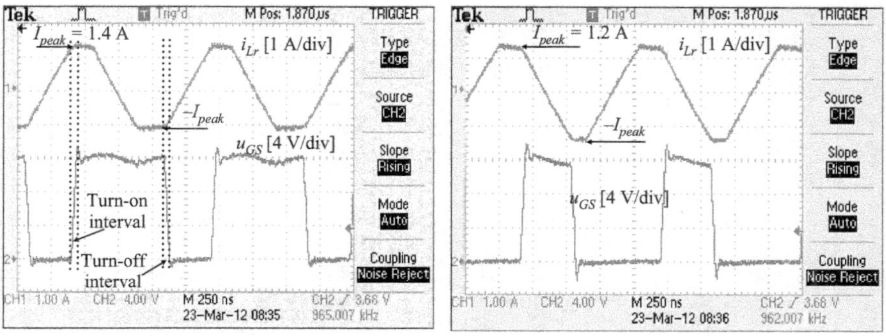

(a) $D > 0.5$ (b) $D < 0.5$

Figure 5.13 CS inductor current i_{Lr} and gate-drive voltage u_{GS} with different duty cycle

Figure 5.14(a) gives the measured efficiency comparison between the CSD and the conventional VSD with different line voltages under full-load condition. Due to the fast switching speed and the switching loss reduction of CSD, the efficiency improvement is achieved throughout the whole input line voltage. Particularly, with 110 Vac input voltage and 300 W output, an efficiency improvement of 3.2% is achieved, while with 220 Vac input voltage, an efficiency improvement of 1.5% is achieved over the conventional VSD. Figure 5.14(b) gives the measured efficiency comparison with different line voltages under half-load condition. Similar efficiency improvements are achieved over the VSD due to fast switching speed.

5.3 Digital adaptive power factor correction

5.3.1 Introduction

The HB adaptive continuous CSD was adopted due to its simple topology and control. The FB adaptive continuous CSD was used to eliminate the blocking capacitor in the HB topology to improve its dynamic performance. However, the continuous adaptive CSD has following disadvantages:

- The drive current is associated with the switching frequency, the duty cycle, and the drive voltage. Therefore the continuous CSD is not suitable to the variable frequency applications;

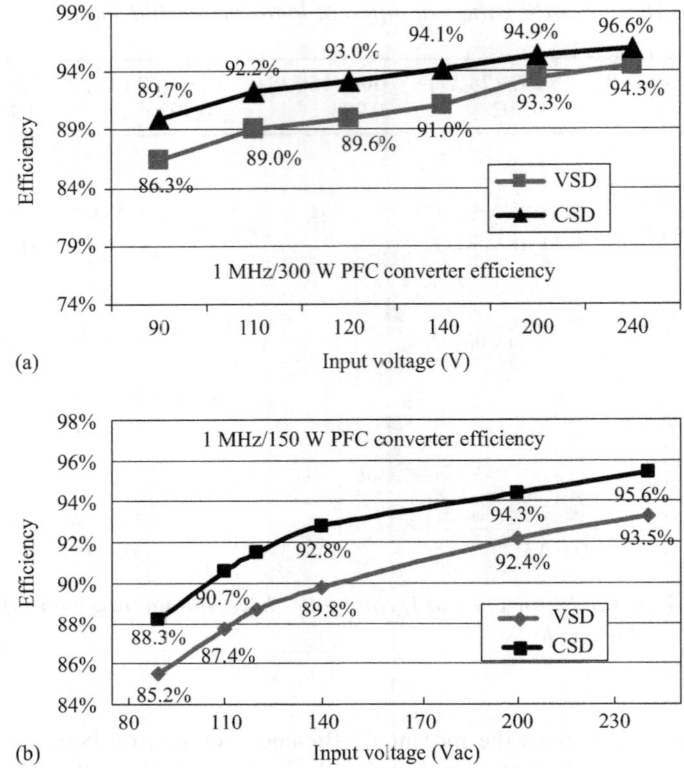

Figure 5.14 Efficiency comparison with different line voltages: top: CSD; bottom: conventional VSD; (a) under full load and (b) under half load

- The drive current strongly depends on the drive voltage and duty cycle. Therefore the optimal design and realization of the adaptive-drive current is complicated;
- The turn-on and turn-off drive current can only be the same which reduces the effectiveness of the efficiency improvement in a wide range;
- The CS inductor current is continuous, which results in high circulating loss in the drive circuit.

To realize the decoupling between the drive current, the duty cycle and the switching frequency, the control strategy of the HB and FB CSDs need to be modified to achieve discontinuous current. The adaptive-drive current is adjusted by the drive voltage that is realized by the analog linear regulator. The drive current is only related to the drive voltage. Therefore the relationship between the adaptive current and the load current is simple and can be easily realized. The CS inductor current is discontinuous, which leads to the reduced circulating loss in the drive

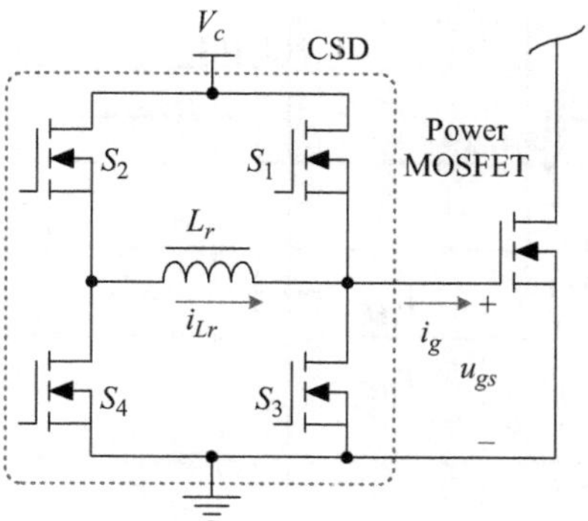

Figure 5.15 Topology of FB CSD

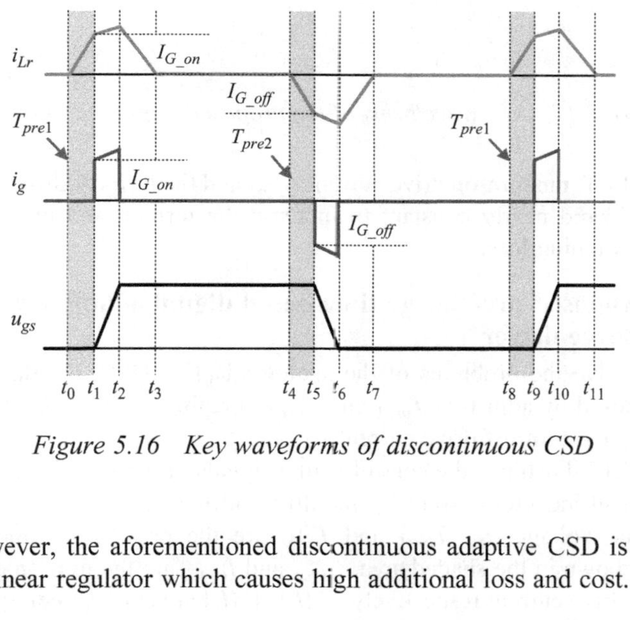

Figure 5.16 Key waveforms of discontinuous CSD

circuit. However, the aforementioned discontinuous adaptive CSD is realized by the analog linear regulator which causes high additional loss and cost.

5.3.2 Principle of the proposed digital adaptive discontinuous current source driver

5.3.2.1 Topology of the full-bridge current source driver

The topology of the digital adaptive CSD is the FB structure as shown in Figure 5.15. It consists of four drive switches, S_1–S_4, and a small resonant inductor, L_r. V_c is the drive voltage.

Figure 5.16 shows the key waveforms of the CSD. The CS inductor current is discontinuous to achieve the lower circulating loss of the drive circuit. From Figure 5.16, S_1–S_4 are respectively controlled to generate the desired CS inductor current i_{Lr} to provide the nearly constant-drive current to turn on and turn off the main power MOSFET. During the turn-on [t_1, t_2] and turn-off [t_5, t_6] interval of the

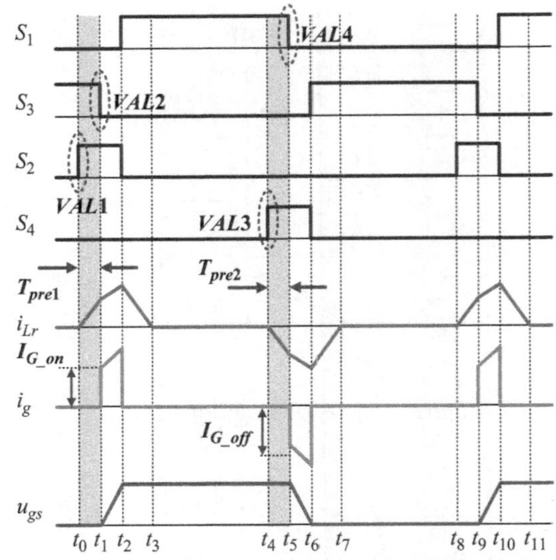

Figure 5.17 Key waveforms of the proposed digital adaptive CSD

power MOSFET, the turn-on drive current I_{G_on} and the turn-off drive current I_{G_off} can be considered nearly constant to increase the turn-on and turn-off speed to reduce the switching loss.

5.3.2.2 Proposed pre-charge time based digital adaptive current source driver

In order to solve the problems of the analog adaptive CSD, the digital adaptive CSD is proposed by adjusting T_{pre1} and T_{pre2} to realize the adaptive turn-on drive current I_{G_on} and turn-off drive current I_{G_off}.

Figure 5.17 illustrates the control gating signals of the four drive switches S_1–S_4, the resonant inductor current i_{Lr}, the drive current i_g, and the power MOSFET gate-to-source voltage u_{gs}. T_{pre1} and T_{pre2} are the pre-charge time of the CS inductor as shown in the shaded area. I_{G_on} and I_{G_off} are the turn-on drive current and turn-off drive current respectively. *VAL*1–*VAL*4 are corresponding to the rising edge of S_2, the falling edge of S_3, the rising edge of S_4, the falling edge of S_1, respectively. *VAL*1–*VAL*4 are the pulse width modulation (PWM) register values in the digital controller which are used to control the duty cycle and period of the PWM signals. The basic idea is to adjust the time interval between *VAL*1 and *VAL*2, and the time interval between *VAL*3 and *VAL*4 in the digital controller, so that the pre-charge time T_{pre1} and T_{pre2} can be adjusted accordingly.

5.3.3 Design procedure and implementation

5.3.3.1 Interleaved boost power factor correction converter under CRM

The proposed digital adaptive CSD was applied to a two-phase interleaved Boost PFC converter under critical conduction mode (CRM) to verify its functionality and

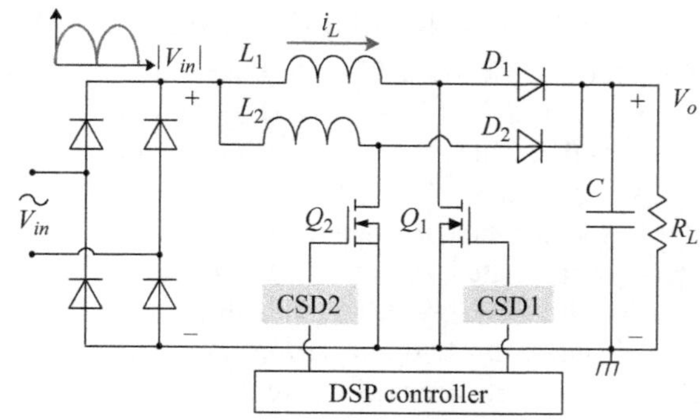

Figure 5.18 Two-phase interleaved boost PFC converter

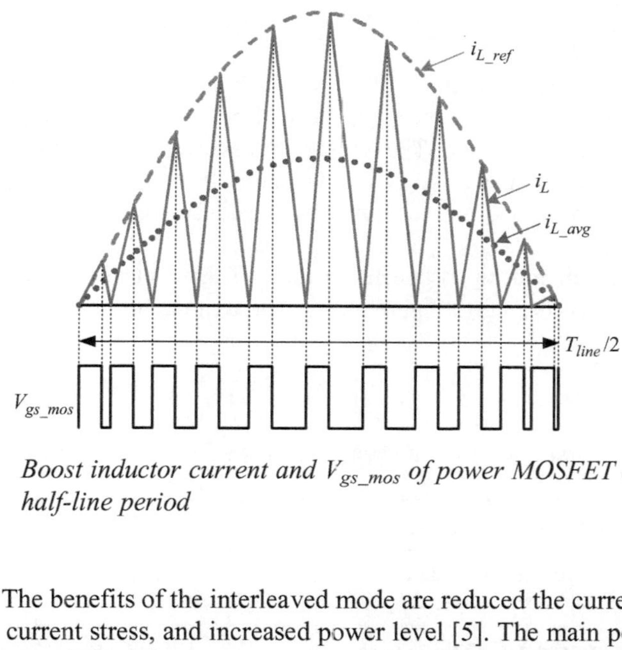

Figure 5.19 Boost inductor current and V_{gs_mos} of power MOSFET during half-line period

advantages. The benefits of the interleaved mode are reduced the current ripple and components current stress, and increased power level [5]. The main power stage is shown in Figure 5.18. The reason to choose the CRM mode is that the power MOSFET owes the different turn-on and turn-off conditions.

Figure 5.19 gives the boost inductor current i_L and gate-to-source voltage V_{gs_mos} during half line period. When the inductor current crosses zero, the switch turns on. When the boost inductor current i_L ramps up till it reaches the reference current i_{L_ref}, the switch turns off at this instant. Since the current through the boost diode decays to zero before the power MOSFET turns on, there is no reverse recovery loss in the boost diode. It is noted that the main switching loss under CRM is the turn-off loss because the power MOSFET turns on at zero current but turns off at a high reference current which is twice of the average current i_{L_avg}. Therefore it is beneficial for the digital adaptive CSD to generate different turn-on and turn-off drive current individually according to different drain currents in this application.

5.3.3.2 Design procedure of the adaptive-drive current

The adaptive-drive current is chosen based on the trade-off between the drive circuit loss and switching loss. The loss analysis provides the design guideline for the proposed digital adaptive CSD (Table 5.4). Although the switching frequency is variable during half-line period, the loss calculation of every switching cycle can be regarded as similar to the constant frequency condition. The basic idea is that the proper drive current is calculated to achieve the lowest total loss during every switching cycle, so that the total loss during half-line period can be minimized.

Since the turn-on and turn-off drain current are different, the turn-on drive current I_{G_on} and turn-off drive current I_{G_off} are designed, respectively. Compared to the turn-off loss, the turn-on loss is much lower. Considering the trade-off between the drive-circuit loss and turn-on delay, the turn-on drive current I_{G_on} is chosen as 2 A in this application.

(1) The tendency of the turn-off drive current I_{G_off} *during half-line period*

Figure 5.20 illustrates the total loss P_{total} as function of I_{G_off} with different turn-off drain current i_D. The arrow curve gives the tendency of the optimized turn-off drive current with different i_D and f_s. The turn-off drain current i_D increases linearly from 1.0 to 2.5 A, the optimized I_{G_off} also increases from 1.2 to 2.2 A accordingly. Therefore, the tendency of the red curve can be considered as a linear relationship approximately.

The linear relationship between the turn-off drive current I_{G_off} and turn-off drain current i_D according to Figure 5.20 is approximately:

$$I_{G\_off} = 0.7 + 0.7 \cdot i_D \tag{5.11}$$

Table 5.4 CSD design parameters

Circuit parameters	
Switching frequency, f_s	90–500 kHz
Gate drive voltage, V_c	12 V
Power MOSFET	**IPP60R099CPA**
Total gate charge@V_{gs} = 12 V	60 nC
Internal gate resistance, R_g	1.3 Ω
Drive switches S_1–S_4	**FDN335N**
On resistance, R_{DS_on}	70 mΩ
Total gate charge@V_{gs} = 5 V	3.5 nC
CS inductor L_r	**1812SMS-R12**
Inductor value	120 nH
CS inductor resistance	17.3 mΩ

In the actual implementation, considering the effect of the turn-off delay, when the turn-off drain current is below the threshold current of 1 A, the turn-off drive current is set constant as 1.4 A. Otherwise, the turn-off drive current is chosen according to (5.11).

Figure 5.21 gives the optimized turn-on and turn-off drive currents during half-line period. The turn-on drive current is a constant value as 2 A as designed. As the envelop of the turn-off drain current is a sinusoidal curve during half-line period, the envelope of the optimized turn-off drive current is also nearly a sinusoidal curve. In particular, when the turn-off drain current i_D is lower than the threshold current of 1 A, the gate drive turn-off current i_{G_off} is chosen as 1.4 A constantly.

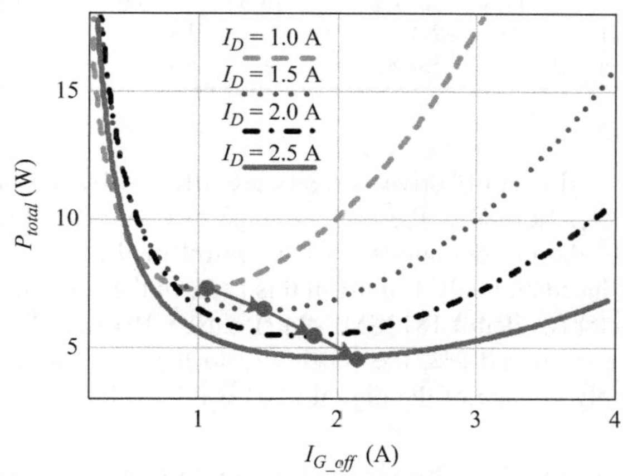

Figure 5.20 Object function P_{total} (I_{G_off}) as a function of drive current I_{G_off} under CRM

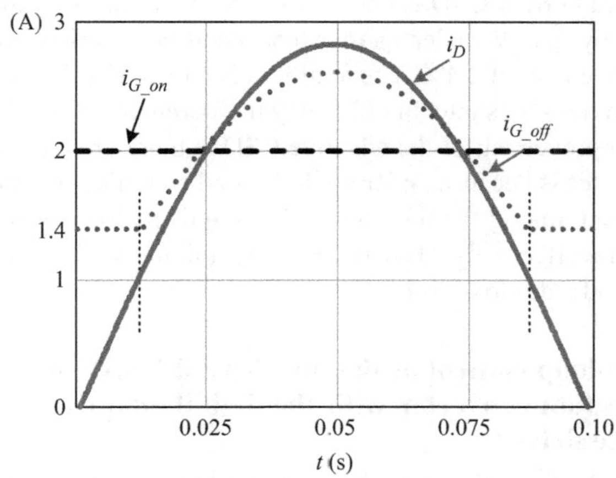

Figure 5.21 Turn-off drain current i_D, drive current i_{G_on} and i_{G_off} during half-line period

Table 5.5 Loss distribution comparison of the boost PFC converter with different drive current types (two channels)

Load	**100 W (25%)**		**200 W (50%)**		**400 W (100%)**	
Drive current type	**Constant**	**Adaptive**	**Constant**	**Adaptive**	**Constant**	**Adaptive**
Switching loss (W)	2.6	2.9	5.1	5.2	11.3	10.9
Drive circuit loss (W)	5.8	3.1	5.8	3.0	5.8	2.6
Conduction loss (W)	0.3	0.3	1.0	1.0	4.0	4.0
Inductor loss (W)	0.3	0.3	1.1	1.1	4.3	4.3
Rectifier bridge loss (W)	1.9	1.9	3.7	3.7	7.2	7.2
Boost diode loss (W)	0.9	0.9	1.9	1.9	3.7	3.7
Total loss (W)	11.8	9.4	18.6	15.9	36.3	32.7
Loss reduction (W)	2.4		2.7		3.6	
Efficiency improvement	2.4%		1.4%		0.9%	

The turn-on and turn-off drive currents are affected by the drive voltage V_c and CS inductor L_r. Normally, the drive voltage can be controlled in tight range (<1%). Owing to the discontinuous current control of the inductor, the inductor value is much reduced and only 120 nH in this case. In the experimental prototype, the air core inductor (Coilcraft 1812SMS-R12) is used. Because the tolerance of the air core inductor is around 2%, the sensitivity to the mismatches caused by the inductor and supply voltage of the digital CSD is relatively low and acceptable.

5.3.3.3 Loss comparison between the constant-drive current and the adaptive-drive current

Table 5.5 gives the loss distribution comparison of the boost PFC converter with the constant- and adaptive-drive currents. The turn-on and turn-off drive currents of the constant-drive current CSD is chosen as 2 A. As shown in Table 5.5, the total loss is reduced by 2.4 W under quarter-load condition, which translates into an efficiency improvement of 2.4%. The total loss is reduced by 3.6 W under full-load condition, which translates into an efficiency improvement of 0.9%. It is noted that the efficiency improvement of the adaptive CSD is more effective under the light-load condition. This is because under the light-load condition, the drive-circuit loss becomes dominant among the total loss. The adaptive CSD can reduce the drive-circuit loss more effectively. The target is to minimize the summation of the switching loss and gate-drive loss.

5.3.3.4 Close-loop control of the interleaved boost power factor correction converter with the digital adaptive current source driver

The digital adaptive discontinuous CSD is applied to a two-phase interleaved boost PFC converter to verify its advantages. The interleaved parallel boost PFC converter under CRM is shown in Figure 5.22. The close-loop control diagram is

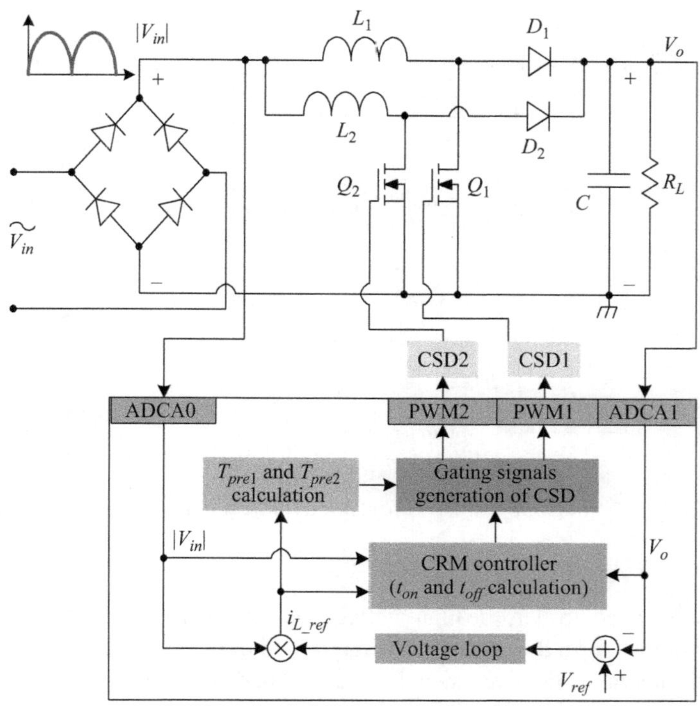

Figure 5.22 Two phase interleaved boost PFC converter

shown in Figure 5.23. The PFC voltage loop is calculated to ensure the stability of the output voltage. The output of the voltage regulator multiplied with the input voltage provides the boost inductor current reference i_{L_ref}.

The CRM controller is designed to ensure the boost inductor current operates under CRM and the input current follows the waveform of the input voltage. In order to reduce the cost, the low-end digital signal processor (DSP) and a simplified control method are used without sampling the boost inductor current directly. In the CRM controller, the on and off interval of the main switch can be calculated as

$$t_{on} = L\frac{i_{L\_ref}(t)}{v_{in}(t)} \tag{5.12}$$

$$t_{off} = L\frac{i_{L\_ref}(t)}{V_o - v_{in}(t)} \tag{5.13}$$

where v_{in} is the input voltage, V_o is the output voltage and i_{L_ref} is the boost inductor current reference, which is the output of the multiplier from the voltage control loop.

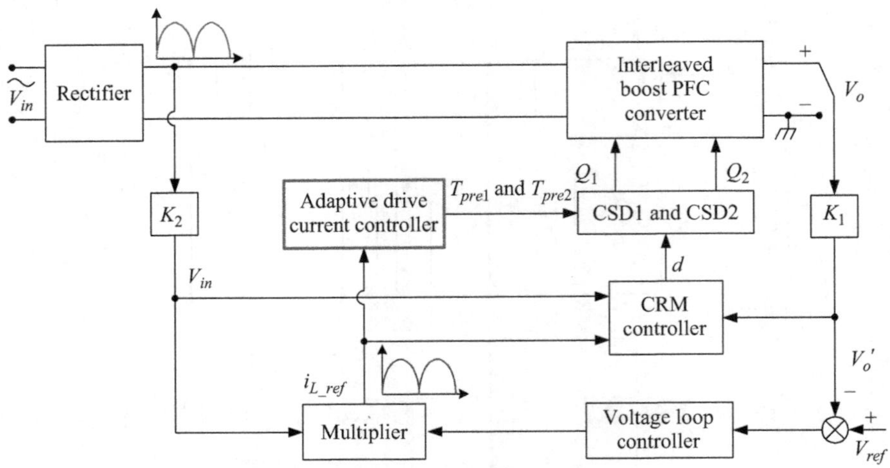

Figure 5.23 Close-loop control diagram of interleaved boost PFC converter

In the CRM controller, the input voltage, output voltage, and the boost inductor current reference from the output of the multiplied of the voltage loop are read. The on and off interval of the main switch are calculated according to (5.12) and (5.13) in the PWM interrupt [6]. Once t_{on} and t_{off} are obtained, the duty cycle and period of the PWM signal of the main switch can be generated.

Since the sampling of the boost inductor current is avoided, the boost inductor current reference i_{L_ref} is chosen as the reference signal of the adaptive turn-off drive current, which represents the turn-off drain current i_D at the turn-off instant. In the adaptive-drive current controller, the turn-on and turn-off pre-charge times T_{pre1}, T_{pre2} can be calculated according to i_{L_ref}. According to t_{on}, t_{off}, T_{pre1}, and T_{pre2} calculated in the PWM interrupt, the gating control signals of the CSD drive switches can be generated in real time before a new modulation period.

Figure 5.24 shows the flowchart of the master phase of the two-phase interleaved boost PFC converter. The PWM output functions properly when the analog to digital converter (ADC) results are in the normal range. Otherwise the PWM output is masked immediately to protect the circuit. In the PFC voltage loop, t_{on} and t_{off} are calculated to realize the PFC function. Meanwhile, the turn-on pre-charge time T_{pre1} and turn-off pre-charge time T_{pre2} are calculated according to i_{L_ref}. Then the PWM register values of the gating signals can be generated. The gating signals of the slave phase are delayed half period of the master slave.

5.3.4 Experimental results and discussion

A 220 Vac input, 380 V/400 W output two-phase interleaved boost PFC converter under CRM was built. The switching frequency varies from 90 to 500 kHz. The

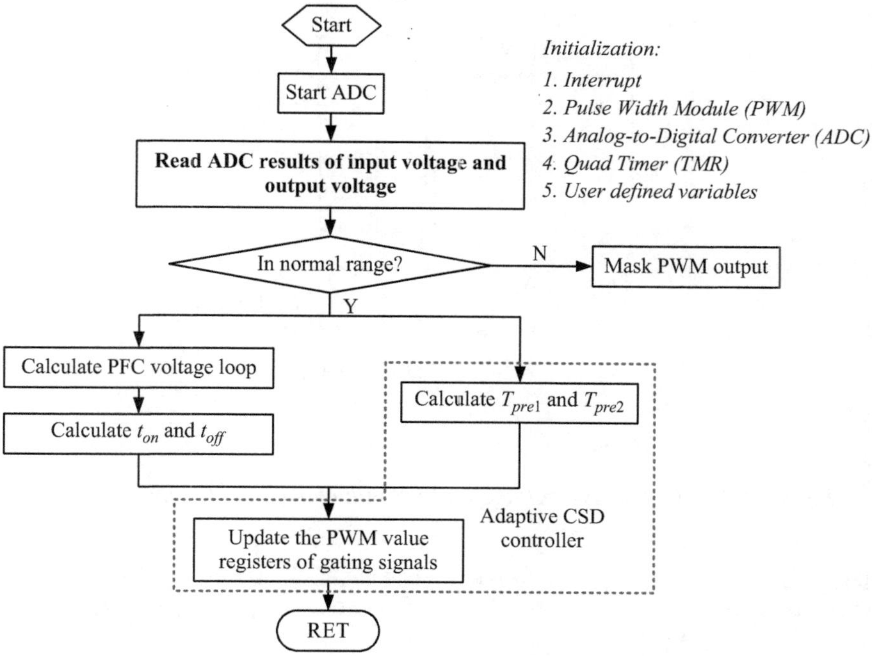

Figure 5.24 Flowchart of boost PFC converter

specifications are: the boost inductor $L = 220$ µH; the output capacitor $C = 440$ µF. The CSD part was built using the discrete components. The FDN335N N-channel MOSFETs were used for the four-drive switches S_1–S_4. The Coilcraft 1812SMS air core inductor (120 nH) was used for the CS inductor. The gate-drive voltage V_c is 12 V. The Freescale MC56F8257 DSP was used for the digital control of the CSD and PFC control loop.

5.3.4.1 Experimental results of the digital adaptive discontinuous current source driver

Figure 5.25 shows the waveforms of the gate-drive voltage and CS inductor current when the turn-off drain current $i_D = 1.5$ A. The amplitude of the gate-drive voltage is 12 V. The peak value of the turn-on drive current and the turn-off current are 2 A and 1.4 A, respectively, which are the optimized values of the drive current.

Figure 5.26(a) and (b) shows the zoomed gate-drive voltage and CS inductor current during the turn-on interval and turn-off interval. It is noted that the turn-on interval time and turn-off interval time are 8 ns and 9 ns, respectively, and fast switching speed is achieved.

Figures 5.27 and 5.28 show the detailed waveforms of the CS inductor current, the gate-drive voltage and the drive switches control signals when the turn-off drain current i_D turns into 2.6 A. The turn-on pre-charge time and the peak CS inductor current are the same. The drive current is 2 A according to the optimal design.

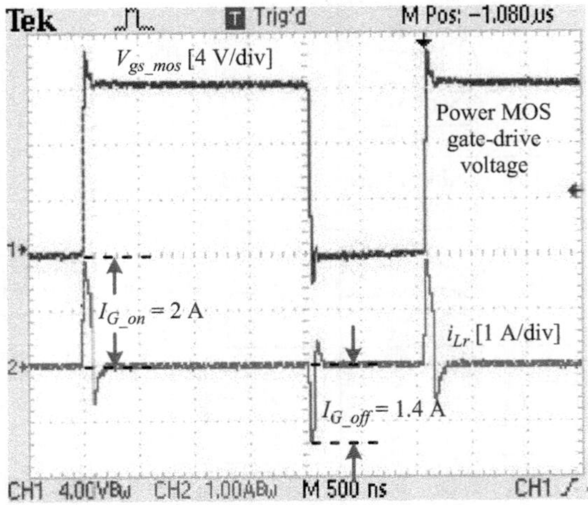

Figure 5.25 Gate-drive voltage and CS inductor current with $i_D = 1.5$ A

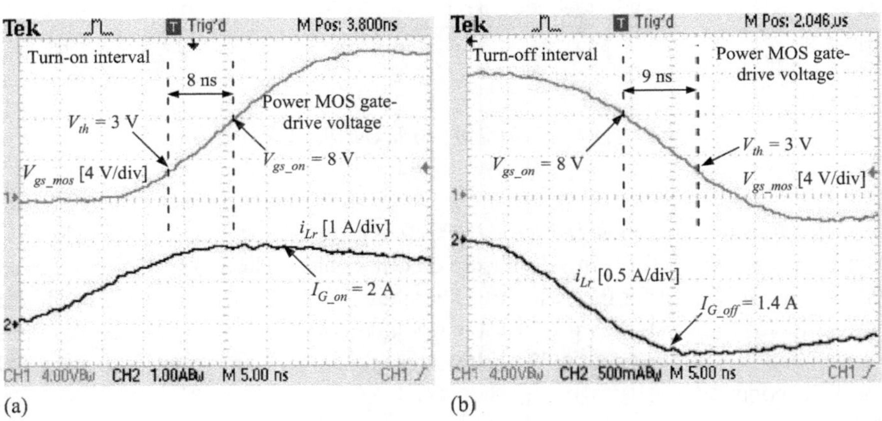

Figure 5.26 Zoomed gate-drive voltage and inductor current with $i_D = 1.5$ A; during (a) turn-on interval and (b) turn-off interval

From Figures 5.26(a) and 5.28(a), it is noted that the turn-on interval time is 8 ns because the turn-on drive current is the same. Compared Figure 5.26(b) to Figure 5.28(a), the turn-off interval time reduces from 9 to 7 ns (a reduction of 22%) to reduce the turn-off loss.

Figure 5.29 shows the drive current and input current during half-line period. As shown in Figure 5.29, the turn-on drive current is constant and the envelope curve of the turn-off drive current is nearly a sinusoidal curve during half-line period according to the optimal design. The turn-off drive current is adjusted dynamically according to different turn-off drain current to realize better trade-off between the turn-off loss and drive-circuit loss.

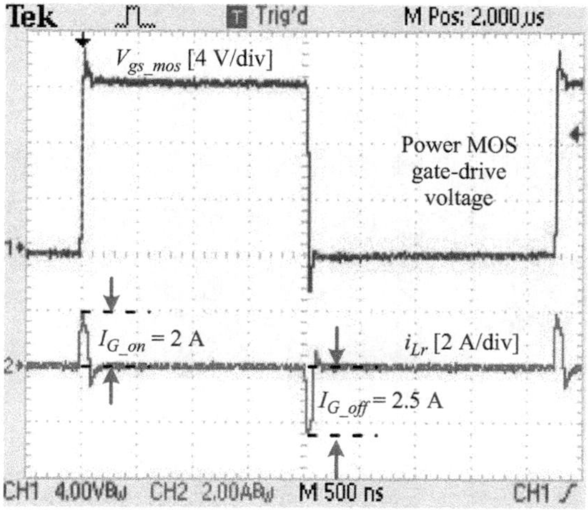

Figure 5.27 Gate-drive voltage and CS inductor current with $i_D = 2.6$ A

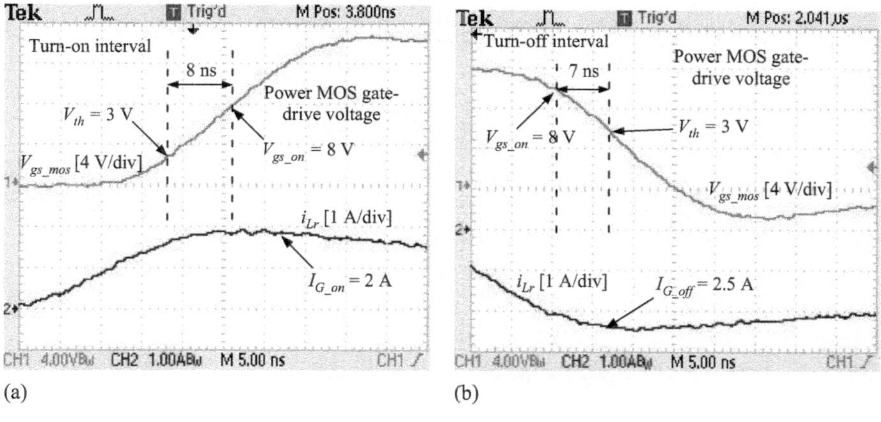

Figure 5.28 Zoomed gate-drive voltage and inductor current with $i_D = 2.6$ A; during (a) turn-on interval and (b) turn-off interval

5.3.4.2 Loss reduction of the digital adaptive discontinuous current source driver

Figure 5.30 shows the measured efficiency comparison between the digital adaptive CSD and constant one. The turn-on and turn-off drive currents of the constant-drive current CSD are chosen as 2 A. As shown in Figure 5.30, the boost PFC with the proposed digital adaptive CSD improves 1.6% over the constant one under the quarter load. An efficiency improvement of 0.7% is achieved over the constant-drive current CSD under the full load. The proposed digital CSD achieves a higher efficiency improvement under the light-load condition over the heavy-load condition.

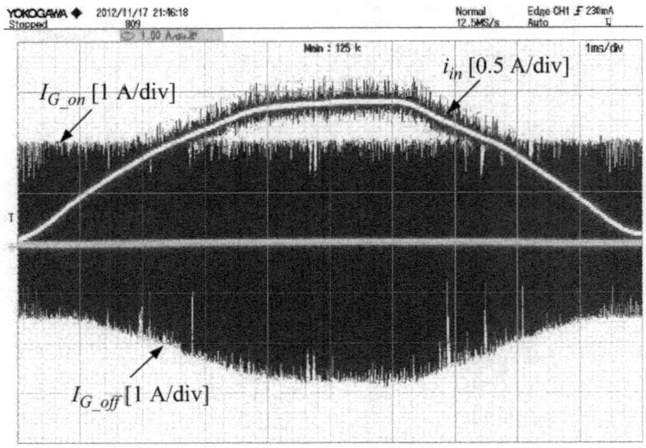

Figure 5.29 Input current of one phase and drive current during half-line period

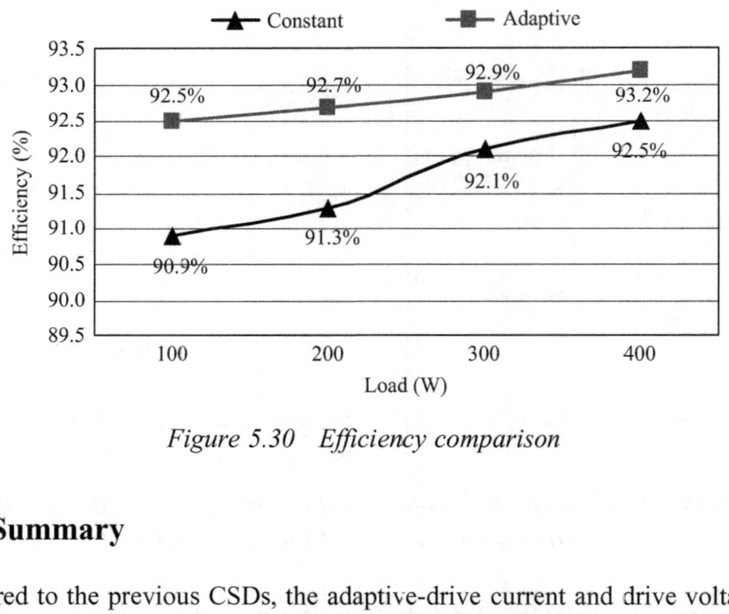

Figure 5.30 Efficiency comparison

5.4 Summary

Compared to the previous CSDs, the adaptive-drive current and drive voltage can be realized to optimize the switching loss reduction and the drive loss reduction at different load currents. It should be noted that the adaptive concept is suitable for both the continuous and discontinuous CSDs regardless of the drive-circuit topologies. The linear regulator can be used to achieve the function of the adaptive voltage and drive current in a cost-effective manner.

The adaptive FB CSD was proposed for boost PFC converters to achieve fast switching speed and significant switching loss reduction. Compared to other CSDs with the constant-drive current, the advantage of the adaptive-drive current can be achieved: further switching loss reduction when the power MOSFET is with higher switching current; while reducing the drive circuit loss when the MOSFET is with lower switching current. This provides better optimal opportunity with the trade-off between the switching loss reduction and CSD drive circuit loss during wide operation range.

The digital CSD is able to achieve the adaptive-drive current in "a freeway" since it has strong compatibility with the digital control technique. It eliminates the additional loss and cost introduced by the linear regulator, reduces the complexity, and cost of the circuit significantly. More importantly, the digital adaptive function can offer a more flexible control algorithm to build the different turn-on and turn-off drive current to achieve the design trade-off between the switching loss and drive circuit loss, and obtain the optimal efficiency in a wide load range.

References

[1] A. Parayandeh and A. Prodic, "Digitally controlled low-power DC–DC converter with segmented output stage and gate charge based instantaneous efficiency optimization," in *Proceedings on IEEE ECCE*, pp. 3870–3875, 2009.

[2] M. M. Jovanovic, "Technology drivers and trends in power supplies for computer/telecom," *Plenary Session Presentation of Applied Power Electronics Conference and Exposition*, 2006.

[3] C. Wang, M. Xu, B. Lu, and F. C. Lee, "New architecture for MHz switching frequency PFC," in *Proceedings on IEEE APEC*, pp. 179–185, 2007.

[4] P. O. Jeannin, D. Frey, J. C. Podvin, *et al.*, "1 MHz power factor correction boost converter with SiC Schottky diode," in *Proceedings on IEEE IAS*, pp. 1267–1272, 2004.

[5] L. Huber, B. T. Irving, C. Adragna, and M. M. Jovanovic, "Implementation of open-loop control for interleaved DCM/CCM boundary boost PFC converters," in *Proceedings on IEEE APEC*, pp. 1010–1016, 2008.

[6] Z. Jiang and H. Li, "DSP control of interleaving critical PFC module for high power application," in *Proceedings on IEEE Industrial Electronics Society Conference (IECON)*, pp. 712–716, 2008.

Chapter 6
Resonant gate drivers

6.1 Resonant gate drivers for multi-megahertz isolated resonant converters

6.1.1 Introduction

Recently, the switching frequency has been pushed up to tens of megahertz (MHz), directly reducing the energy storage requirements of the passive components, shrinking the size and weight, and allowing faster transient response [1,2]. Most of the previous research on the multi-MHz resonant converters focus on the non-isolated applications. For example, the resonant boost converters in [3,4], and the resonant single-ended primary-inductor converters (SEPICs) in [5] have achieved high efficiency and fast dynamic performance. However, the state-of-the-art monolithic power integrated circuits (ICs) have realized the non-isolated power conversion with extremely high power density, such as EN23F0QI from ENPIRION, but the isolation still remains a challenge to the power IC solution due to the difficulty of the integration of the magnetic components. This motivates research efforts in the isolated multi-MHz resonant converters, such as the 75-MHz isolated resonant converter proposed in [6]. The converter manages to achieve very high power density. However, the efficiency needs to be further improved, especially the high loss in the diodes, which is 35% of the total power loss.

It is pointed out that the performance of the diodes degrades seriously as the switching frequency increases to tens of MHz because of the forward recovery [7,8]. There is a large transient voltage across the diode at the turn-on instant, causing high power loss especially at multi-MHz. Therefore, the multiple diodes are paralleled in [9–12] to minimize the conduction loss, resulting in increased component count and cost. However, due to the high-frequency pulse current, the high conduction loss dissipated in the rectifier diodes can still account for 30%–35% of the total power loss [6,7]. In order to reduce the high conduction loss and the frequency-dependent forward recovery loss, the synchronous rectification (SR) technique is strongly desired for the multi-MHz resonant converters, especially in the application of high output current.

Since the conventional hard-switched gate drivers are too inefficient in multi-MHz low-power applications, the resonant gate drivers (RGDs) are normally needed to reduce high-frequency gate-drive loss. However, the RGDs in the previous literatures mostly focus on the control field effect transistors (FETs) in the multi-MHz resonant converters instead of the SR FETs. The multi-stage RGD is widely used in the multi-MHz resonant converters [3,5,6]. It provides the reliable resonant gate voltage and reduces the high-frequency gate loss by storing the drive energy in the resonant tank. However, since the drive signal is produced by the totem pole, the gate voltage averages a fixed value, typically zero or half the supply voltage V_{CC} [3,5,6]. The duty ratio of the metal oxide semiconductor field effect transistor (MOSFET) is also fixed. When applied to the SR, it can hardly provide precise switching timing for the SR FET and induce additional conduction time of the body diode, causing high forward voltage drop and high conduction loss. More importantly, the gate voltage of the RGD is sinusoidal and the slow change of the gate voltage will cause high on-state resistance $R_{DS(on)}$, especially at turn-on and turn-off instants. Hence the average $R_{DS(on)}$ is much higher than that provided in the datasheet. This characteristic of the RGD becomes more apparent when employed to the SR and the high average $R_{DS(on)}$ offsets the benefits of the SR technique seriously. A resonant second harmonic class-E gate driver is proposed in [12], which is a half-wave RGD. It provides a d.c. bias voltage to shift the gate voltage up and down to control the switch duty ratio. However, it consists of as many as seven passive components, which seriously reduce the power density. Besides, the auxiliary switch also needs additional drive circuit (a complementary metal oxide semiconductor (CMOS) hard-switched inverter driver). The RGDs with tunable d.c. offset are also proposed in [13,14]. However, when applied to the SR FET, the driving logic of the drive signals for the control FET and the SR FET still needs to be solved. Furthermore, the complexity of the drive circuit consequently increases the converter size and power loss.

6.1.2 Challenges for the SR drive in the multi-MHz isolated resonant converters

6.1.2.1 The multi-MHz isolated synchronous class-Φ_2 converter

Figure 6.1 gives the schematic of the isolated synchronous class-Φ_2 resonant converter and Table 6.1 lists the specifications. Because the forward recovery causes high conduction loss in the diodes at multi-MHz, the SR technique is employed to the converter. However, since the SR FETs normally have large total gate charge compared to the control FETs, the gate-drive loss at multi-MHz with the conventional hard driver is extremely high and reduces the efficiency seriously.

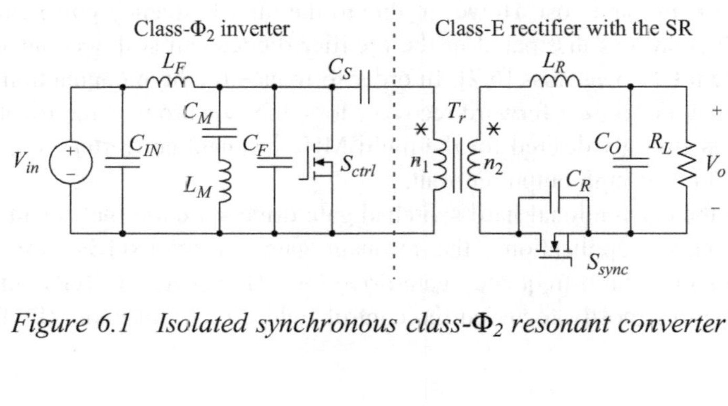

Figure 6.1 Isolated synchronous class-Φ_2 resonant converter

6.1.2.2 Challenges for the SR drive and problems of the conventional RGD

The challenge of the high-frequency SR drive is that the switching cycle is only 100 ns in this case, which causes problem to obtain the desired phase-shift between the drive signals for the control and the SR FETs. The isolation of the drive signal is another problem to be solved.

The multistage RGD is considered for the SR because of its simplicity and efficient gate drive. However, the drive signal of the RGD in previous research is generated by the totem pole. When applied to the SR in this case, the external-driven scheme will induce the propagation delay of the driver ICs (normally tens of ns), which may cause ill-effect on the drive signal timing during the switching cycle of 100 ns. Meanwhile, the self-driven SR is a simple drive scheme because the drive signals for the SR FETs are provided by the circuit node voltage inherently without additional controllers and hardware delay problems [15–17]. Therefore, as shown in Figure 6.2, an auxiliary winding is chosen to provide the drive signal and isolation. Figure 6.3 shows the sinusoidal gate voltage of the conventional RGD. With the drive signal provided by the auxiliary winding of the transformer, the resonant gate voltage averages zero. There are two main drawbacks of the conventional RGD.

1. Figure 6.4 shows the typical curve of the on-state resistance $R_{DS(on)}$ as a function of the gate-to-source voltage v_{gs}, based on the datasheet of a Si MOSFET from Vishay (SiS862DN). It is observed that the $R_{DS(on)}$ changes significantly with v_{gs}. When v_{gs} just exceeds the threshold voltage V_{th} or declines close to V_{th}, especially at the turn-on and turn-off instants, $R_{DS(on)}$ is much high. Conversely, when v_{gs} exceeds the critical voltage V_{cv} (typically 5–6 V), $R_{DS(on)}$ is much lower and reduces smoothly.

Table 6.1 Converter specifications

V_{in}	V_o	I_o	P_o	f_s
16–21 V	5 V	2 A	10 W	10 MHz

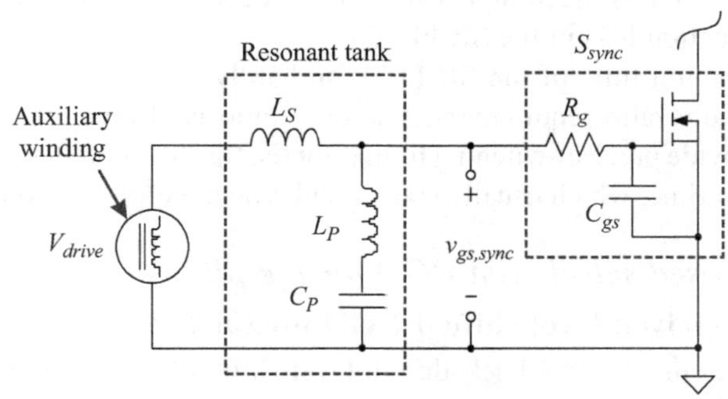

Figure 6.2 Conventional RGD with auxiliary winding

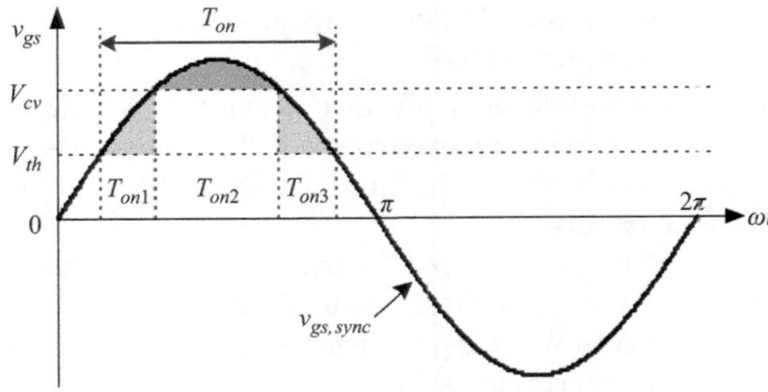

Figure 6.3 Resonant drive voltage of the conventional RGD

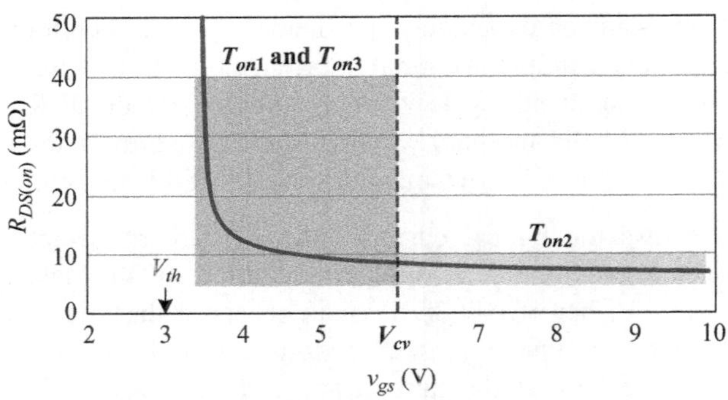

Figure 6.4 On-resistance $R_{DS(on)}$ vs. gate voltage v_{gs}

Based on the curve of $R_{DS(on)}$ vs. v_{gs}, the conduction time in Figure 6.3 is divided into three intervals: the conduction time with a low $R_{DS(on)}$, T_{on2}, and the conduction time with a high $R_{DS(on)}$, T_{on1} and T_{on3}. The problem is that T_{on2} is quite short (less than 50% of the total conduction time T_{on}), while T_{on1} and T_{on3} are relatively long (more than 50% of T_{on}). As a result, the average $R_{DS(on)}$ of the MOSFET is much higher than the rated value in the datasheet, causing high conduction loss in the SR FET.

2. The conduction time of the SR FET can hardly be adjusted flexibly to meet different duty ratio requirement. The duty ratio is always less than 50% and cannot provide precise switching timing for the SR. It will induce the body diode conduction time, which usually causes high voltage drop and conduction loss.

6.1.3 Proposed self-driven RGD for the SR

6.1.3.1 Self-driven level-shifted RGD for the SR

Figure 6.5 gives the proposed self-driven level-shifted RGD for the SR FET. The auxiliary winding, n_3, provides the drive signal. The resonant tank provides the desired voltage gain and phase-shift from n_3 to the gate of the SR FET.

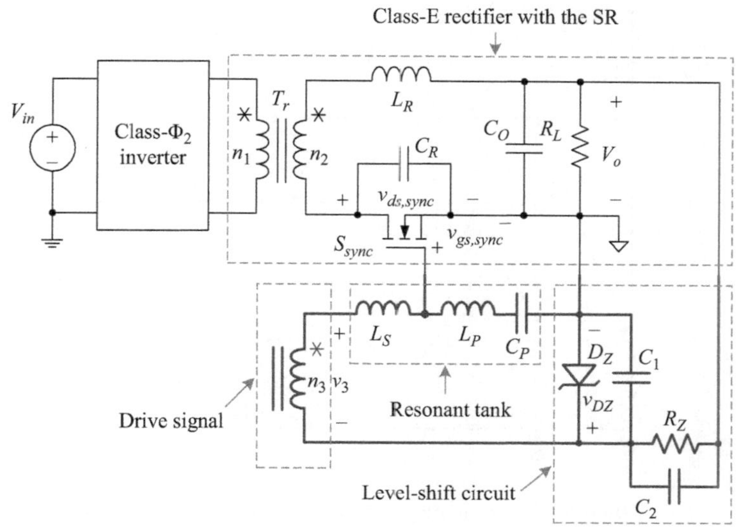

Figure 6.5 Proposed self-driven level-shifted RGD

The level-shift circuit provides a tunable d.c. bias to increase the peak gate voltage, so that the conduction time with the optimal $R_{DS(on)}$ can be extended and the average $R_{DS(on)}$ of the MOSFET can be reduced. Moreover, by tuning the phase-shift of the resonant tank and the d.c. bias, the proposed RGD can provide precise switching timing for the SR FET.

Figure 6.6 shows the key waveforms. The voltage across the auxiliary winding, v_3, is transferred through the resonant tank to the a.c. sine component of the gate voltage $v_{gs,sync}$. The voltage across D_Z, v_{DZ}, is kept constant at its Zener voltage V_B. So the DC bias of $v_{gs,sync}$ also equals V_B. At t_0, the drain voltage $v_{ds,sync}$ resonates to zero, and $v_{gs,sync}$ increases in a resonant manner and just exceeds the threshold voltage V_{th}, so the SR FET S_{sync} turns on immediately. At t_1, $v_{gs,sync}$ decreases below V_{th}, so S_{sync} turns off and $v_{ds,sync}$ begins to rise. After t_1, $v_{gs,sync}$ resonates always below V_{th} until the next switching cycle.

6.1.3.2 Problems of the level-shifted RGD under ON-OFF control

The ON-OFF control is normally used in the multi-MHz resonant converters owing to its simple structure and fast transient response [3,18]. Figure 6.7 shows the block diagram and the key waveforms. The output voltage V_o is compared to the reference V_{ref} by the ON-OFF controller. The output of the controller, V_{ctrl}, modulates the converter ON and OFF. The ON-OFF modulation frequency f_{mod} is much lower than the switching frequency f_s. To ensure fast transient response, the ON-OFF control scheme requires that the converter turns ON and OFF quickly. Therefore, for the driver, shown as v_{gs} in Figure 6.7, the gate voltage is supposed to achieve fast shutdown and buildup under ON-OFF operation.

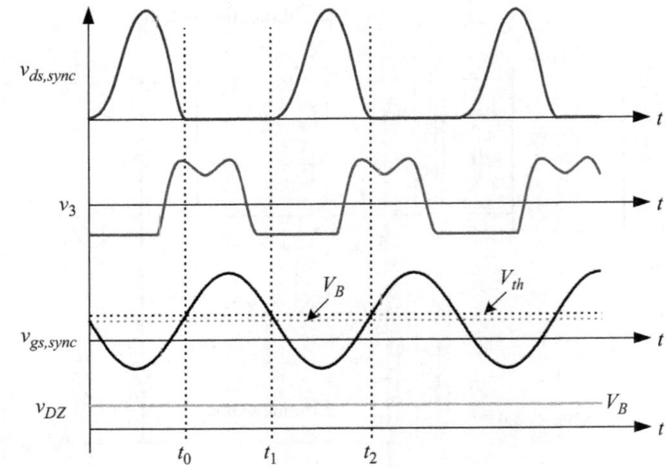

Figure 6.6 Key waveforms of the level-shifted RGD

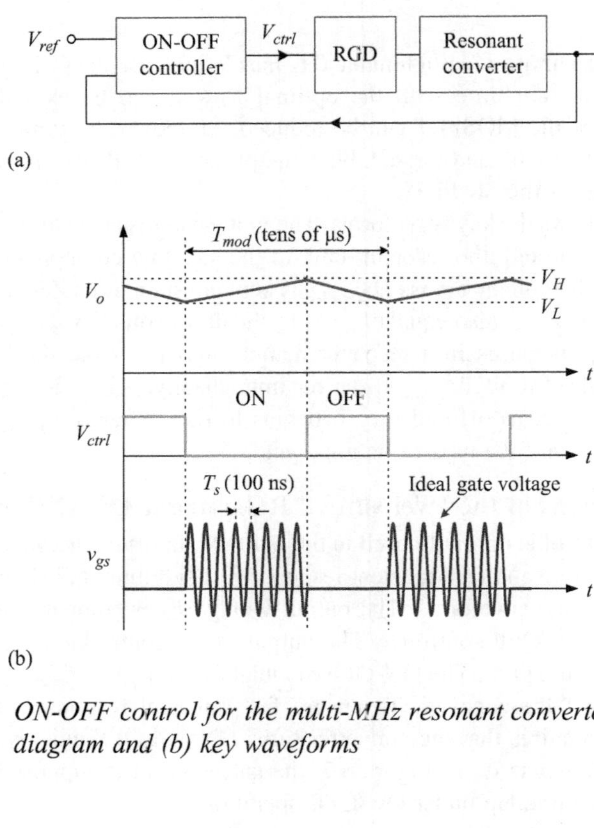

Figure 6.7 ON-OFF control for the multi-MHz resonant converters; (a) block diagram and (b) key waveforms

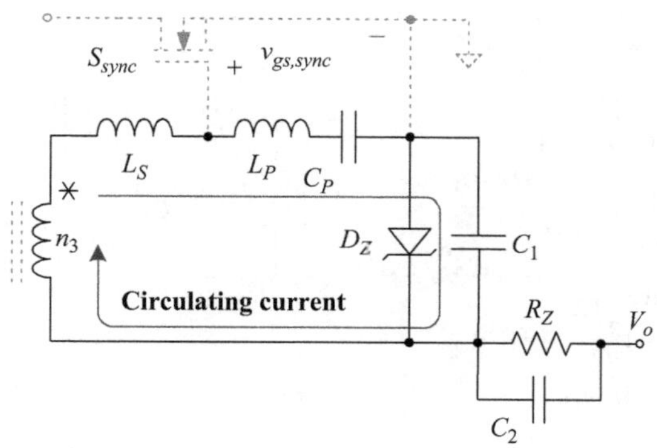

Figure 6.8 Circulating current when converter turns OFF

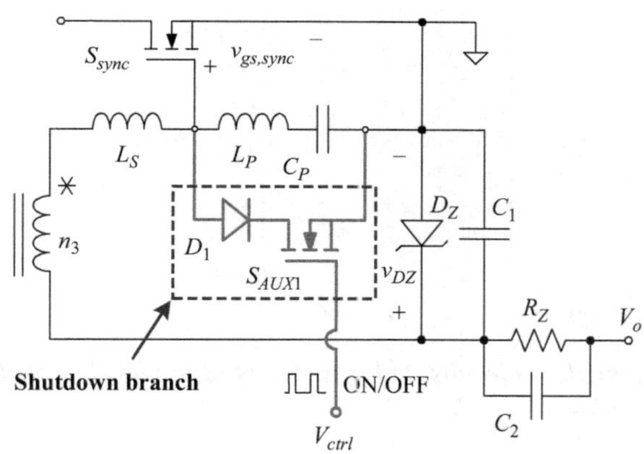

Figure 6.9 Self-driven level-shifted RGD with a shutdown branch

The problem of the proposed RGD in Figure 6.5 is that there is a circulating current and gate energy resonant oscillation in the driver when the converter turns OFF. As shown in Figure 6.8, the voltage across the auxiliary winding n_3 drops to zero, and the drive energy stored in the resonant tank induces a circulating current in the LC network in the drive circuit, until the drive energy is dissipated by the parasitic resistances. This circulating current will cause the gate voltage resonant oscillation, which may lead to undesired false turn-on of the SR FET and high circulating loss.

To block the circulating current, in Figure 6.9, a shutdown branch is added to the drive circuit in parallel with the gate and source. The branch comprises a diode D_1 and an auxiliary switch S_{AUX1} controlled by the control signal V_{ctrl} [3,5,6]. However, although the shutdown branch blocks the drive circulating current when the converter turns OFF, it causes another problem when the converter turns ON. There are three operation intervals of the RGD and Figure 6.10 gives the equivalent circuits and the operation waveforms.

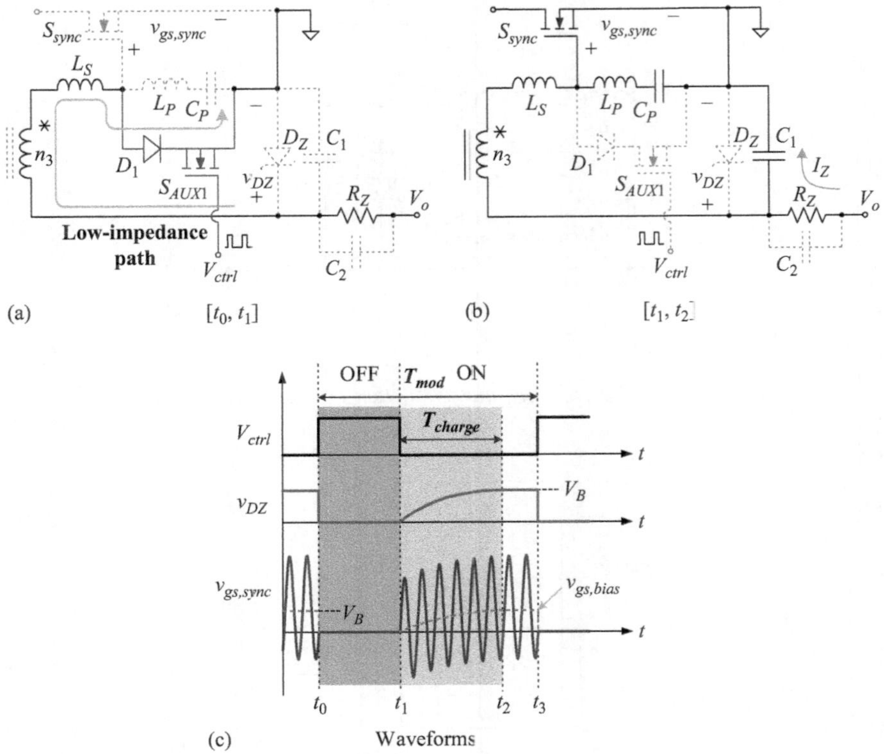

Figure 6.10 Equivalent circuits and waveforms of the level-shifted RGD with the shutdown branch

1. Interval 1 $[t_0, t_1]$: V_{ctrl} sets high and the converter turns OFF. In Figure 6.10(a), S_{AUX1} turns on and D_1 conducts to enable the shutdown branch. The auxiliary winding n_3, the inductance L_S and the shutdown branch form a low-impedance current path. The capacitance C_1 is discharged to zero quickly by this low-impedance path and cannot be charged again because D_1 blocks the reverse charging current. So in Figure 6.10(c), $v_{gs,sync}$ is clamped to zero from t_0 to t_1. v_{DZ} falls to zero at t_0 and keeps constant at zero from t_0 to t_1.
2. Interval 2 $[t_1, t_2]$: V_{ctrl} sets low and the converter turns ON. S_{AUX1} turns off and the shutdown branch is disabled. So the passive components begin to resonate and C_1 is charged through R_Z by V_o. The voltage across C_1, v_{DZ}, (which is actually the bias voltage of $v_{gs,sync}$, $v_{gs,bias}$) increases as

$$v_{DZ}(t) = v_{gs,bias}(t) = V_B[1 - \exp(-(t - t_1)/R_Z C_1)] \tag{6.1}$$

3. Interval 3 $[t_2, t_3]$: v_{DZ} reaches V_B and keeps constant. Note that if v_{DZ} cannot reach V_B from t_1 to t_3, this interval does not exist.

During Interval 2, the actual bias voltage $v_{gs,bias}$ is always lower than the desired bias V_B. From Figure 6.10(c), it should be noted that the period of Interval 2,

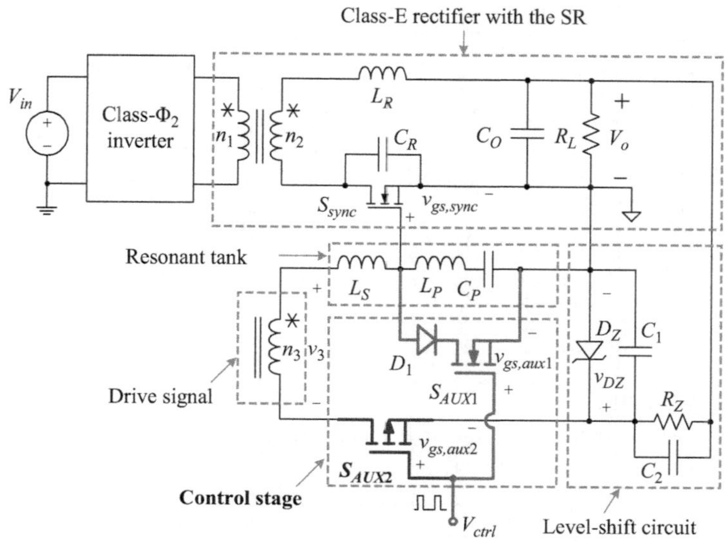

Figure 6.11 Proposed ON-OFF controlled level-shifted RGD for the SR

$T_{charge} = t_2 - t_1$, is too long compared to the modulation cycle T_{mod}. Normally, T_{mod} is several μs to tens of μs, while the time constant of R_Z and C_1 ($\tau_{RC} = R_Z C_1$) is also in the microsecond range. It means that the bias voltage can hardly reach the desired voltage level during the ON status, seriously offsetting the benefits of the d.c. bias in the level-shifted RGD under close-loop.

6.1.3.3 Proposed ON-OFF controlled level-shifted RGD for the SR

To solve the charging and discharging problem of v_{DZ}, another auxiliary switch S_{AUX2} (P-Channel) is introduced to the driver, as shown in Figure 6.11. S_{AUX2} is connected in series between the auxiliary winding n_3 and the level-shift circuit. S_{AUX2} and D_1, S_{AUX1} form the control stage of the proposed ON-OFF controlled level-shifted RGD. Since the control signal V_{ctrl} is a square wave between 0 and V_{CC} (the supply voltage), the gate voltage of S_{AUX2}, $v_{gs,aux2}$, is a square wave between $-V_B$ and $V_{CC} - V_B$. Therefore, S_{AUX1} and S_{AUX2} can share V_{ctrl} as the drive voltage as long as $-V_B$ is lower than the threshold of S_{AUX2}.

Figure 6.12 shows the equivalent circuits and the key waveforms of the proposed RGD. In Figure 6.12(a), at t_0, V_{ctrl} sets high and S_{AUX1} turns on to shut down the gate voltage. $v_{gs,aux2}$ increases to $V_{CC} - V_B$ to turn off S_{AUX2}. The low-impedance path is blocked and the voltage across D_Z will not be discharged. So in Figure 6.12(c), from t_0 to t_1, $v_{gs,sync}$ is clamped to zero and v_{DZ} is kept stable at V_B. In Figure 6.12(b), at t_1, V_{ctrl} sets low and S_{AUX1} turns off to disable the shutdown branch. $v_{gs,aux2}$ falls to $-V_B$ so that S_{AUX2} turns on to conduct the drive current. The resonant components in the RGD begin to resonate to provide switching-frequency drive voltage. Since v_{DZ} is not discharged before t_1, there is no charging process from t_1 to t_2. So in Figure 6.12(c), from t_1 to t_2, v_{DZ} still keeps stable at V_B and

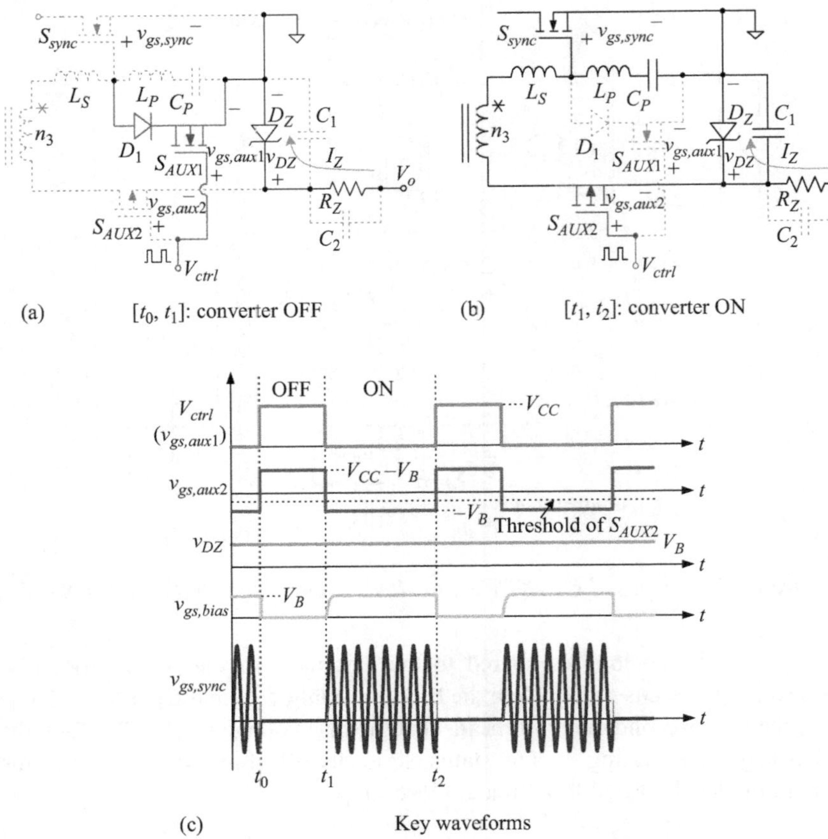

Figure 6.12 Equivalent circuits and waveforms of the ON-OFF controlled level-shifted RGD

$v_{gs,bias}$ increases to V_B as soon as the converter turns ON. Therefore, the benefits of the level-shifted RGD can be maintained under ON-OFF operation.

6.1.3.4 Design of the level-shifted RGD

Auxiliary winding

The voltage across the winding n_3 is determined by the output of the inverter and the turns ratio of the primary and the auxiliary windings n_1:n_3. The output of the class-Φ_2 inverter is approximately a square wave between V_{in} and $-V_{in}$. So the voltage across n_3, v_3, swings between n_3V_{in}/n_1 and $-n_3V_{in}/n_1$. A step-down voltage is preferred so that it is easier to design the resonant tank to generate the desired phase-shift and voltage gain. Since the turns ratio of the primary and the secondary windings, n_1:n_2, is 4:1 to balance the voltage and current stress on the SR FET, there are three possible options for n_1:n_3. The final design of n_1:n_3 is 4:1 because if it is set to 4:2 or 4:3, the peak gate voltage would easily be high or even exceed the limits of the device (±20 V), which results in the over voltage damage of the semiconductor device.

Resonant tank

The capacitance C_P blocks the d.c. bias voltage. The inductances, L_P and L_S, set the winding to gate (ac component) transfer function, which is

$$H(s) = \frac{V_{gs,ac}(s)}{V_3(s)} = \frac{sR_gL_PC_{iss} + L_P}{s^2L_PL_SC_{iss} + sR_gC_{iss}(L_P + L_S) + L_P + L_S} \tag{6.2}$$

So the frequency characteristics are

$$A(\omega) = |H(j\omega)| \tag{6.3}$$

$$\varphi(\omega) = \arctan\left(\frac{\mathrm{Im}(H(j\omega))}{\mathrm{Re}(H(j\omega))}\right) \tag{6.4}$$

The inductances need to be designed to obtain the desired voltage gain and phase-shift. The phase-shift of the drive signal v_3 to the gate voltage $v_{gs,sync}$, φ, can be obtained by the simulation of the powertrain. Replacing the SR FET with an ideal diode, the phase angle difference between the output of the inverter and the ideal conduction time of the diode are the desired φ. In this case, $\varphi \approx -26°$. Substituting φ into (6.4), the relation between L_P and L_S can be described as $L_S = f(L_P)$.

The resonant frequency of L_P and C_{iss} is below f_s, so an initial point for L_P can be selected as

$$L_P > 1/4\pi^2 f_s^2 C_{iss} \tag{6.5}$$

So the value of L_S can be determined. Then the voltage gain $A(\omega) = V_{gsac}/V_3$ needs to be checked, which is supposed to ensure a proper amplitude for the drive voltage, that is, $V_{th} < V_{gs,ac} < V_{gs,max}$, where $V_{gs,max}$ is the maximum gate voltage of the device.

And the voltage across n_3 is

$$V_3 = \frac{n_3}{n_1} V_{in} \tag{6.6}$$

So,

$$\frac{n_1 V_{th}}{n_3 V_{in,min}} < A(\omega_s) < \frac{n_1 V_{gs,max}}{n_3 V_{in,max}} \tag{6.7}$$

After the voltage gain is checked, the values of L_P and L_S can be determined. Some trial-and-error tuning is still required to make the final design.

Level-shift circuit

The level-shift circuit provides the desired d.c. bias V_B for the RGD. The value of V_B is determined by the duty ratio and the threshold voltage of the SR FET. The duty ratio is set to 0.5 to balance the voltage and current stress on the device. So the d.c. bias voltage V_B is supposed to be around the threshold voltage, that is, $V_B = V_{th}$. Since the switch selected for the SR is SiS862DN from Vishay and the threshold voltage is around 2 V, the Zener diode TZS4679 with a Zener voltage of 2 V is selected as D_Z.

6.1.3.5 Loss analysis of the SR with the proposed RGD

Drive loss of the SR

Figure 6.13 gives the loss model of the proposed RGD when the converter is ON, where R_g is the gate resistance of S_{sync}, R_3 is the ESR of n_3, R_{LS} and R_{LP} are the ESRs of L_S and L_P, respectively, R_{AUX2} is the on-state resistance of S_{AUX2}. To avoid complexity, the ESRs and ESLs of the capacitance are not considered.

In Figure 6.13, the gate voltage of the RGD, $v_{gs,sync}$, is

$$v_{gs,sync}(t) = V_B + V_{gs,ac}\sin(2\pi f_s t + \theta) \tag{6.8}$$

The phase shift between $v_{gs,sync}$ and the voltage across the input capacitance, v_{Ciss}, is

$$\Delta\theta = \arctan 2\pi f_s C_{iss}\cdot R_g \approx 0 \tag{6.9}$$

So the voltage across C_{iss} is

$$v_{Ciss}(t) = v_{gs,sync}(t)\cdot\cos\Delta\theta \approx v_{gs,sync}(t) \tag{6.10}$$

The current through R_g is

$$i_g(t) = i_{Ciss}(t) = C_{iss}\frac{dv_{Ciss}(t)}{dt} = 2\pi f_s V_{gs,ac} C_{iss}\cos(2\pi f_s t + \theta) \tag{6.11}$$

Therefore, the loss dissipation in R_g is

$$P_{Rg} = I_g^2 R_g = 2\pi^2 f_s^2 V_{gs,ac}^2 C_{iss}^2 R_g \tag{6.12}$$

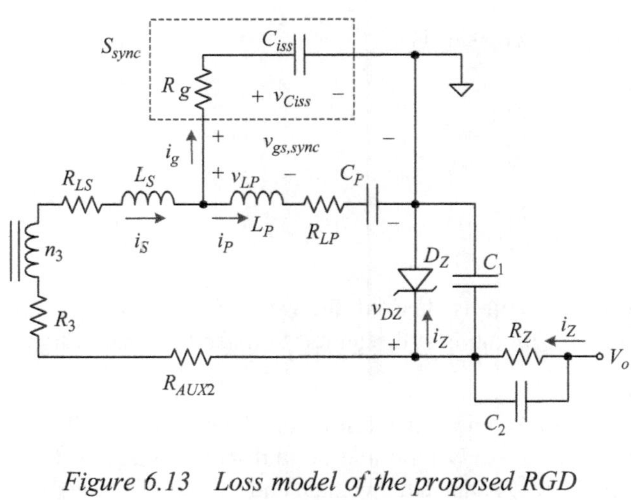

Figure 6.13 Loss model of the proposed RGD

The voltage across C_P is V_B. Since the impedance of R_{LP} is quite small compared to L_P, the phase shift between $v_{gs,sync}$ and v_{LP} can be ignored. So v_{LP} is

$$v_{LP}(t) = V_{gs,ac}\sin(2\pi f_s t + \theta) \tag{6.13}$$

With the integral of v_{LP}, the current through L_P is

$$i_P(t) = \frac{1}{L_P}\int v_{LP}(t)dt = -\frac{V_{gs,ac}}{2\pi f_s L_P}\cos(2\pi f_s t + \theta) \tag{6.14}$$

The power loss in the ESR of L_P can be calculated as

$$P_{LP} = I_P^2 \cdot R_{LP} = \frac{V_{gs,ac}^2}{8\pi^2 f_s^2 L_P^2} R_{LP} \tag{6.15}$$

Combining (6.11) and (6.14), the current through R_3 is

$$i_S(t) = i_g(t) + i_P(t) = \left(2\pi f_s V_{gs,ac} C_{iss} - \frac{V_{gs,ac}}{2\pi f_s L_P}\right)\cos(2\pi f_s t + \theta) \tag{6.16}$$

So the loss dissipated in R_3, R_{LS}, and R_{AUX2} can be calculated as

$$P_{R3} = I_S^2 \cdot R_3 \tag{6.17}$$

$$P_{LS} = I_S^2 \cdot R_{LS} \tag{6.18}$$

$$P_{SAUX2} = I_S^2 \cdot R_{AUX2} \tag{6.19}$$

$$I_S = \sqrt{2}\pi f_s V_{gs,ac} C_{iss} - \frac{V_{gs,ac}}{2\sqrt{2}\pi f_s L_P} \tag{6.20}$$

The current through R_Z is

$$i_Z(t) = \frac{V_o - V_B}{R_Z} \tag{6.21}$$

Owing to the high resistance of R_Z, the power loss in the level-shift circuit is negligible.

Combining (6.12), (6.15), and (6.17)–(6.19), the total drive loss of the proposed RGD is

$$P_D = P_{R3} + P_{Rg} + P_{LS} + P_{LP} + P_{SAUX2} \tag{6.22}$$

Conduction loss of the SR

The conduction loss of the SR consists of the conduction loss in the MOSFET, $P_{CON,SR}$, and the conduction loss in the body diode, $P_{CON,BD}$, which can be calculated as

$$P_{CON,SR} = I_{SR,rms}^2 \cdot R_{DS(on),avg} \tag{6.23}$$

$$P_{CON,BD} = V_{F,BD} \cdot I_{BD,avg} \tag{6.24}$$

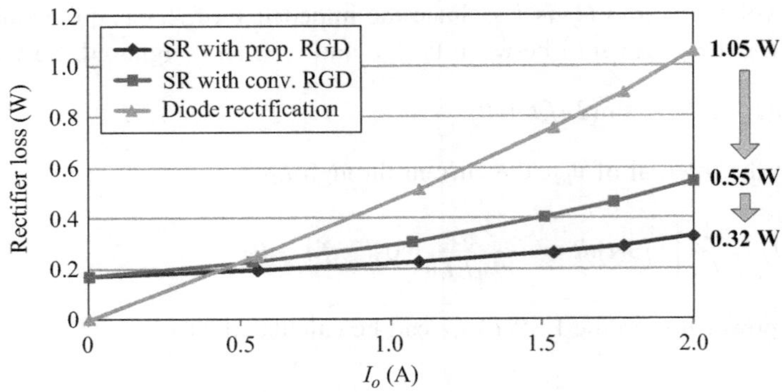

Figure 6.14 Rectifier device loss comparison (RGD drive loss included)

where $I_{SR,rms}$ is the RMS current of the MOSFET, $V_{F,BD}$ is the forward voltage of the body diode (normally 0.7–0.8 V) and can be found in the datasheet, $I_{BD,avg}$ is the average current of the body diode.

Rectifier loss comparison among different rectification schemes
Figure 6.14 gives the calculated rectifier loss comparison among three different rectification schemes. The loss of the rectifier diode is calculated based on the recommended formula for STPS3L60 (60 V/3 A, ST) as

$$P_{diode} = 0.44 \times I_{avg} + 0.05 \times I_{rms}^2 \tag{6.25}$$

From Figure 6.14, it is observed that when the output current I_o is below 0.5 A, the power loss of the SR FET is higher than the diode rectification because of the drive loss. When I_o is over 0.5 A, the SR schemes become more efficient than the diode scheme. More importantly, the conduction loss of the SR with the proposed RGD is reduced compared to the conventional RGD. At 5 V/2 A output, the loss of the SR with the proposed RGD is 0.32 W (3.2% of P_o), a reduction of 42% (0.23 W) compared to the SR with the conventional RGD (0.55 W), and a reduction of 72% (0.73 W) compared to the diode rectification (1.05 W).

6.1.4 Experimental verification and discussion

To verify the proposed circuit, an experimental prototype operating at 10 MHz, 18 V input and 5 V/2 A output was built. Figure 6.15 gives the photograph of the prototype. Tables 6.2 and 6.3 give the component values and part numbers in the power stage and the proposed RGD respectively. The inductance in the inverter stage, L_B, is provided by the leakage inductance of the transformer. The values of C_F and C_R in Table 6.2 are the external capacitances in parallel with the output capacitance of the MOSFETs.

Figure 6.16 shows the drain and gate voltage waveforms of the control FET. It is observed that the drain voltage is approximately half-wave symmetric. The peak drain voltage is 47 V at 18 V input (2.61 times of the input voltage), and the

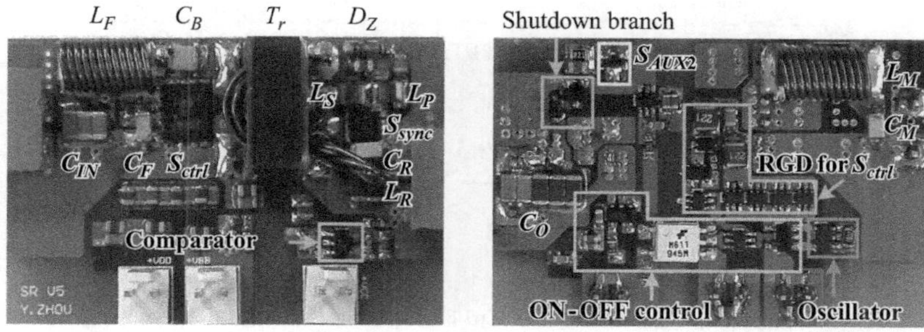

(a) (b)

Figure 6.15 Photograph of the prototype; (a) top and (b) bottom

Table 6.2 Components in the power stage

S_{ctrl}	**Si7898DP (Vishay) (150 V, N-channel)**	S_{sync}	**SiS862DN (Vishay) (60 V, N-channel)**
L_F, L_M	2222SQ-221 (Coilcraft)	L_R	A04T (Coilcraft)
C_F	268 pF	C_R	2,940 pF
C_M	247 pF	C_B	15 nF
C_{IN}	20 μF	C_O	235 μF
n_1:n_2:n_3	4:1:1	Core	ER 14.5/3/7 (4F1)

Table 6.3 Components in the proposed RGD

L_S	680 nH	D_1	BAS170WS (Vishay)
L_P	150 nH	S_{AUX1}	FDV303N (Fairchild)
C_P	0.1 μF	S_{AUX2}	NTA4151P (ON Semi)
R_Z	470 Ω	D_Z	TZS4679 (Vishay)

converter provides good zero-voltage switching (ZVS) achievement for the control FET.

Figure 6.17 shows the drain and gate voltage waveforms of the SR FET. It is observed that the proposed RGD provides a sinusoidal drive voltage with a DC bias of 2 V and the amplitude of the sine component is 5.8 V, agreeing well with the theoretical analysis. The drain and gate oscillation is caused by the parasitic package inductance and the nonlinearity of the parasitic capacitance of the device. The duty ratio of the SR FET is 0.5 and the peak drain voltage is 20 V.

Figure 6.18 shows the close-loop gate voltage of the proposed RGD and the control signal. The modulation frequency is 55 kHz at full load. It is observed that when the converter turns OFF, the gate voltage is clamped to zero quickly. When

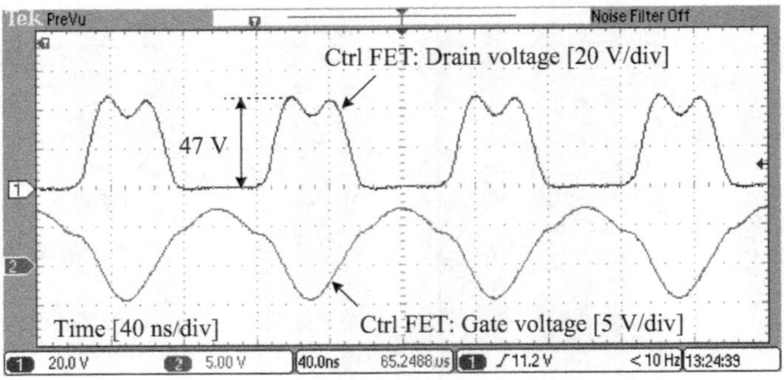

Figure 6.16 Waveforms of the Ctrl FET: $V_{in} = 18$ V, $V_o = 5$ V, and $f_s = 10$ MHz

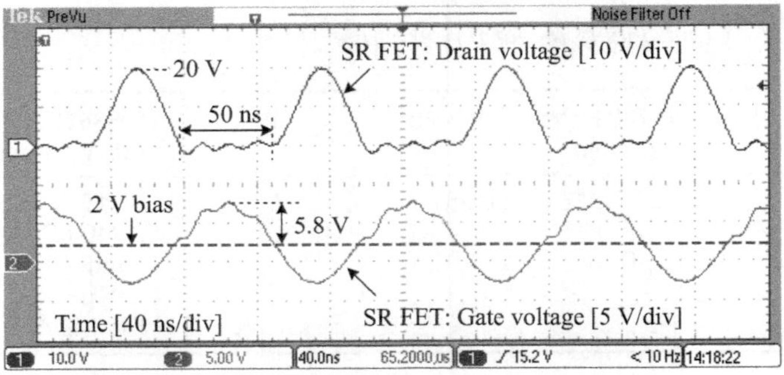

Figure 6.17 Waveforms of the SR FET: $V_{in} = 18$ V, $V_o = 5$ V, and $f_s = 10$ MHz

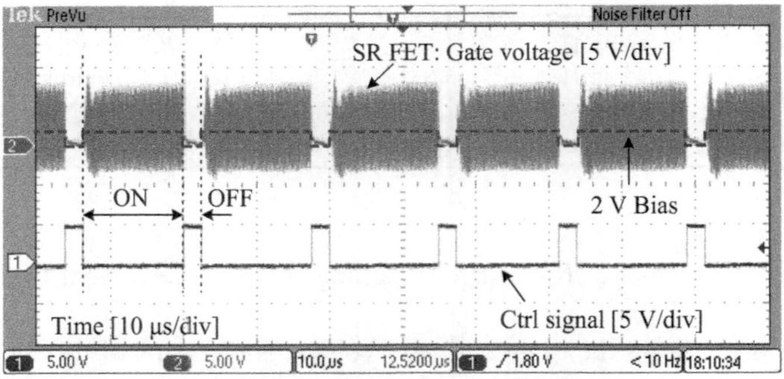

Figure 6.18 Gate voltage of the SR and the control signal: $V_{in} = 18$ V, $V_o = 5$ V and $P_o = 10$ W

the converter turns ON, the gate voltage begins to resonate, and the bias voltage of the gate voltage reaches the desired Zener voltage (2 V in this case) immediately.

Figure 6.19 shows the voltage across the Zener diode D_Z in the proposed RGD. It can be seen that there is no charging and discharging process of v_{DZ} during the

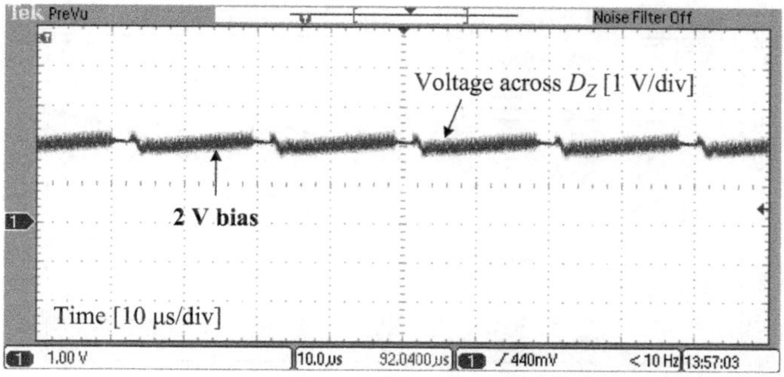

Figure 6.19 Voltage across D_Z: $V_{in} = 18$ V, $V_o = 5$ V, and $P_o = 10$ W

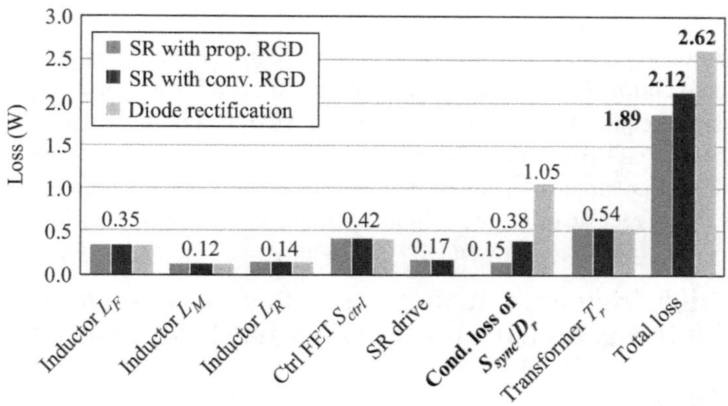

Figure 6.20 Loss breakdown: $V_{in} = 18$ V, $V_o = 5$ V, and $P_o = 10$ W

ON-OFF operation, and it remains stable at 2 V (the Zener voltage of D_Z), agreeing well with the theoretical analysis in Section 6.1.3.

Two additional prototypes rectified by the SR with the conventional RGD and the diodes were built respectively. For fair comparison, other parameters in the power stage are the same. Two Schottky diodes (PMEG4020ER, 40 V/2 A, NXP) are used in parallel. Since $R_{DS(on)}$ with the RGD varies all the time during one switching cycle and the switching period is only 100 ns, it is difficult to measure $R_{DS(on)}$ directly. However, the conduction loss reduction owing to the reduction of $R_{DS(on)}$ was verified by the temperature rise reduction and the efficiency improvement.

Figure 6.20 gives the loss breakdown comparison under the nominal condition. Owing to the reduction in the conduction loss, the total power loss of the converter with the proposed RGD is 1.89 W, a reduction of 0.23 W compared with the conventional RGD (2.12 W), and a reduction of 0.73 W compared to the diode rectification (2.62 W).

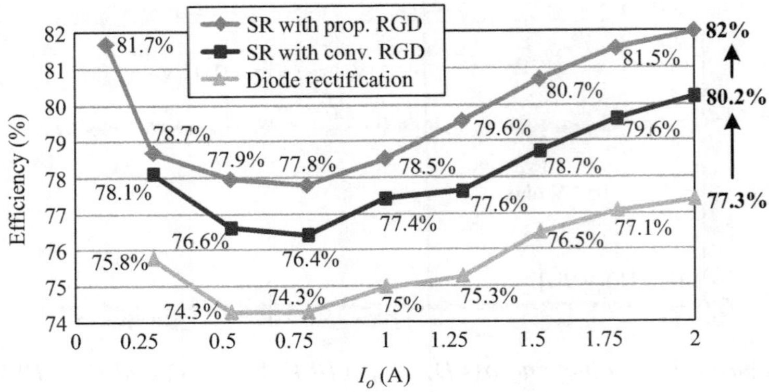

Figure 6.21 Efficiency comparison: $V_{in} = 18$ V and $V_o = 5$ V

Figure 6.21 gives the close-loop efficiency comparison. Compared to the SR with the conventional RGD and the rectifier diodes, the SR with the proposed RGD achieves significant efficiency improvement in wide load range. At full load of 2 A, the proposed RGD improves the efficiency from 80.2% using the conventional RGD to 82% (an improvement of 1.8%). Compared to the efficiency of 77.3% using the diode rectification, the efficiency improvement is 4.7% at full load.

6.2 A high-frequency dual-channel isolated resonant gate driver with low gate-drive loss for ZVS full-bridge converters

6.2.1 Introduction

To increase the switching frequency of the power converters can help to reduce the size and volume of the passive components, such as the inductors, transformers, and capacitors, so that the power density can be increased. However, as the switching frequency increases, the switching loss and gate-drive loss increase proportionally. In recent years, a lot of work has been done to reduce, or eliminate the switching loss. Soft switching technique is one of the effective solutions to reduce the switching loss at high switching frequency. With ZVS or zero-current switching (ZCS), the full-bridge (FB) converters exhibit lower switching loss and are widely used in many applications [19,20]. However, the excessive gate drive loss may still exist and cause lower efficiency and reliability as the switching frequency increases.

In the last 20 years, RGDs have been proposed with the objective of recovering gate energy loss in the conventional voltage source drivers (VSDs) [21–23]. The basic idea of the RGDs is to utilize the L-C resonance to recover the gate-drive energy stored in the gate capacitance of the power MOSFET. The review on the RGD topologies concentrating on gate energy recovery efficiency has been reported in [24]. The general circuit model and design method of the RGDs are proposed in [25]. A high-speed RGD with controlled peak gate voltage is proposed for the silicon carbide (SiC) MOSFETs in [26]. An assessment of the RGD techniques for use in low power d.c./d.c. converters is presented in [27], and the mathematical model is built to estimate the power loss of the drive circuit in [28]. However, there was little work reported on the RGDs suitable for the FB converters to reduce the excessive gate drive loss with the ZVS technique at high switching frequency.

Most of the reported RGDs can be classified into three categories when applied to a ZVS FB converter.

1. The RGDs designed for one single power MOSFET [29–35]. To drive an FB converter, four sets of these RGDs are needed, which increases the complexity and cost significantly. The RGDs in [33–35] can provide the negative gate-drive voltage to prevent the false-trigger problem caused by fast *dv/dt* during the turn-off transition. Though the RGD in [33] can provide the negative gate-drive voltage, the amplitude of the gate-to-source voltage is affected by the resonant components seriously. Compared to [34], the RGD proposed in [35] is more concise and can recover more gate-drive energy.
2. The RGDs designed for two ground-sharing power MOSFETs [36]. Basically, these RGDs are only able to drive two power MOSFETs sharing the same ground and not applicable to the power MOSFETs in one bridge leg. These RGDs are not suitable for the FB converter with the floating ground.
3. The RGDs designed for two power MOSFETs in one bridge leg [37]. These RGDs can reduce the complexity of the drive circuits in the FB converters. However, the level-shift circuit in the RGD is prone to noise and may lead to the shoot-through problem in the FB converter [37]. In [38], the power MOSFET gate terminal is not actively clamped high or low, which results in lower noise immunity. In the CSD proposed in Section 3.2, the resonant inductor of the RGD suffers the input voltage. When the input voltage becomes high, the resonant inductance should have to increase significantly, which is not suitable for the FB converters. Furthermore, this type of the RGDs cannot provide the negative gate-drive voltage to prevent the false trigger in the FB converters.

Based on the above analysis, the desired RGD for an FB converter should have the following characteristics: (1) recover the gate-drive energy with high efficiency; (2) is able to drive two power MOSFETs in one bridge leg as dual-channel drivers with low components count and low cost; and (3) provide the negative gate-drive voltage to prevent the false trigger with high reliability.

6.2.2 Review of gate-drive circuits

6.2.2.1 The conventional voltage source drivers for the FB converters

For the FB converters, the conventional transformer coupled VSD shown in Figure 6.22 is widely used. S_1–S_4 are the drive switches, D_1–D_4 the freewheel diodes, R_{ext1} and R_{ext2} are the external resistors, R_{g1} and R_{g2} are the gate mesh resistance. This VSD has the ability to drive two power MOSFETs in one bridge leg with isolation, leading to the low complexity and high reliability.

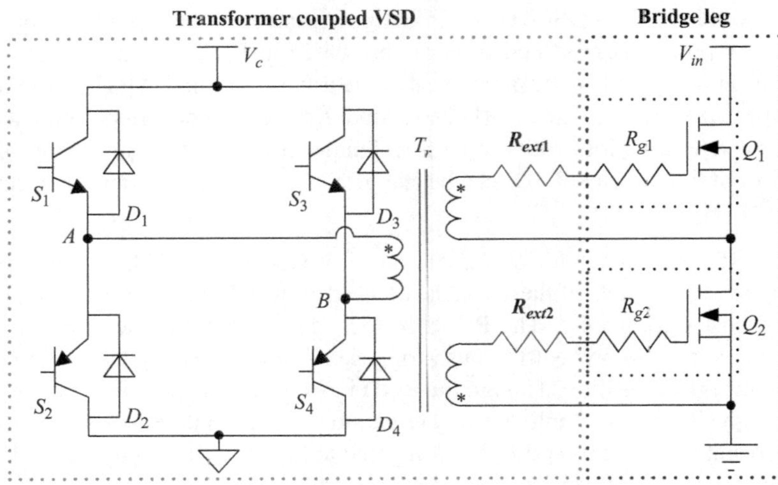

Figure 6.22 The transformer coupled VSD for two MOSFETs in one bridge leg

Unfortunately, in this VSD, the energy provided to drive the power MOSFETs still dissipates totally in the resistors R_{g1}, R_{ext1}, R_{g2}, and R_{ext2}.

6.2.2.2 Candidates of the RGDs for the FB converters

Most of the reported RGD topologies are for non-isolated application. However, the FB converters require isolated gate drivers. Among these topologies, the RGDs proposed in [35] has respective advantages in the FB converter applications. It should be interesting to investigate these two RGDs particularly.

A high dual-channel high-side low-side CSD proposed in Chapter 3 is able to drive two power MOSFETs in one leg. However, the disadvantages of this RGD are as follows: (1) the input voltage is applied across the resonant inductance entirely. When the input voltage increases, the inductance has to be increased to guarantee the normal operation of the RGD. Therefore, this dual-channel RGD is only suitable for the low-voltage applications; (2) the dual-channel RGD cannot provide negative gate-drive voltage, which may cause false-trigger problem in the FB converters; (3) this dual-channel RGD only provides two non-isolated drive signals, so it is not suitable to drive the half-bridge (HB) and FB converters in high power applications; and (4) the current of the resonant inductance L_r is continuous, which increases the conduction loss of the drive circuit itself.

6.2.3 Proposed resonant gate driver for FB converters and principle of operation

The proposed isolated RGD is shown in Figure 6.23. The circuit consists of four drive switches S_1–S_4, the drive transformer T_r, two resonant inductors L_{r1} and L_{r2}. L_{r1} and L_{r2} can be the leakage inductance of the transformer. Q_1 is the high-side switch and Q_2 is the low-side switch in one bridge leg.

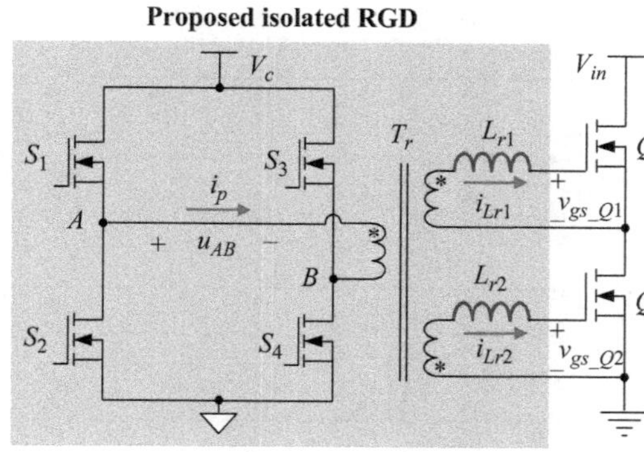

Figure 6.23 The proposed isolated RGD

The gate waveforms of four drive switches, S_1–S_4, the voltage between A and B, u_{AB}, the primary-side current, i_p, along with the resonant inductor current, the gate-to-source voltage v_{gs_Q1}, v_{gs_Q2} are illustrated in Figure 6.24. S_1–S_2 and S_3–S_4 are switched out of phase with the complementary control, respectively. It is observed that the pulse width of u_{AB} is controlled by the switching status of the four drive switches S_1–S_4. i_{L1} and i_{L2} are the resonant current through L_{r1} and L_{r2}, respectively. The value of i_p is the summation of the absolute value of i_{L1} and i_{L2}. The inductor current i_{L1}, i_{L2}, i.e., the gate current of power MOSFET is sinusoidal and discontinuous, which reduces the additional conduction loss. The proposed RGD turns the MOSFETs on and off by using the resonant current injected into the input capacitance as shaded during $[t_1, t_2]$ and $[t_3, t_4]$. Moreover, the gate-to-source voltage v_{gs_Q1} and v_{gs_Q2} are complementary and sinusoidal during the switching transition.

The operation principle of the proposed RGD includes four operating modes during one period as illustrated in Figure 6.25(a)–(d). C_{g_Q1} and C_{g_Q2} represent the input capacitance of power MOSFETs Q_1 and Q_2, respectively. Initially, it is assumed that Q_1 is off and Q_2 is on before t_0 as shown in Figure 6.25(a).

1. Mode 1 $[t_0, t_1]$: S_2 and S_3 conduct during this interval as illustrated in Figure 6.25(a). The voltage between A and B, u_{AB} is clamped to $-V_c$. As the ratio of the drive transformer is 1:1, the value of the gate-to-source voltage v_{gs_Q1} is initially charged at $-V_c$. This negative gate-drive voltage prevents the false trigger due to fast dv/dt problem in the FB converter. As the polarities of the secondary windings are opposite for the power MOSFETs Q_1 and Q_2, the gate-to-source voltage v_{gs_Q2} is initially charged at V_c, and Q_2 is on during this interval.
2. Mode 2 $[t_1, t_2]$: At t_1, since the resonant current starts from $i_{Lr1} = i_{Lr2} = i_p = 0$, S_3 turns off with ZCS and S_4 turns on with ZCS. The voltage $u_{AB} = 0$ and the voltage of the secondary windings is clamped to zero. With the initial electrostatic energy stored in the input capacitance, the resonance between the C_{g_Q1} and L_{r1}, C_{g_Q2}, and L_{r2} starts respectively as illustrated in Figure 6.25(b).

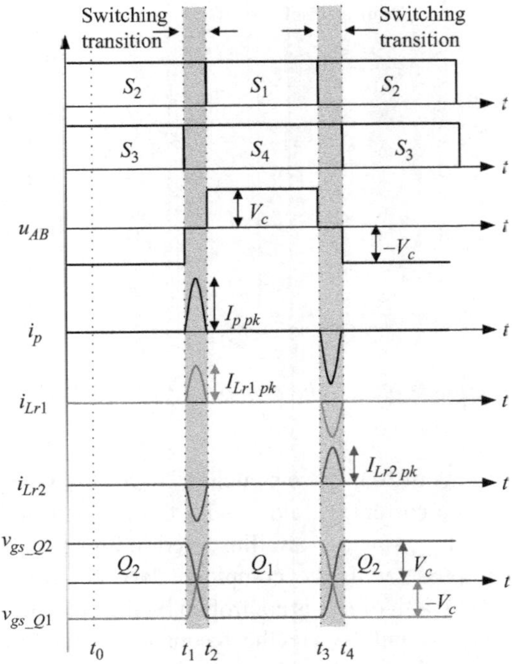

Figure 6.24 Key waveforms of the proposed RGD

It is observed that the drive current circulates among S_2 and S_4, L_{r1} and C_{g_Q1}, L_{r2} and C_{g_Q2}, and no current flows through the d.c. power supply V_c. Therefore, the power supply provides no electric power to the gate-drive circuit theoretically during the switching transition, which indicates that the gate-drive loss is reduced significantly by the proposed RGD. During this interval, the power MOSFETs Q_1 and Q_2 operate in the complementary mode. At t_2, v_{gs_Q1} reaches V_c and v_{gs_Q2} reaches $-V_c$, Q_1 turns on and Q_2 turns off at the end of this interval.

3. Mode 3 [t_2, t_3]: At t_2, since the resonant transition ends when $i_{Lr1}=i_{Lr2}=i_p=0$, S_2 turns off with ZCS and S_1 turns on with ZCS. As the voltage u_{AB} is clamped to V_c, the voltage v_{gs_Q1} is actively clamped to V_c, and v_{gs_Q2} is actively clamped to $-V_c$ as illustrated in Figure 6.25(c). Therefore, $v_{gs\_Q1}=V_c$ and $v_{gs\_Q2}=-V_c$ remain unchanged during this interval. Therefore, this interval can be set to adjust the duty cycle.
4. Mode 4 [t_3, t_4]: At t_3, the resonant current starts from $i_{Lr1}=i_{Lr2}=i_p=0$. S_1 turns off with ZCS and S_2 turns on with ZCS. This interval as illustrated in Figure 6.25(d) is similar to operation Mode 2 as shown in Figure 6.25(b). However, the direction of the resonant current i_{Lr1}, i_{Lr2} is opposite. At t_4, v_{gs_Q1} reaches $-V_c$ and v_{gs_Q2} reaches V_c. So Q_1 turns off and Q_2 turns on at the end of this interval.

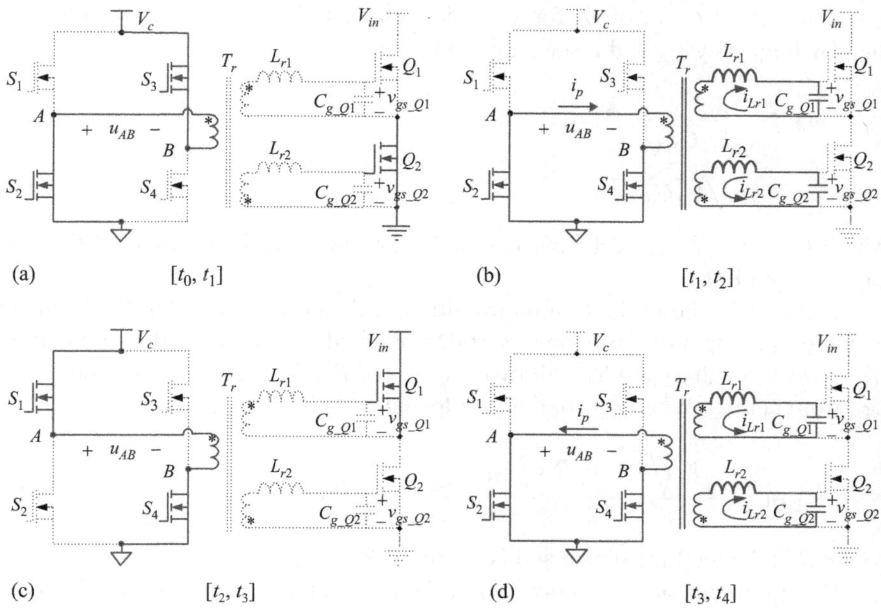

Figure 6.25 Equivalent circuits during the switching period

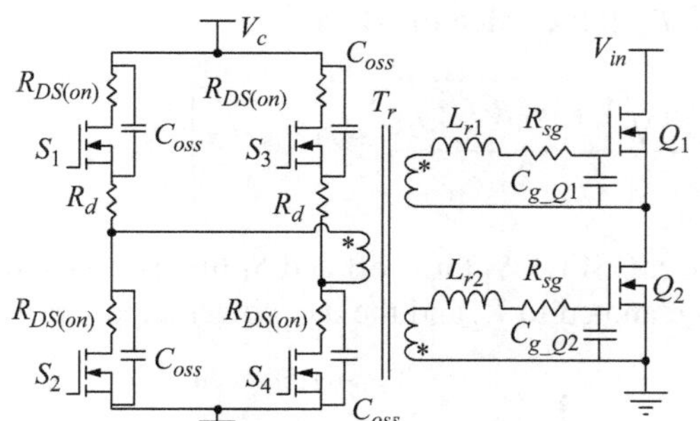

Figure 6.26 Loss analysis model of the proposed RGD

6.2.4 Loss analysis and optimal design

6.2.4.1 The gate drive loss with the proposed RGD

Figure 6.26 shows the equivalent loss model of the proposed RGD, which includes the circuit elements producing the power consumption in the practical gate drive circuit. $R_{DS(on)}$ is the on-state resistance and C_{oss} is the output capacitance in the drive MOSFETs; S_1–S_4. R_{sg} is the summation of the winding resistance in resonant inductor and transformer, and the gate-pattern resistance inside the power MOSFET. The damping resistance R_d is connected in series with S_1 and S_3.

As the two power MOSFETs Q_1 and Q_2 are controlled in the complementary mode, the gate drive loss of Q_1 is chosen to analyze. The resonant inductor L_{r1} and gate capacitance C_{g_Q1} of Q_1 form a series resonant circuit, in which the resonant angular frequency ω_R and resonant period T_R are

$$\omega_R = \frac{1}{\sqrt{L_{r1} C_{g\_Q1}}} \tag{6.26}$$

$$T_R = 2\pi\sqrt{L_{r1} C_{g\_Q1}} \tag{6.27}$$

where L_{r1} is the value of the resonant inductor and C_{g_Q1} is the value of the gate capacitance of Q_1.

Figure 6.27 shows the turn-on transition of the power MOSFET Q_1. With the operation principle of the proposed RGD presented in Section 3, the gate current i_{Lr1} flows through S_2 and S_4 with two $R_{DS(on)}$ and R_{sg} during the turn-on transition, i.e., from t_0 to t_2. When $t_0 = 0$, the gate-to-source voltage v_{gs_Q1} is

$$v_{gs\_Q1} = -\frac{V_c\sqrt{4 + \omega_R^2 R^2 C_{g\_Q1}^2}}{2} e^{-(RC_{g\_Q1}/2)\omega_R t} \cos \omega_R t \tag{6.28}$$

where R is the total resistance and is given by $R = 2R_{DS(on)} + R_{sg}$.

Due to the resistance energy loss of the resonant circuit, the amplitude of the gate-to-source voltage decreases and there is a voltage drop after each switching transition. The voltage drop is defined as ΔV as shown in Figure 6.27. When $t_2 = T_R/2$, v_{gs_Q1} reaches the maximum value during the turn-on transition. From 6.51 and 6.52, with $t_2 = T_R/2$, the value of ΔV is

$$\Delta V = V_c \left[1 - \frac{\sqrt{4 + \omega_R^2 R^2 C_{g\_Q1}^2}}{2} e^{-(RC_{g\_Q1}/2)\pi}\right] \tag{6.29}$$

At t_2, the drive MOSFET S_2 turns off and S_1 turns on as shown in Figure 6.27. Then, the v_{gs_Q1} is clamped to V_c and the d.c. power supply provides an amount of

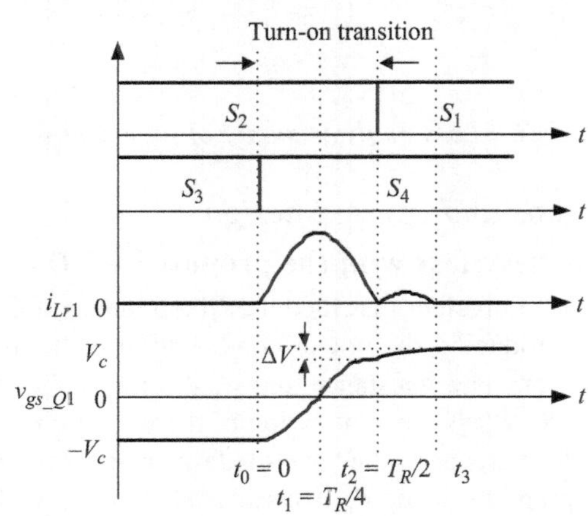

Figure 6.27 The turn-on transition of the power MOSFET Q_1

electric energy, which is equal to $C_{g\_Q1}V_c\Delta V$ for the turn-on procedure. In the same way, $C_{g\_Q1}V_c\Delta V$ is also provided for the turn-off procedure. Therefore, the electric power to compensate for the voltage drop ΔV is

$$P_{c\_RGD} = 2f_{sw}C_{g\_Q1}V_c\Delta V \tag{6.30}$$

Although the total gate charge of S_1–S_4 is low, it may still cause some losses at high switching frequency. The switching frequency of these four MOSFETs is the same as the f_{sw} of the power MOSFETs. The total gate drive loss of S_1–S_4 is

$$P_s = 4Q_{g\_s}V_{gs\_s}f_{sw} \tag{6.31}$$

where Q_{g_s} is the total gate charge of drive MOSFETs, V_{gs_s} is the drive voltage of S_1–S_4.

In addition, the power loss P_r of the output capacitance C_{oss} in the drive MOSFETs S_1–S_4 is

$$P_r = 4C_{oss}V_c^2f_{sw} \tag{6.32}$$

The transformer loss consists of the magnetic hysteresis loss, eddy-current loss and residual loss. The total transformer loss is

$$P_t = \eta f_s^{\alpha}B_m^{\beta}V \tag{6.33}$$

where η is the loss coefficient, α is the frequency index (>1), β is the magnetic induction index (>1), B_m is the magnetic induction, and V is the volume of the magnetic core.

From (6.30) to (6.33), the power consumption P_{d_RGD} for driving two MOSFETs in one bridge leg with the proposed RGD is

$$P_{d\_RGD} = 2P_{c\_RGD} + P_s + P_r + P_t \tag{6.34}$$

6.2.4.2 The gate drive loss comparison between the proposed RGD and conventional VSD

For comparison, the conventional transformer coupled VSD in Figure 6.22 is analyzed. The gate-drive loss for one power MOSFET with the conventional VSD is calculated. Take account of the negative gate voltage, the d.c. power supply has to provide an electric energy as large as $2C_{g\_Q1}V_c^2$ to turn the power MOSFETs on or off. This electric energy provided by the dc power supply is consumed in the gate resistor R_g. Therefore, the power loss P_{c_VSD} in the gate resistor is

$$P_{c\_VSD} = 4f_{sw}C_{g\_Q1}V_c^2 \tag{6.35}$$

From (6.31) to (6.33) and (6.35), the power consumption for driving two MOSFETs in one bridge leg with the conventional transformer coupled VSD is

$$P_{d\_VSD} = 2P_{c\_VSD} + P_s + P_r + P_t \tag{6.36}$$

To demonstrate the gate-drive loss reduction with the proposed RGD, the design parameters are given in Table 6.4. The design parameters provided in this part agree with the experimental parameters.

Based upon the design parameters above, the calculated results are provided in the Table 6.5. With the proposed RGD, the gate-drive loss of two MOSFETs in one bridge leg is reduced from 3.2 to 0.72 W (a reduction of 78%). As a result, the gate-drive loss of four MOSFETs in an FB converter is reduced from 6.4 to 1.44 W (a reduction of 78%), translating into an efficiency improvement of 0.5% in a 1-kW 500-kHz FB converter.

Table 6.4 Design parameters of the proposed RGD

Switching frequency, f_{sw}	500 kHz
Gate drive voltage, V_c	15 V
Resonant inductor, L_{r1}, L_{r2}	246 nH
Power MOSFET, Q_1, Q_2	IPP50R199CP
Total gate charge, Q_g	50 nC
Internal gate resistance, R_g	2.2 Ω
Drive MOSFETs, S_1–S_4	FDN335N
Total gate charge, Q_{g_s}	3.7 nC
On resistance, $R_{DS(on)}$	0.07 Ω
Output capacitance, C_{oss}	80 pF

Table 6.5 Gate-drive loss calculation comparison between the proposed RGD and conventional VSD

	Conventional VSD	**Proposed RGD**	**Loss reduction**
Power consumption for the resistance P_{c_VSD}, P_{c_RGD}	3 W (P_{c_VSD})	0.53 W (P_{c_RGD})	2.47 W (a reduction of 82%)
Gate drive loss of S_1–S_4, P_s	0.037 W	0.037 W	–
Power loss of the output capacitance in S_1–S_4, P_r	0.036 W	0.036 W	–
Transformer loss, P_t	0.12 W	0.12 W	–
Gate drive loss of two MOSFETs in one bridge leg	**3.2 W** (P_{d_VSD})	**0.72 W** (P_{d_RGD})	**2.48 W** (a reduction of 78%)
Gate drive loss of four MOSFETs in an FB converter	**6.4 W**	**1.44 W**	**4.96 W** (a reduction of 78%)

6.2.4.3 The switching loss comparison between the proposed RGD and conventional VSD

This section analyzes the potential switching loss reduction with the proposed RGD. In the ZVS FB converter, the turn-on loss is significantly reduced due to ZVS turn-on condition. However, there is still some turn-off loss normally. In the conventional VSD, the gate current decays significantly from its peak value due to the RC type charging and discharging as shown in Figure 6.28. On the other hand, the gate current in the proposed RGD is sinusoidal as shown in Figure 6.28.

For the conventional VSD, the turn-off switching loss, P_{off_VSD} is given approximated by (6.37). In (6.37), V_{ds} represents the voltage across the switch, I_{off_pk} is the peak current through the switch at turn off and t_{f_VSD} is the fall time given by (6.38). In (6.38), I_{th_off} is the gate current when the gate voltage is at the threshold and is given by (6.39). Also in (6.38), I_{pl_off} is the gate current when the gate voltage is at the plateau and is given by (6.40). t_{f_VSD}, I_{th_off} and I_{pl_off} are clearly shown in Figure 6.28. Other parameters include the power MOSFET total gate charge at the beginning of the plateau, Q_{pl}, the total gate charge at the threshold, Q_{th}, the gate to drain charge, Q_{gd}, the MOSFET internal gate resistance, R_g, and the external gate resistance in the drive circuit, R_{ext}:

$$P_{off\_VSD} = \frac{1}{2} f_s V_{ds} I_{off\_pk} t_{f\_VSD} \tag{6.37}$$

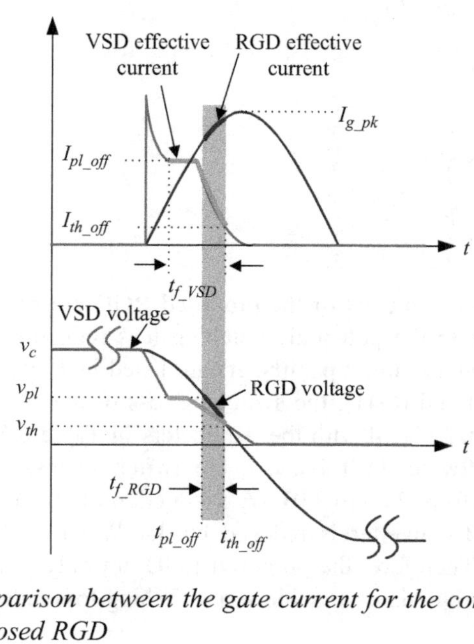

Figure 6.28 Comparison between the gate current for the conventional VSD and proposed RGD

$$t_{f\_VSD} = \frac{Q_{pl} - Q_{th}}{\frac{1}{2}\left|I_{th\_off} + I_{pl\_off}\right|} + \frac{Q_{gd}}{\left|I_{pl\_off}\right|} \tag{6.38}$$

$$I_{th\_off} = \frac{V_{th}}{R_{ext} + R_g} \tag{6.39}$$

$$I_{pl\_off} = \frac{V_{pl}}{R_{ext} + R_g} \tag{6.40}$$

On the other hand, for the proposed RGD with the sinusoidal gate current, the turn-off switching loss, P_{off_RGD} is approximated by (6.41), where I_{g_avg} is the average gate current during the turn-off transition, t_{f_RGD}. In (6.42), I_{g_pk} is the maximal gate current given by (6.43), t_{pl_off} is the time when the gate voltage is at the plateau and is given by (6.44). Also in (6.42), t_{th_off} is the time when the gate voltage is at the threshold and is given by (6.45). I_{g_pk}, t_{pl_off}, and t_{th_off} are clearly shown in Figure 6.28. The shaded area between t_{pl_off} and t_{th_off} is the transition produces the switching loss in the proposed RGD. It is observed that the magnitude of the RGD current in this shaded area is greater than I_{pl_off} and I_{th_off}, respectively. The result is that the turn-off time and the switching loss decrease. Other parameters noted include the input capacitance of the power MOSFET, C_{iss}, the resonant inductor L_r and the gate drive voltage V_c:

$$P_{off\_RGD} = \frac{1}{2} f_s V_{ds} I_{off\_pk} \frac{Q_{pl} - Q_{th} + Q_{gd}}{\left|I_{g\_avg}\right|} \tag{6.41}$$

$$\left|I_{g\_avg}\right| = \frac{\int_{t_{pl\_off}}^{t_{th\_off}} I_{g\_pk} \sin t dt}{t_{th\_off} - t_{pl\_off}} = \frac{\cos t_{pl\_off} - \cos t_{th\_off}}{t_{th\_off} - t_{pl\_off}} I_{g\_pk} \tag{6.42}$$

$$I_{g\_pk} = \sqrt{\frac{C_{iss}}{L_r}} V_c \tag{6.43}$$

$$t_{pl\_off} = \arccos \frac{V_{pl}}{V_c} \tag{6.44}$$

$$t_{th\_off} = \arccos \frac{V_{th}}{V_c} \tag{6.45}$$

The key design parameters of the proposed RGD are presented in Table 6.4 above. To demonstrate the potential switching loss reduction with the proposed RGD, the supplementary circuit parameters are listed in Table 6.6.

Based on (6.37) and (6.41), the switching loss of the conventional VSD and proposed RGD are calculated with the parameters in Table 6.6 respectively using the MATHCAD software. In Table 6.7, the switching loss of one MOSFET is reduced by 0.85 W (from 2.86 to 2.01 W, a reduction of 30%). The total switching loss in the ZVS FB converter is reduced by 3.4 W (from 11.44 to 8.04 W, a reduction of 30%). Therefore, the proposed RGD not only has a significant gate-drive loss reduction, but also can achieve a switching loss reduction.

Table 6.6 Circuit parameters for switching loss calculation comparison

Converter topology	**ZVS FB**
Voltage across the MOSFET, V_{ds}	200 V
MOSFET turn-off current, I_{off_pk}	5 A
External impedance, R_{ext}	2 Ω
Power MOSFET	**IPP50R199CP**
Q_{gd}	11 nC
Q_{th}	5 nC
Q_{pl}	7.5 nC
V_{th}	3 V
V_{pl}	5.2 V
C_{iss}	3300 pF

Table 6.7 Switching loss calculation comparison between the proposed RGD and conventional VSD

	Switching loss of one MOSFET in the ZVS FB converter	**Switching loss of four MOSFETs in the ZVS FB converter**
Conventional VSD	2.86 W	11.44 W
Proposed RGD	2.01 W	8.04 W
Switching loss reduction	**0.85 W (a reduction of 30%)**	**3.4 W (a reduction of 30%)**

6.2.4.4 Optimal design for the proposed RGD

In the proposed RGD, the resonant inductance is the crucial component. The inductance value of L_{r1} and L_{r2} are assumed equal as $L_{r1} = L_{r2} = L_r$. When the switching frequency f_{sw} is fixed, i.e., 500 kHz, the value of the resonant inductance L_r remains to be determined. Therefore, three design rules need to be followed: (1) to ensure the resonant manner of the proposed RGD; (2) fast driving speed; and (3) to minimize the overall loss of the gate-drive loss and switching loss.

(1) Rule 1: To ensure the resonance of L_r–C_{iss}

In the proposed RGD, to ensure that the total resistance R in the drive circuit has no effect on the natural resonance between the resonant inductor L_r and input capacitance C_{iss}, the characteristic impedance R_{LC} of L_r and C_{iss} has to be large enough and should meet:

$$R_{LC} = \sqrt{\frac{L_r}{C_{iss}}} = (2 \sim 3) \times R \tag{6.46}$$

where R is the summation of the winding resistance in resonant inductor and transformer, the gate-pattern resistance inside the power MOSFET and the on-resistance of the drive MOSFETs.

From (6.46), the value of the resonant inductor L_r can be calculated as

$$L_r = [(2 \sim 3) \times R]^2 C_{iss} \tag{6.47}$$

From (6.47), based on the circuit parameters from Table 6.4, the value of the resonant inductor L_r can be solved to be larger than 240 nH ($L_r \geq 240$ nH).

(2) Rule 2: Fast driving speed

The value of L_r affects the rising time t_r and falling time t_f of driving a power MOSFET, i.e., the switching speed. In the proposed RGD, as the characteristic impedance R_{LC} is much larger than the total resistance R in the resonant circuit, the gate drive voltage of power MOSFET Q_1 or Q_2 can be regarded as sinusoidal during the turn-on transition. As shown in Figure 6.29, the rising time t_r defined as the interval for v_{gs} to rise from zero to V_c is

$$t_r = \frac{1}{2}\pi\sqrt{L_r C_{iss}} \tag{6.48}$$

Similarly, the falling time t_f defined as the interval for V_{gs} to fall from V_c to zero is

$$t_f = \frac{1}{2}\pi\sqrt{L_r C_{iss}} \tag{6.49}$$

From (6.48) and (6.49), it is noted that the larger L_r is, the slower the power MOSFET can be driven. Considering ZVS operation of one bridge leg, less than 5% of the switching period is usually allowed as (6.50):

$$t_d = t_r + t_f = \pi\sqrt{L_r C_{iss}} \leq \frac{5\%}{f_s} \tag{6.50}$$

where t_d is the total driving time of the power MOSFET.

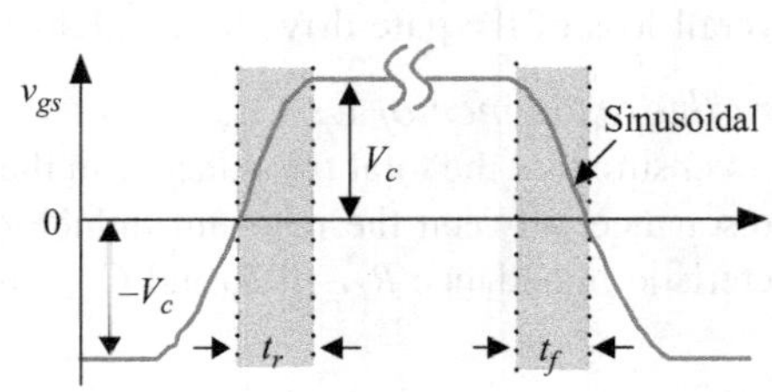

Figure 6.29 t_r and t_f during the switching period

From (6.50), the maximal value of L_r is solved:

$$L_r \leq \frac{\left(\frac{5\%}{\pi f_s}\right)^2}{C_{iss}} \tag{6.51}$$

Based upon the given design parameters in Table 6.4, the value of L_r can be calculated to be less than 300 nH ($L_r \leq 300$ nH).

In summary, from the design Rule 1 and Rule 2, the accepted design range for L_r is from 240 to 300 nH as illustrated in Figure 6.30.

(3) Rule 3: To minimize the overall loss of the gate drive loss and switching loss

From (6.34), the power consumption P_{d_RGD} is the gate drive loss for driving two MOSFETs in one bridge leg. From (6.41), the power consumption P_{off_RGD} is the turn-off switching loss, i.e., the switching loss for one MOSFET. Then, the switching loss P_{sw_RGD} for two power MOSFETs in one bridge leg is

$$P_{sw\_RGD} = 2P_{off\_RGD} \tag{6.52}$$

Then, the overall loss of the gate-drive loss and switching loss in one bridge leg, P_{sum} is

$$P_{sum} = P_{d\_RGD} + P_{sw\_RGD} \tag{6.53}$$

From (6.34), (6.41), (6.48), and (6.49), P_{d_RGD}, P_{sw_RGD}, and P_{sum} as function of the resonant inductance L_r are illustrated in Figure 6.31. From Figure 6.31, it is observed that with the increase of the resonant inductance L_r, the gate-drive loss P_{d_RGD} is reduced. The reduction slew rate of the gate-drive loss tends to be slow when L_r becomes large. With the increase of L_r, the switching loss P_{sw_RGD} increase. At the same time, the overall loss P_{sum} of the gate drive loss and switching loss also increase since the switching loss is the dominate element compared to the gate-drive loss.

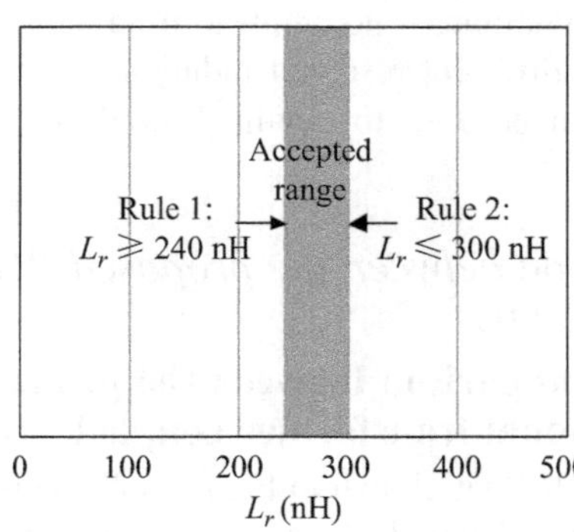

Figure 6.30 The design requirement of Rule 1 and Rule 2

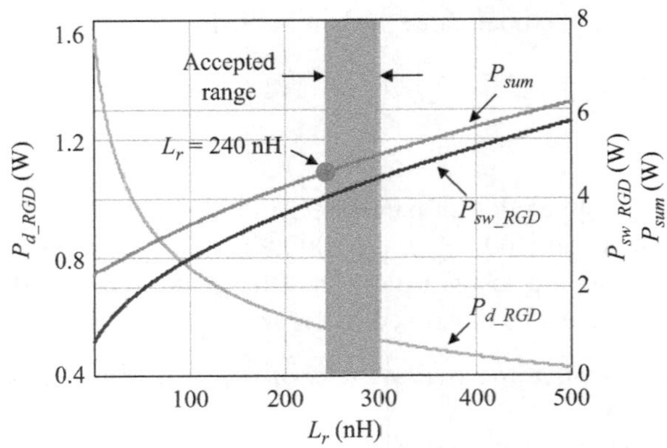

Figure 6.31 P_{d_RGD}, P_{sw_RGD}, and P_{sum} *as functions of the resonant inductance* L_r

As a conclusion, in order to minimize the overall loss of the gate-drive loss and the switching loss, the resonant inductance L_r should to be chosen as small as possible in the accepted range. Therefore, the value of L_r is about 240 nH for the optimum design in the proposed RGD with the desired specifications.

6.2.4.5 Dead time control in the proposed RGD

Since the proposed RGD is used to drive the HB leg, the dead time should be inserted to avoid the short-through. The waveforms of the gate-drive voltage of Q_1 and Q_2 are shown in Figure 6.24. It is observed that the waveforms of gate-drive voltage of Q_1 and Q_2 are complementary. When v_{gs_Q1} changes from the positive maximum value to the negative maximum value, v_{gs_Q2} changes from the negative maximum value to the positive maximum value simultaneously. As a result, there is no voltage overlap between two drive voltage signals, which prevents the short-through during the dead time. Considering the tolerance of the input capacitance of Q_1 and Q_2 in practical applications, the value of two resonant inductors in each RGD can be adjusted to guarantee enough dead time between two complementary signals. Moreover, since different resonant inductor values lead to different resonant periods and this can be used to obtain desired dead time according to the applications.

6.2.5 The comparison between the proposed RGD and previous gate-drive circuits

6.2.5.1 Operation comparison between the proposed RGD and conventional transformer coupled VSD

The transformer coupled VSD is shown in Figure 6.22. This VSD can be used with either coreless or core-based transformer. The coreless transformer in [39] simplifies the VSD and performs virtually as well as its core-based counterpart; however, it is still a VSD-based driver and has no capability to recovery high-frequency gate-drive loss. The proposed isolated RGD is shown in Figure 6.23. The two

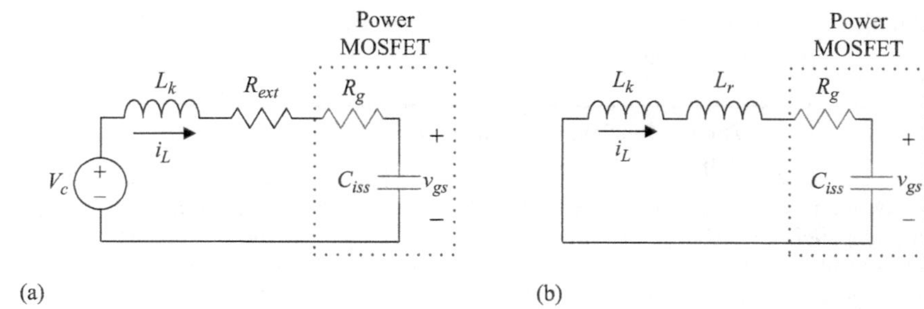

Figure 6.32 The simplified equivalent circuits during the turn-on transition; (a) conventional transformer coupled VSD and (b) proposed RGD

gate-drive circuits are both able to drive two power MOSFETs. Considering the leakage inductance of the transformer, the simplified equivalent circuit of the transformer coupled gate driver during the turn-on transition is shown in Figure 6.32(a). For comparison, the simplified equivalent circuit of the proposed RGD during the turn-on transition is shown in Figure 6.32(b).

For comparison, in the transformer coupled VSD as shown in Figure 6.32(a), C_{iss} represents the input capacitance of the power MOSFET. The inductance L_k consists of the leakage inductance and stray inductance, which is as a disadvantageous element. R_g is the gate mesh resistance of the power MOSFET and R_{ext} is the external damping resistance. Due to the existence of the inductance, the gate-to-source voltage v_{gs} is oscillating during the charging and discharging intervals. Therefore, the external resistance R_{ext} has to be inserted in the gate drive circuit to damp the oscillation and make the *R-L-C* second-order system under the overdamping condition. But the problem is that the external resistance slows down the driving speed and increases the switching loss.

In the proposed RGD as shown in Figure 6.32(b), the resistance is only the gate mesh resistance of the power MOSFET. The leakage inductance of the drive transformer and stray inductance no longer exist as the disadvantageous elements. The proposed RGD utilizes the *L-C* resonance to recover the charge energy of the input capacitance. An additional resonant inductance L_r may also be applied to adjust the value of the total inductance for the optimum design.

The operation principles of the transformer coupled VSD and the proposed RGD are quite different. The key point of the gate-drive energy recovery in the proposed RGD is the different control strategy from the transformer coupled VSD. Unlike the VSD, the proposed RGD can provide an additional interval in the switching transition, when the primary-side voltage u_{AB} of the drive transformer is zero as shaded area in Figure 6.24. During this interval, the proposed RGD utilizes the energy stored in the gate capacitance instead of the drive voltage supply to switch the polarities of the gate voltage through the energy exchange in the resonance manner. However, the VSD operates in the overdamping mode as an *R-L-C* second-order system during the charging and discharging period. The energy provided by the d.c. power supply V_c is entirely dissipated in the damping resistance,

Table 6.8 Comparison between conventional transformer coupled VSD and proposed RGD

	The conventional transformer coupled VSD	The proposed RGD
Control strategy in the switching transition	Utilize the d.c. power supply to change the gate-to-source voltage	Provide an additional interval for the *L-C* resonance to recover gate energy
Energy stored in C_{iss}	Dissipated in the damping resistance totally	Recovered by utilizing the *L-C* resonance, reduce gate-drive loss significantly
Leakage inductance of the transformer, L_k	A disadvantageous element, make the drive voltage to be oscillating	Be utilized to recover the gate-drive energy
External damping resistance, R_{ext}	Should be inserted to damp the oscillation, slow down the driving speed, increase switching loss	Not needed

which increases the gate-drive loss. Based on the above analysis, the difference between the conventional VSD and the proposed RGD is summarized in Table 6.8.

6.2.5.2 Comparison between the proposed RGD and conventional VSD with gate charge

The NMOSFET IPP50R199CP (V_{DS} = 550 V, I_D = 17 A, Q_g = 50 nC) from Infineon is chosen as the power MOSFET in the experimental prototype in this paper. It is noted that not only this power MOSFET, other power MOSFETs can be driven by the proposed RGD in different applications. In high current applications, to reduce the on-state resistance of the power MOSFET, the super junction structure is introduced in MOSFETs, which increase the gate charge Q_g and the input capacitance C_{iss} of the power MOSFET. Figure 6.33 shows the gate-drive loss P_{d_leg} for one bridge leg as a function of the gate charge Q_g. The gate-drive loss for one bridge leg with the proposed RGD is given in (6.34) and the gate-drive loss for one bridge leg with the conventional VSD is given in (6.36). It is observed that as the gate charge increases, the gate-drive loss in both gate drivers is increasing correspondingly. In addition, under larger gate charge situations, the more gate-drive loss can be reduced by the proposed RGD. For example, the gate charge Q_g of the CoolMOS™ SPW52N50C3 (V_{DS} = 560 V, I_D = 52 A) from Infineon is 290 nC. From (6.34) and (6.36), about 11.4 W can be reduced by the proposed RGD as shown in Figure 6.33. Therefore, with the proposed RGD, a more significant gate-drive loss reduction can be achieved in larger gate charge situations.

6.2.5.3 Applications extension

The proposed RGD is able to provide two isolated complementary drive signals for a pair of power MOSFETs. Therefore, it is not only suitable for the traditional ZVS

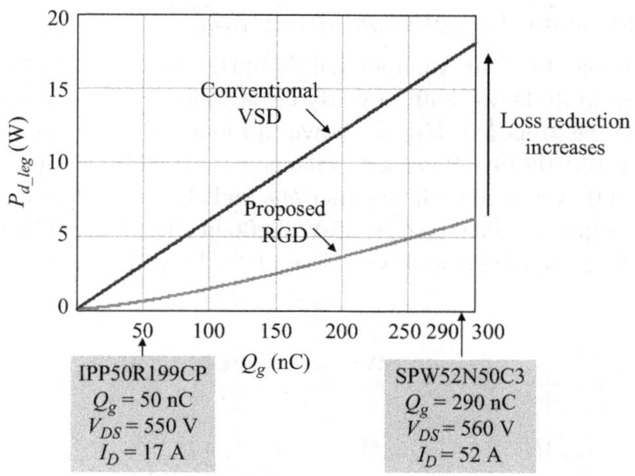

Figure 6.33 The relationship between the gate charge and gate-drive loss

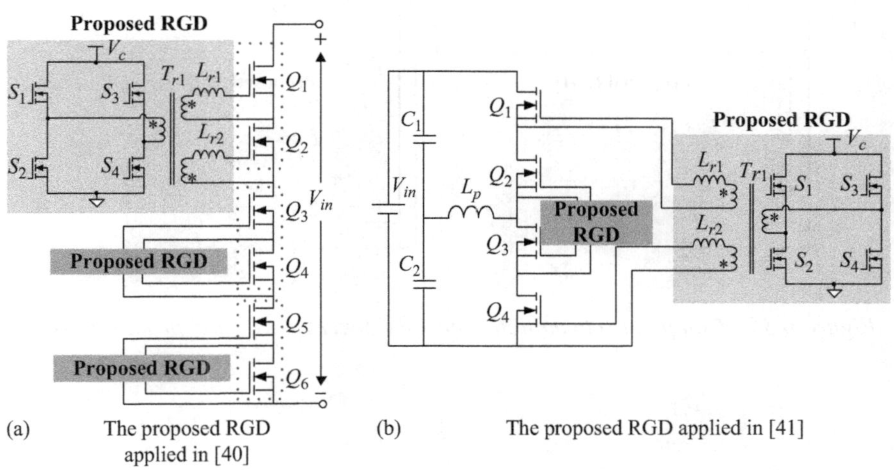

Figure 6.34 The ZVZCS converters in [40,41] with the proposed RGD

FB converters, but also can be applied to other ZVZCS converters at a higher power level. Two structures of ZVZCS converters presented in [40,41] operate in a higher power level and have a good efficiency. Figure 6.34(a) shows the proposed RGD applied to the ZVZCS converter in [40], in which each pair of power MOSFETs Q_1 and Q_2, Q_3 and Q_4, Q_5 and Q_6 requires two complementary drive signals. Similarly, Figure 6.34(b) shows the proposed RGD applied to the converter in [41]. The proposed RGD can be applied to these ZVZCS converters to reduce the gate driver loss and turn-off loss at high switching frequency.

6.2.6 Experimental results and discussion

A 200 VDC input, 48 V/20 A output and 500-kHz phase-shift ZVS FB converter with the proposed RGD was built to verify the advantage. The schematic of the FB converter with the proposed RGD is shown in Figure 6.35. The components used are as follows: IPP50R199CP is used as the power MOSFET Q_1–Q_4; the Schottky diode DSSK60-02A is used as the rectifier D_{R1} and D_{R2}; the output inductance L_f is 6 uH and the output capacitance C_f is 63 V, 3,000 uF. The FDN335N is used as the drive switch S_1–S_8 and the gate drive voltage $V_c = 15$ V. The magnetic core of the

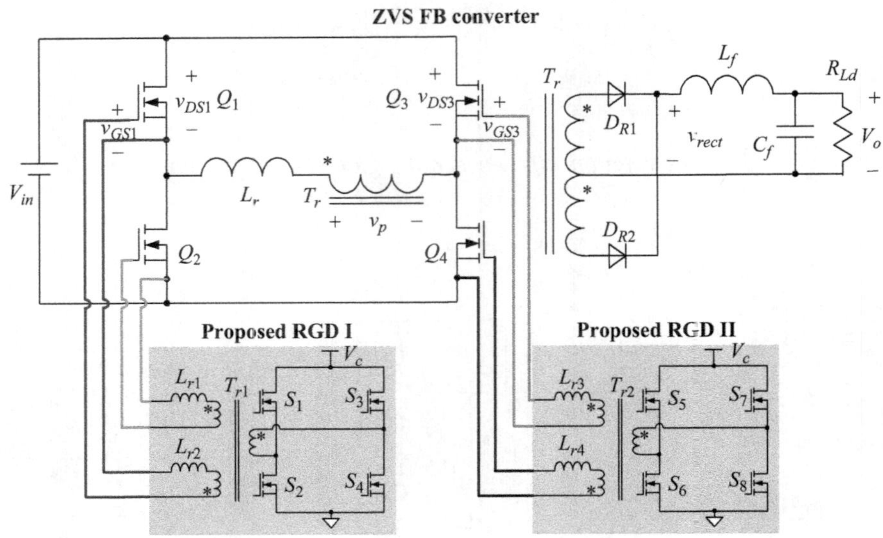

Figure 6.35 Complete schematic of the FB converter with the proposed RGD

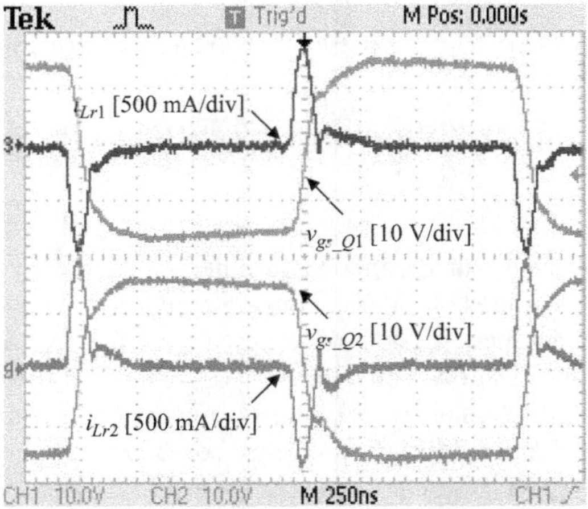

Figure 6.36 Two complementary drive voltage and resonant inductor current

drive transformer T_{r1} and T_{r2} is EE 13. The turn ratio of T_{r1} and T_{r2} is 1:1. The magnetic inductance is 66.5 uH and leakage inductance is 246 nH, which are measured by the HEWLETT PACKARD 4284A precision LCR meter. Because the value of the leakage inductance fits the optimum design range, they can be used as the resonant inductance L_{r1} and L_{r2} in this case.

Figure 6.36 shows the resonant inductor current i_{Lr1}, i_{Lr2} and the gate-drive voltage v_{gs_Q1}, v_{gs_Q2}, which agrees with the theoretic waveforms in Figure 6.24. It is observed that the gate-drive voltages, v_{gs_Q1} and v_{gs_Q2}, are complementary; the directions of the resonant inductor currents, i_{Lr1} and i_{Lr2}, are also opposite. This experimental waveform indicates that the proposed RGD has the ability to drive two power MOSFETs in one bridge leg, leading to the low complexity and cost in the FB converters.

Figure 6.37(a) illustrates the relationship between v_{gs_Q1}, i_{Lr1} and the voltage between A and B, v_{AB} during the turn-on interval. It is observed that the resonance

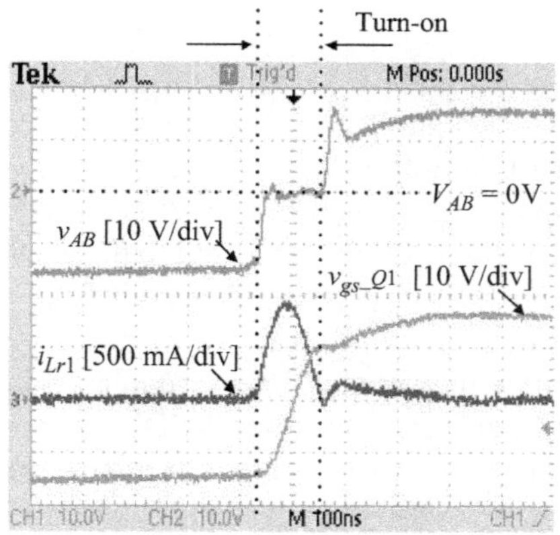

(a)

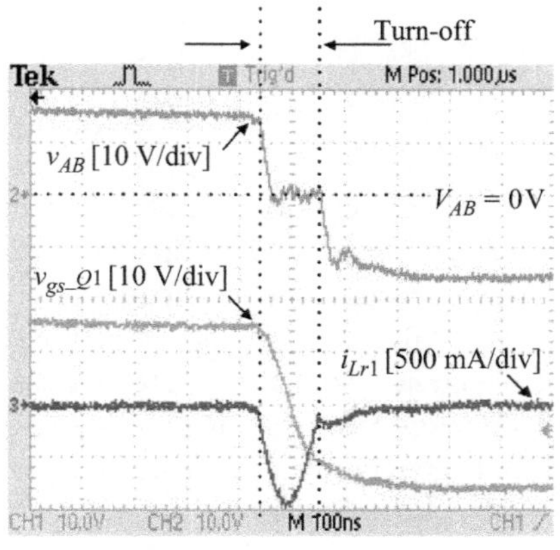

(b)

Figure 6.37 Resonant switching transition: v_{gs_Q1}, i_{Lr1}, and v_{AB}; (a) turn-on interval and (b) turn-off interval

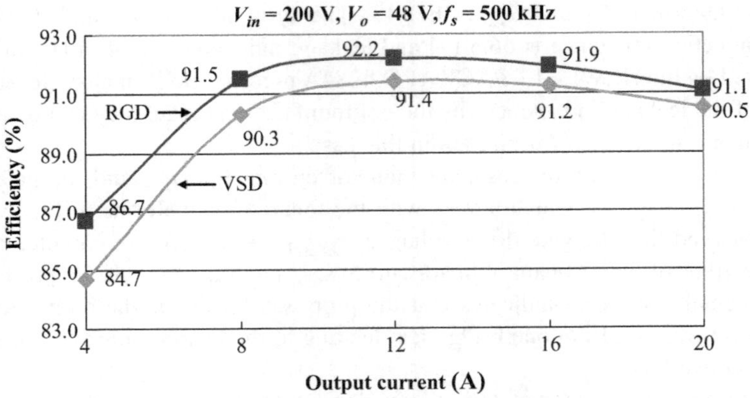

Figure 6.38 Efficiency comparison

only happens during the stage of $v_{AB} = 0$. This interval, i.e., $v_{AB} = 0$, is the key difference between the conventional VSDs from the control strategy. During this interval, the proposed RGD utilizes the *L-C* resonance to recover the gate-drive energy stored in the input capacitance of the power MOSFET. Therefore, the gate-drive loss can be reduced significantly compared to the conventional VSDs. For the turn-off interval, the detailed waveforms are provided in Figure 6.37(b), and they are similar to the turn-on interval.

Figure 6.38 shows the measured efficiency comparison between the proposed RGD and conventional transformer coupled VSD. It is observed that at 4 A, the efficiency is improved from 84.7% to 86.7% (an improvement of 2.0%). At 12 A, the efficiency is improved from 91.4% to 92.2% (an improvement of 0.8%). It should be also noted that because of the low gate-drive loss, the proposed RGD improves the efficiency effectively in full-load range over the conventional transformer coupled VSD.

6.3 Summary

The forward recovery of the diodes causes poor performance and high conduction loss in the multi-MHz resonant converters. A simple and efficient self-driven RGD is presented. A control stage comprised of a shutdown branch and an auxiliary switch is introduced to the RGD to block the circulating current and the low-impedance path in the drive circuit, so that the gate voltage can achieve fast shutdown and buildup under ON-OFF operation to ensure fast transient response. Moreover, the proposed RGD generates a tunable d.c. bias to increase the peak gate voltage and extend the conduction time with the optimal $R_{DS(on)}$, so that the average $R_{DS(on)}$ and the associated conduction loss in the SR FET can be reduced. It also provides precise switching timing for the SR to minimize the body diode conduction loss.

An isolated RGD for two MOSFETs in one bridge-leg is presented. The proposed RGD can provide two complementary drive signals to drive two MOSFETs, which can be used to drive the HB leg in FB converters. Moreover, with the negative drive voltage capability, the proposed RGD ensures high reliability in the FB converters over the previously proposed RGDs.

References

[1] D. J. Perreault, J. Hu, J. M. Rivas, *et al.*, "Opportunities and challenges in very high frequency power conversion," in *Proceedings on IEEE APEC*, pp. 1–14, 2009.

[2] A. Knott, T. M. Andersen, P. Kamby, *et al.*, "Evolution of very high frequency power supplies," *IEEE Journal of Emerging and Selected Topics in Power Electronics*, vol. 2, no. 3, pp. 386–394, 2014.

[3] R. C. N. Pilawa-Podgurski, A. D. Sagneri, J. M. Rivas, D. I. Aderson, and D. J. Perreault, "Very high frequency resonant boost converters," *IEEE Transactions on Power Electronics*, vol. 24, no. 6, pp. 1654–1665, 2009.

[4] J. M. Burkhart, R. Korsunsky, and D. J. Perreault, "Design methodology for a very high frequency resonant boost converter," *IEEE Transactions on Power Electronics*, vol. 28, no. 4, pp. 1929–1937, 2013.

[5] J. Hu, A. D. Sagneri, J. M. Rivas, Y. Han, S. M. Davis, and D. J. Perreault, "High frequency resonant SEPIC converter with wide input and output voltage ranges," *IEEE Transactions on Power Electronics*, vol. 27, no. 1, pp. 189–200, 2012.

[6] A. D. Sagneri, D. I. Anderson, and D. J. Perreault, "Transformer synthesis for VHF converters," in *Proceedings on IEEE IPEC*, pp. 2347–2353, 2010.

[7] L. C. Raymond, W. Liang, and J. M. Rivas-Davila, "Performance evaluation of diodes in 27.12 MHz Class-D resonant rectifiers under high voltage and high slew rate conditions," in *Proceedings on COMPEL*, 2014.

[8] J. A. Santiago-Gonzalez, K. M. Elbaggari, K. K. Afridi, and D. J. Perreault, "Design of class E resonant rectifiers and diode evaluation for VHF power conversion," *IEEE Transactions on Power Electronics*, vol. 30, no. 9, pp. 4960–4972, 2015.

[9] M. Madsen, A. Knott, and M. A. E. Andersen, "Low power very high frequency switch-mode power supply with 50 V input and 5 V output," *IEEE Transactions on Power Electronics*, vol. 29, no. 12, pp. 6569–6580, 2014.

[10] M. Madsen, A. Knott, and M. A. E. Andersen, "Very high frequency half bridge DC/DC converters," in *Proceedings on IEEE APEC*, pp. 1409–1414, 2014.

[11] M. P. Madsen, J. A. Pedersen, A. Knott, and M. A. E. Andersen, "Self-oscillating resonant gate drive for resonant inverters and rectifiers composed solely of passive components," in *Proceedings on IEEE APEC*, pp. 2029–2035, 2014.

[12] J. M. Rivas, O. Leitermann, Y. Han, and D. J. Perreault, "A very high frequency DC–DC converter based on a class-Φ_2 resonant inverter," *IEEE Transactions on Power Electronics*, vol. 26, no. 10, pp. 2980–2992, 2011.

[13] Y. Tang and A. Khaligh, "Miniaturized bridgeless high-frequency resonant AC–DC step-up/step-down converters," *IEEE Transactions on Power Electronics*, vol. 29, no. 12, pp. 6518–6533, 2014.

[14] P. Shamsi and B. Fabak, "Design and development of very high frequency resonant DC–DC boost converters," *IEEE Transactions on Power Electronics*, vol. 27, no. 8, pp. 3725–3733, 2012.

[15] M. Arias, M. F. Diaz, D. G. Lamar, D. Balocco, A. A. Diallo, and J. Sebastian, "High-efficiency asymmetrical half-bridge converter without electrolytic capacitor for low-output-voltage AC–DC LED drivers," *IEEE Transactions on Power Electronics*, vol. 28, no. 5, pp. 2539–2550, 2013.

[16] A. Fernandez, D. G. Lamar, M. Rodriguez, M. M. Hernando, J. Sebastian, and M. Arias, "Self-driven synchronous rectification with input voltage tracking for converters with a symmetrically driven transformer," *IEEE Transactions on Industrial Electronics*, vol. 56, no. 5, pp. 1440–1445, 2009.

[17] J. Yang and H. Do, "High-efficiency ZVS AC–DC LED diver using a self-driven synchronous rectifier," *IEEE Transactions on Circuits and Systems*, vol. 61, no. 8, pp. 2505–2512, 2014.

[18] M. Kovacevic, A. Knott, and M. A. E. Andersen, "A VHF interleaved self-oscillating resonant SEPIC converter with phase-shift burst-mode control," in *Proceedings on IEEE APEC*, pp. 1402–1408, 2014.

[19] M. Pahlevaninezhad, P. Das, J. Drobnik, P. K. Jain, and A. Bakhshai, "A novel ZVZCS full-bridge dc–dc converter used for electric vehicles," *IEEE Transactions on Power Electronics*, vol. 27, no. 6, pp. 2752–2769, 2012.

[20] A. F. Bakan, N. Altintaş, and I. Aksoy, "An improved PSFB PWM DC–DC converter for high-power and frequency applications," *IEEE Transactions on Power Electronics*, vol. 28, no. 1, pp. 64–74, 2013.

[21] N. Teerakawanich and C. M. Johnson, "A new resonant gate driver with bipolar gate voltage and gate energy recovery," in *Proceedings on IEEE APEC*, pp. 2424–2428, 2013.

[22] H. Fujita, "A resonant gate-drive circuit with optically isolated control signal and power supply for fast-switching and high-voltage power semiconductor devices," *IEEE Transactions on Power Electronics*, vol. 28, no. 11, pp. 5423–5430, 2013.

[23] J. T. Strydom, M. A. de Rooij, and J. D. van Wyk, "A comparison of fundamental gate-driver topologies for high frequency applications," in *Proceedings on IEEE APEC*, pp. 1045–1052, 2004.

[24] M. M. Swamy, T. Kume, and N. Takada, "An efficient resonant gate-drive scheme for high-frequency applications," *IEEE Transactions on Power Electronics*, vol. 48, no. 4, pp. 1418–1431, 2012.

[25] P. Anthony, N. McNeill, and D. Holliday, "A first approach to a design method for resonant gate driver architectures," *IEEE Transactions on Power Electronics*, vol. 27, no. 8, pp. 3855–3868, 2012.

[26] P. Anthony, N. McNeill, and D. Holliday, "High-speed resonant gate driver with controlled peak gate voltage for silicon carbide MOSFETs," in *Proceedings on IEEE ECCE*, pp. 2961–2968, 2012.

[27] P. Dwane, D. O' Sullivan and M. G. Egan, "An assessment of resonant gate drive techniques for use in modern low power dc–dc converters," in *Proceedings on IEEE APEC*, vol. 3, pp. 1572–1580, 2005.

[28] T. López, G. Sauerlaender, T. Duerbaum, and T. Tolle, "A detailed analysis of a resonant gate driver for PWM applications," in *Proceedings on IEEE APEC*, pp. 873–878, 2003.

[29] X. Zhou, Z. Liang, and A. Huang, "A high-dynamic range current source gate driver for switching-loss reduction of high-side switch in buck converter," *IEEE Transactions on Power Electronics*, vol. 25, no. 6, pp. 1439–1443, 2010.

[30] D. Maksimovic, "A MOS gate drive with resonant transitions," in *Proceedings on IEEE Power Electronics Specialists Conference (PESC)*, pp. 527–532, 1991.

[31] Y. Chen, F. Lee, L. Amoroso, and H. P. Wu, "A resonant MOSFET gate driver with efficient energy recovery," *IEEE Transactions on Power Electronics*, vol. 19, no. 2, pp. 470–477, 2004.

[32] W. Eberle, Y. F. Liu, and P. C. Sen, "A new resonant gate-drive circuit with efficient energy recovery and low conduction loss," *IEEE Transactions on Industrial Electronics*, vol. 55, no. 5, pp. 2213–2221, 2008.

[33] I. D. de Vries, "A resonant power MOSFET/IGBT gate driver," in *Proceedings on IEEE APEC*, pp. 179–185, 2002.

[34] S. Pan and P. K. Jain, "A new pulse resonant MOSFET gate driver with efficient energy recovery," in *Proceedings on IEEE PESC*, pp. 1–5, 2006.

[35] H. Fujita, "A resonant gate-drive circuit capable of high-frequency and high-efficiency operation," *IEEE Transactions on Power Electronics*, vol. 25, no. 4, pp. 962–969, 2010.

[36] Q. Li and P. Wolfs, "The power loss optimization of a current fed ZVS two-inductor boost converter with a resonant transition gate drive," *IEEE Transactions on Power Electronics*, vol. 21, no. 5, pp. 1253–1263, 2006.

[37] Y. Ren, M. Xu, Y. Meng, and F. C. Lee, "12 V VR efficiency improvement based on two-stage approach and a novel gate driver," in *Proceedings on IEEE PESC*, pp. 2635–2641, 2005.

[38] K. Yao and F. C. Lee, "A novel resonant gate driver for high frequency synchronous buck converters," *IEEE Transactions on Power Electronics*, vol. 17, no. 2, pp. 180–186, 2002.

[39] S. C. Tang, S. Y. (R.) Hui, and H. S.-H. Chung, "Coreless printed circuit board (PCB) transformers with multiple secondary windings for complementary gate drive circuits," *IEEE Transactions on Power Electronics*, vol. 14, no. 3, pp. 431–437, 1999.

[40] T. T. Song, H. Wang, H. S. H. Chung, S. Tapuhi, and A. Ioinovici, "A high-voltage ZVZCS DC–DC converter with low voltage stress," *IEEE Transactions on Power Electronics*, vol. 23, no. 6, pp. 2630–2647, 2008.

[41] T. T. Song, N. Huang, and A. Ioinovici, "A zero-voltage and zero-current switching three-level DC–DC converter with reduced rectifier voltage stress and soft-switching-oriented optimized design," *IEEE Transactions on Power Electronics*, vol. 21, no. 5, pp. 1204–1212, 2006.

Chapter 7
eGaN HEMTs gate drivers

7.1 Three-level gate drivers

7.1.1 Introduction

With the excellent Figure of Merit, enhancement mode gallium nitride (eGaN) high-electron mobility transistors (HEMTs) are expected to improve power converters to higher efficiency and power density. When they are used, the considerations are as follows: (1) the gate voltage cannot exceed the maximum rating of 6 V; (2) prevent the false turn-on due to the low threshold voltage V_{th} (typically 1.4 V); (3) there is no body-diode, so the reverse current has to flow via the channel by the reverse conduction mechanism, resulting in high reverse conduction loss.

A three-level driving method is proposed to reduce the eGaN synchronous rectifier (SR) reverse conduction loss in synchronous buck converters in [1]. However, the driving method is limited to the eGaN SRs in half-bridge structures with complementary driving signals. There is no reported analysis considering the eGaN control HEMT reverse conduction loss before zero-voltage switching (ZVS) turn-on interval (normally tens of ns). The eGaN HEMT reverse conduction voltage (>1.4 V) is much higher than that of a Si metal oxide semiconductor field effect transistor (MOSFET) (around 0.7 V) due to the reverse conduction mechanism. More importantly, with multi-megahertz (MHz) (the switching period of hundreds of ns), the reverse conduction loss becomes severe and has to be minimized. Moreover, the turn-off ringing and high *dv/dt* event due to the parasitics may cause the false turn-on problem [2]. For wide band gap devices with fast switching frequency, the false turn-on problem is more serious. A negative gate bias is applied to SiC MOSFETs in [3] to prevent the false turn-on problem. The result shows that the negative bias solution is not suitable for SiC MOSFETs due to the lower threshold voltage and small margin for negative gate bias. The negative voltage bias is applied when the eGaN HEMT turns off to improve the *dv/dt* immunity in [4]. However, whether a negative gate bias has to be applied to improve the *dv/dt* immunity or how to select the appropriate negative bias value is not discussed.

On the other hand, SR techniques are strongly desired because diodes have low performance at multi-MHz due to forward recovery [5]. Conventional Si SR MOSFET driving schemes can be classified into three categories: (1) the control-based, (2) voltage-based and (3) current-based methods. The control-based method detects the SR body-diode turn-on and turn-off instants and generates the gate-driving signal by anticipating the turn-on and turn-off instants based on the previous switching pattern, limiting multi-MHz applications [6]. The voltage-based and current-based methods, such as [7–9], determine the turn-on and turn-off instants based on sampling the SR instantaneous drain-to-source voltage and current, respectively. However, to meet the strict voltage requirements, driving integrated circuits (ICs) are normally needed with multi-MHz. The driving IC propagation delay makes generated driving signals lag to the desired one, which results in high reverse conduction loss during propagation time. Therefore, these driving methods for Si SR MOSFETs could hardly be transplanted to multi-MHz eGaN SRs directly, especially with resonant gate drivers since the resonant voltage can hardly be controlled under the maximum safe gate voltage [10]. For eGaN SR HEMTs, the driving signal timing is achieved with a RC delay circuit in [11]. The delay circuit is tuned iteratively to match the simulation results. However, the driving signal timing problem has not been analyzed quantitatively yet. A voltage following control is proposed in [12]. It is derived that the switches gate timing relationship is fixed when the output voltage is proportional to the input voltage. But this control scheme is limited to d.c. transformer (DCX) applications while not suitable for regulated output converters.

7.1.2 Driving requirements for eGaN HEMTs in resonant SEPIC converters

7.1.2.1 Multi-MHz isolated resonant SEPIC converter

Figure 7.1 shows the eGaN-isolated resonant single-ended primary-inductor converter (SEPIC) SR converter. It consists of class Φ_2 inverter, transformer, and class E rectifier. V_{in} is from 18 to 24 V. V_{out} and I_{out} are 5 V and 2 A, respectively. The detailed power-stage design procedure can be found in [13–16]. Table 7.1 gives the parameters. The following analysis and experimental verification are regarding to the converter in Figure 7.1 with the ON/OFF control [16].

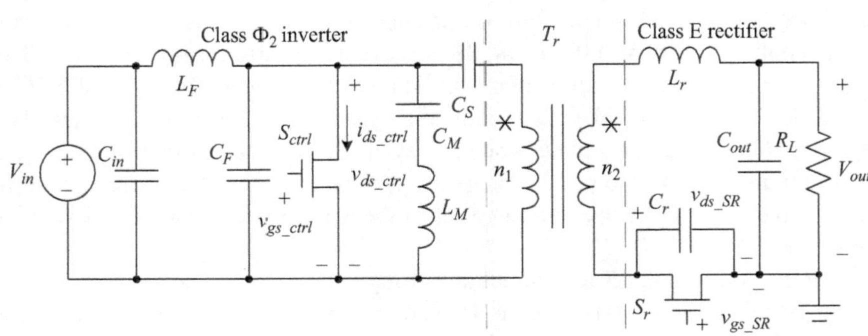

Figure 7.1 Isolated resonant SR SEPIC converter

7.1.2.2 Reverse conduction mechanism

Figure 7.2 shows eGaN HEMT equivalent operation circuits and waveforms during the reverse conduction mechanism. The reverse conduction procedure before ZVS turn-on interval is presented as follows.

1. Stage 1 [t_0, t_1] (Figure 7.2(a)):
   At $t = t_0$, v_{sd} resonates to zero and the eGaN HEMT is ready to realize ZVS turn-on. During this stage, $v_{gs} = 0$, which means that C_{gs} is short circuit and

Table 7.1 Power-stage component values

L_F	270 nH (Coilcraft)	C_F	450 pF
L_M	270 nH (Coilcraft)	C_M	470 pF
L_r	33 nH (Coilcraft)	C_r	4 nF
C_{in}	20 μF	C_S	15 nF
C_{out}	47 μF	n_1/n_2	4/1
S_{ctrl}	EPC2001 (EPC, 100 V, 25 A)	S_r	EPC2015 (EPC, 40 V, 33 A)

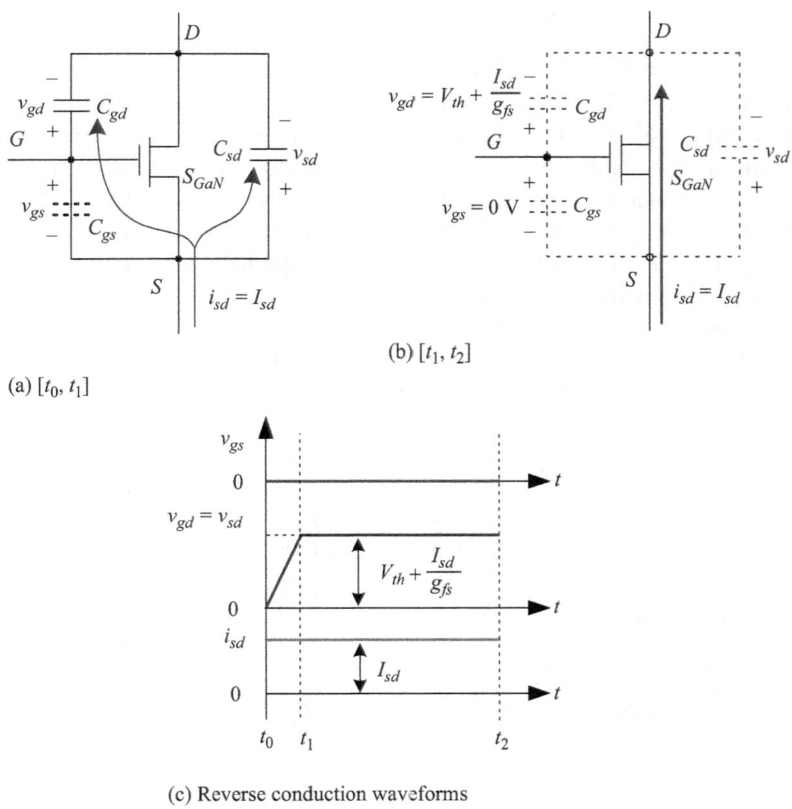

Figure 7.2 Reverse conduction equivalent circuits

C_{gd} is paralleled with C_{sd}. The reverse current i_{sd} flows into the source terminal charging C_{gd} and C_{sd}. Until the capacitance voltage $v_{gd} = v_{sd}$ approaches threshold voltage V_{th}, the channel begins to build up. The relationship between v_{gd} and i_{sd_ch} flows through the channel is

$$g_{fs}(v_{gd} - V_{th}) = i_{sd\_ch} \tag{7.1}$$

where g_{fs} is the transconductance. To make $i_{sd\_ch} = I_{sd}$ at $t = t_1$, v_{gd} should be

$$v_{gd} = V_{th} + \frac{I_{sd}}{g_{fs}} \tag{7.2}$$

2. Stage 2 [t_1, t_2] (Figure 7.2(b)):
During this stage, the reverse current $i_{sd} = I_{sd}$ flows via the channel and the reverse conduction voltage is

$$v_{sd} = v_{gd} - v_{gs} = V_{th} + \frac{I_{sd}}{g_{fs}} - v_{gs} \tag{7.3}$$

From (7.3), with $v_{gs} = 0$ V, $v_{sd} = v_{gd} = V_{th} + I_{sd}/g_{fs} \geq 1.4$ V and as v_{gs} increases, v_{sd} decreases.

7.1.2.3 Control HEMT driving consideration

For eGaN control HEMT, high reverse conduction loss before ZVS turn-on needs special attention. Normally, this resonant converter can be tuned at optimal operating point. For example, in Figure 7.3(a), the control HEMT happens to turn on when v_{ds_ctrl} resonates to zero at $t = t_0$ under 18 V input voltage. Due to ON/OFF control, the output power is controlled by the converter ON/OFF duty ratio while the control HEMT drive signal is fixed [16]. The optimal operating condition can be maintained under 18 V input voltage in full-load range. However, in Figure 7.3(b), when the input voltage increases to 24 V, v_{ds_ctrl} resonates to zero at $t = t_0$ and the control HEMT turns on under ZVS at $t = t_1$. Then, the reverse conduction mechanism is triggered, and reverse conduction time t_{rc} is 20 ns during [t_0, t_1].

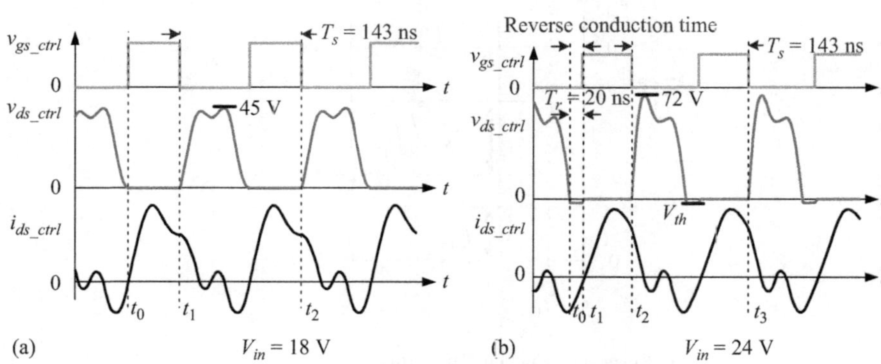

Figure 7.3 Control HEMT under different ZVS condition

The reverse conduction loss P_R before ZVS turn-on with the ON/OFF control is

$$P_R = V_{SD} \cdot I_{SD} \cdot t_{rc} \cdot f_s \cdot D_c \tag{7.4}$$

where V_{SD} is the reverse conduction voltage drop, I_{SD} is the reverse conduction current, t_{rc} is the reverse conduction time, f_s is the switching frequency, and D_c is the power-stage duty ratio. With $f_s = 7$ MHz, $V_{SD} = 1.6$ V, $I_{SD} = 1.5$ A, and $D_c = 0.5$, the reverse conduction loss is 0.17 W (1.7% of the output power) and should be minimized.

7.1.2.4 SR HEMT driving consideration

Compared with fixed control HEMT driving signal, the SR driving signal timing is more challenging. In resonant converters, driving signals of control and SR HEMTs are not complementary. On the other hand, to meet the eGaN strict driving voltage requirements, commercial driving ICs, such as LM5114, are recommended. However, in multi-MHz applications, the driving IC propagation delay has to be considered. For example, the driving IC LM5114 propagation delay is 14 ns typically. This is around 10% of the period $T_S = 143$ ns with 7 MHz. So the driving IC propagation delay is important to satisfy the accurate driving signal timing to ensure ZVS and minimize the reverse conduction loss and circulating loss.

7.1.3 Proposed three-level gate drivers for eGaN HEMTs

Figure 7.4 shows proposed gate-driver circuits. For control HEMTs with fixed driving signal, a three-level driver is proposed to reduce the reverse conduction loss by constructing the driving voltage. Figure 7.5 shows optimal control HEMT driving waveforms. For eGaN SRs, a driver compensating the driving IC propagation delay is proposed. Figure 7.6 shows the corresponding eGaN SR waveforms.

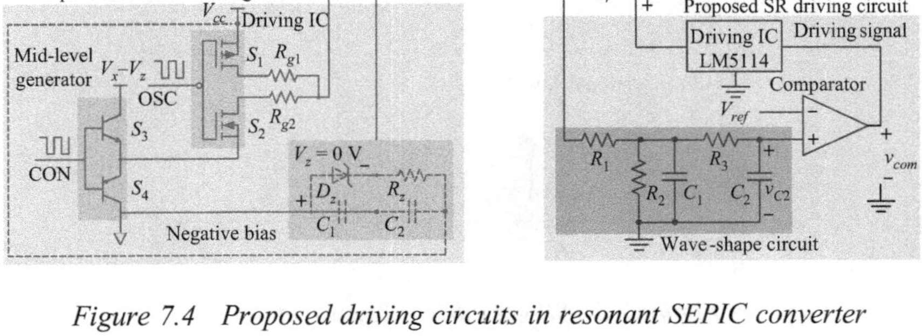

Figure 7.4 Proposed driving circuits in resonant SEPIC converter

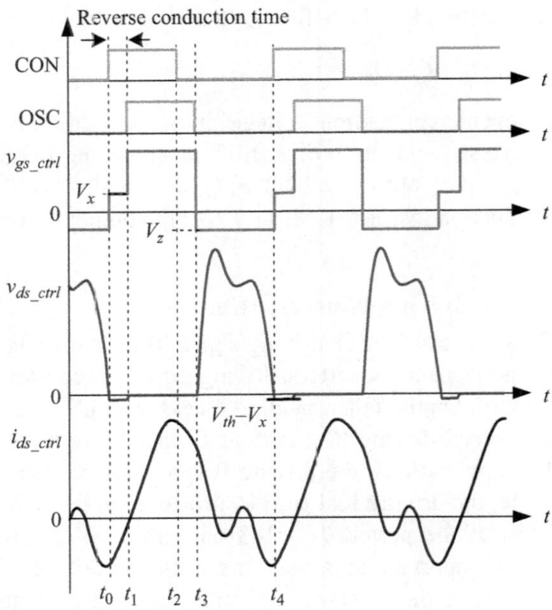

Figure 7.5 Control HEMT mid-level driving waveform

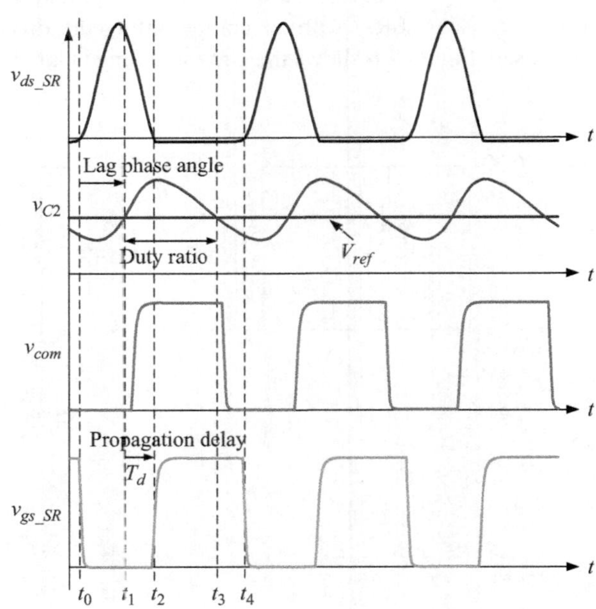

Figure 7.6 SR gate driver key waveforms

7.1.3.1 Proposed eGaN control HEMTs three-level driving circuit

In Figure 7.5, a positive gate bias V_x ($V_x < V_{th}$) is provided by the mid-level generator to reduce the reverse conduction voltage from t_0 to t_1, since V_{SD} decreases when V_{GS} increased. The reverse conduction voltage reduces from around (V_{th}–V_z) (with the negative bias V_z) or V_{th} (without the negative bias) to (V_{th}–V_x) since $v_{sd} = v_{gd} - v_{gs}$ from (7.3). If the negative bias V_z is applied and V_z and V_x are selected to be −2 V and 1 V, respectively, the reverse conduction voltage can be reduced significantly from 3.6 to 0.6 V (a reduction of 83%). Then the reverse conduction loss would be reduced from 0.38 to 0.06 W during before ZVS turn-on interval. From t_3 to t_4, V_z can be provided by the Zener diode D_z to improve *dv/dt* immunity if needed. The signal CON leads to the signal OSC and the leading phase angle represents the mid-level voltage duty ratio.

7.1.3.2 Proposed driving circuit for eGaN SR HEMTs

In Figure 7.6, the driving IC is used to meet the strict voltage requirement. The driving IC propagation delay is considered in the proposed driving circuit. v_{ds_SR} is sensed and filtered to obtain v_{C2}. The lag phase angle considers the driving IC propagation delay, which is accurately designed by the HEMT rectifier mathematic model. Then v_{C2} is compared with a reference voltage V_{ref} to obtain v_{com}. Depending on V_{ref}, the appropriate duty ratio of v_{com} can be determined. Therefore, the accurate SR driving signal timing requirements can be guaranteed by designing the lag phase angle and duty ratio to ensure ZVS turn-on, meanwhile, minimize the reverse conduction loss and circulating loss.

7.1.4 Operation principle of three-level gate driver

7.1.4.1 Operation principle

The equivalent operation circuits are illustrated in Figure 7.7.

1. Stage 1 [t_0, t_1] (Figure 7.7(a)):
   v_{ds_ctrl} resonates to zero at $t = t_0$, and the reverse current flows into the source terminal, triggering the reverse conduction mechanism. The signal CON sets and the signal OSC is low at $t = t_0$. During this stage, V_x reduces the reverse conduction voltage from (V_{th}–V_z) to (V_{th}–V_x) since $v_{sd\_ctrl} = v_{gd\_ctrl} - v_{gs\_ctrl}$ from (7.3). To select V_x value, the following constraint is considered:

   $$0 < V_x < V_{th} \tag{7.5}$$

   V_x should be smaller than V_{th} to avoid the short through issue. The larger V_x is, the smaller the reverse conduction voltage would be. Suppose the reverse conduction mechanism is not triggered before ZVS turn-on under the optimal condition while the reverse conduction time under the worst condition is t_{rc_wst}. Then the interval to ensure ZVS turn-on in the entire operation range should be determined as:

   $$t_1 - t_0 = t_{rc\_wst} \tag{7.6}$$

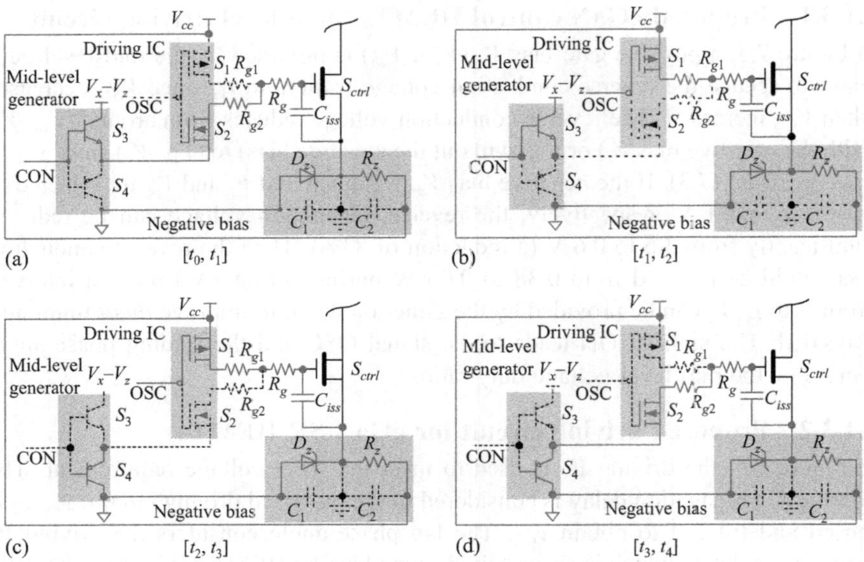

Figure 7.7 Equivalent operation circuits

This stage ends at $t = t_1$ when the control HEMT turns on.

2. Stage 2 $[t_1, t_2]$ (Figure 7.7(b)):
The signal OSC sets and control HEMT turns on under ZVS at $t = t_1$. With R_{g1} damping the input capacitance charge circuit, the gate voltage ringing is much limited:

$$R_{g1} = 2\sqrt{\frac{L_{g\_tot}}{C_{iss}}} - R_{S1} - R_g \tag{7.7}$$

where L_{g_tot} is the total gate loop inductance, R_{S1} is the switch S_1 on-resistance, and R_g is the inner resistance. For EPC2001C, R_g is typically 0.3 Ω.

3. Stage 3 $[t_2, t_3]$ (Figure 7.7(c)):
The eGaN HEMT state during this stage is the same as the state in Stage 2. The difference is that the signal CON resets at $t = t_2$, preparing for the eGaN HEMT to turn off. The eGaN HEMT is in the conduction status until the stage ends at $t = t_3$.

4. Stage 4 $[t_3, t_4]$ (Figure 7.7(d)):
The signal OSC resets and control HEMT turns off at $t = t_3$. During this stage, V_z can be applied to improve *dv/dt* immunity when necessary. An appropriate resistance R_{g2} is connected in series in the discharge loop to damp the gate voltage ringing, which prevents the false turn-on problem when the negative bias is not applied and prevents the voltage from exceeding the minimum rating of −5 V when the negative bias is applied. This stage ends at $t = t_4$ when the control HEMT is about to conduct reversely.

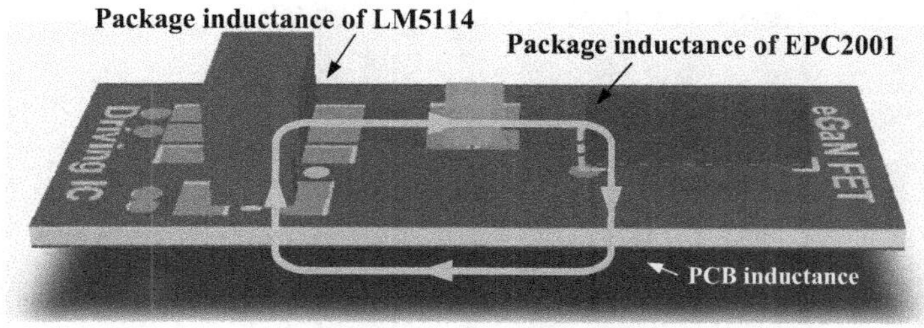

Figure 7.8 Optimized eGaN HEMT driving loop layout

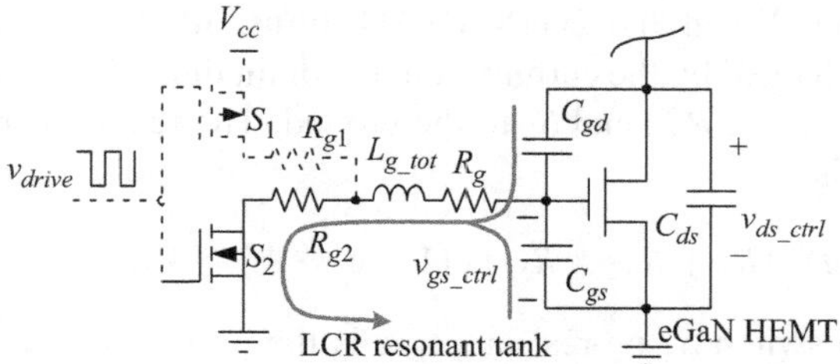

Figure 7.9 Ringing during turn-off transition

7.1.4.2 Negative gate bias during turn-off transition

Driving loop layout

The gate voltage ringing, caused by the gate loop inductance, requires to be damped by an appropriate resistance. Figure 7.8 shows the optimized eGaN driving loop layout. The printed circuit board (PCB) inductance is minimized by: (1) placing the driving IC close to the eGaN HEMT; (2) placing the driving path and the return path on top of each other; and (3) widening the traces between the driving IC and the eGaN HEMT gate. In our design prototype, the PCB inductance is extracted by the Ansoft Q3D extractor and the total gate loop inductance L_{g_tot} is 3.7 nH.

Gate resistance to damp ringing

Figure 7.9 shows the driving circuit during the turns-off transition. C_{iss} ($= C_{gs} + C_{gd}$) is discharged by the current travelling from the gate through L_{g_tot} to R_{g2} and S_2 to the ground. When the LCR resonant tank is under-damped, the ringing happens and the gate voltage amplitude may exceed V_{th}, causing the false turn-on. Therefore, it is recommended to select R_{g2} to ensure the LCR resonant tank critical-damped:

$$R_{g2} = 2\sqrt{\frac{L_{g\_tot}}{C_{iss}}} - R_{S2} - R_g \tag{7.8}$$

Negative bias requirement

With R_{g2} to damp the ringing, another issue of high *dv/dt* event should be considered to prevent the false turn-on. Figure 7.10 shows the *dv/dt* event during the

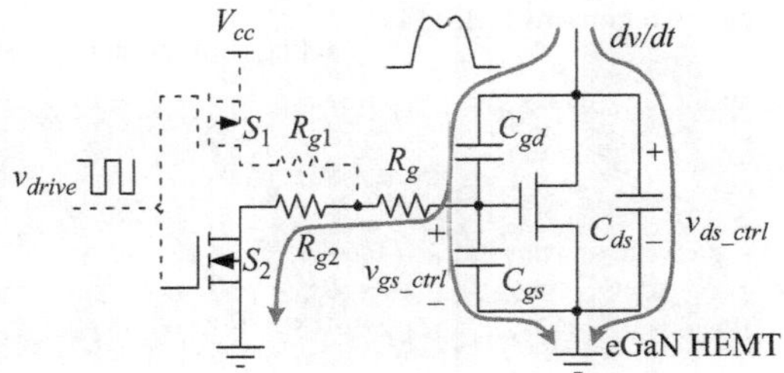

Figure 7.10 High dv/dt event during turn-off transition

turn-off transition. When the eGaN HEMT turns off, v_{ds_ctrl} rises, the Miller capacitor C_{gd} is charged by the current from the drain through C_{gd}. Then the current travels through C_{gs}, R_g, R_{g2}, and S_2 to the ground. The requirement to avoid *dv/dt* (Miller) turn-on is

$$C_{gd} \cdot (dv/dt) \cdot (R_g + R_{g2} + R_{S2}) \cdot (1 - e^{-dt_S/\alpha}) < V_{th} \tag{7.9}$$

where R_{S2} is the switch S_2 on-resistance, α is the passive network time constant $(R_g + R_{g2} + R_{S2}) \cdot (C_{gd} + C_{gs})$ and dt_s is the *dv/dt* switching time [2,3].

Therefore, with R_{g2} and R_{S2} to ground, the maximum drain node allowable *dv/dt* is:

$$(dv/dt)_{max\_allow} \approx \frac{V_{th}}{(R_g + R_{g2} + R_{S2}) \cdot C_{gd}} \tag{7.10}$$

$(dv/dt)_{max\_allow}$ is calculated to compare with $(dv/dt)_{max\_actual}$, which is the maximum drain node *dv/dt*, to determine whether the negative voltage bias is needed. The $(dv/dt)_{max\_actual}$ value can be obtained from the simulation. If $(dv/dt)_{max\_actual}$ exceeds $(dv/dt)_{max\_allow}$, a negative bias V_z should be added to increase $(dv/dt)_{max\_allow}$. The increased maximum *dv/dt* with the negative bias is $(dv/dt)'_{max\_allow}$ in (7.11):

$$(dv/dt)'_{max\_allow} \approx \frac{V_{th} + |V_z|}{(R_g + R_{g2} + R_{S2}) \cdot C_{gd}} \tag{7.11}$$

V_z can be derived from (7.11),

$$|V_z| \approx (R_g + R_{g2} + R_{S2}) \cdot C_{gd} \cdot (dv/dt)'_{max\_allow} - V_{th} \tag{7.12}$$

7.1.4.3 Design example of control HEMT driving circuit

EPC2001 is selected as the control HEMT, and its C_{gd} and C_{gs} are 850 pF and 20 pF, respectively. The design steps are as follows.

1. Step 1: Select R_{g2}
   R_{g2} is selected to ensure the driving circuit discharge loop critical-damped by (7.8). With the driving circuit layout in Figure 7.8, L_{g_tot} is extracted to be

3.7 nH. $R_{S1} = 2.2\ \Omega$, $R_{S2} = 0.2\ \Omega$, so R_{g1} and R_{g2} are calculated to be 1.7 Ω and 3.7 Ω, respectively, by (7.7) and (7.8). The nominal value of 2 Ω and 4 Ω are used, respectively.

2. Step 2: Determine the negative bias voltage
   $(dv/dt)_{max\_allow}$ is calculated to compare with $(dv/dt)_{max\_actual}$ to determine if the negative gate bias is required. From (7.10), the maximum dv/dt is calculated as 16.3 V/ns with $R_{g2} = 4\ \Omega$, $R_{S2} = 0.2\ \Omega$, $R_g = 0.3\ \Omega$, $V_{th} = 1.4$ V, and $C_{gd} = 20$ pF.

In the isolated resonant SEPIC converter, the maximum dv/dt of $(dv/dt)_{max\_actual}$ increases with the input voltage increased. $(dv/dt)_{max\_actual}$ is 3.1 V/ns under the maximum 24 V input voltage, which is smaller than $(dv/dt)_{max\_allow}$. Therefore, the negative voltage bias is not required in this case.

Finally, the designed eGaN control HEMT driving circuit parameters are listed in Table 7.2.

Figure 7.11 shows the loss distribution comparison with and without the mid-level voltage under full-load condition when $V_{in} = 24$ V. The calculated power loss using driving circuits with and without the mid-level voltage are 3.24 W and 3.35 W, which correspond to the efficiency of 75.5% and 74.9%, respectively.

7.1.5 Rectifier mathematic modeling and design

7.1.5.1 Determine component initial value in wave-shape circuit

The wave-shape circuit functions in Figure 7.5 are as follows: (1) reduce the sampled voltage amplitude into the comparator input range, and (2) phase shift

Table 7.2 Driving circuit component values

R_{g1}/R_{g2}	2/4 Ω	Driving IC	LM5114 (TI)
S_3	MMBT2222A (NXP)	S_4	PMBT2907A (NXP)

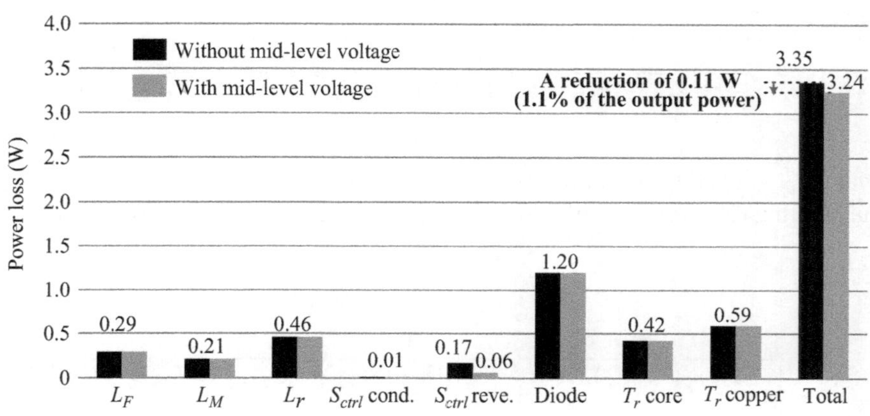

Figure 7.11 Loss distribution comparison

v_{ds_SR} to satisfy the accurate SR signal timing requirement. A simple second-order RC filter is a good choice to reduce the amplitude and shift v_{ds_SR} phase to obtain v_{C2} in Figure 7.6 and the transfer function is

$$H(i\omega) = \frac{v_{C2}(i\omega)}{v_{ds\_SR}(i\omega)} = \frac{R_2}{R_1 + R_2} \cdot \frac{1}{(i\omega)^2 (R_1 \| R_2) R_3 C_1 C_2 + (i\omega)(R_1 \| R_2)(C_1 + C_2) + (i\omega) R_3 C_2 + 1} \tag{7.13}$$

Since the second-order RC filter transfer function is mostly determined by RC network cutoff frequency, R_1, R_2, R_3, C_1, and C_2 play the similar role of reducing the amplitude and shifting the phase angle. Therefore, tuning C_2 solely can vary the reduced amplitude and phase shift angle of v_{ds_SR}. It is recommended to remain the other components value unchanged and select C_2 to simplify the design. To decide the other components initial value, the following constraint should be considered:

$$v_{C2\_max} = v_{C2\_avg} + \frac{1}{2} v_{C2\_ac} < V_{dd} \tag{7.14}$$

where V_{dd} is the comparator supply voltage, v_{C2_avg} is the average voltage of v_{C2} and v_{C2_ac} is the swing voltage amplitude. The formula means that the voltage across C_2 should be smaller than the comparator supply voltage. In Figure 7.5, the average voltage across L_r is zero according to voltage-second balance, so v_{ds_SR} average voltage is V_{out}. v_{ds_SR} average voltage is divided by R_1 and R_2, so (7.15) can be obtained:

$$v_{C2\_avg} = \frac{R_2}{R_1 + R_2} \cdot V_{out} \tag{7.15}$$

The v_{C2} swing voltage amplitude in (7.16) can be derived from (7.13).

$$v_{C2\_ac} = \frac{R_2}{R_1 + R_2} \cdot \left| \frac{1}{-\omega_s^2 (R_1 \| R_2) R_3 C_1 C_2 + i\omega_s (R_1 \| R_2)(C_1 + C_2) + i\omega_s R_3 C_2 + 1} \right| \cdot V_{ds\_SR1} \tag{7.16}$$

where V_{ds_SR1} is the v_{ds_SR} fundamental amplitude and ω_s is the angular frequency. Substituting (7.15) and (7.16) into (7.14), the constraint can be expressed in (7.17).

$$\frac{R_2}{R_1 + R_2} \cdot \left(V_{out} + \frac{1}{2} \left| \frac{1}{-\omega_s^2 (R_1 \| R_2) R_3 C_1 C_2 + i\omega_s (R_1 \| R_2)(C_1 + C_2) + i\omega_s R_3 C_2 + 1} \right| \cdot V_{ds\_SR1} \right) < V_{dd} \tag{7.17}$$

However, with the above constraint, there are still a large number of possible solutions. A set of the initial values of R_1, R_2, R_3, and C_1 is determined as shown in Table 7.3.

After determining the initial values of R_1, R_2, R_3, and C_1, the next step is to select the value of C_2 and V_{ref}. They can be determined to satisfy: (1) the SR gate should turn on simultaneously at the instant when the SR drain voltage resonates to zero, and (2) the driving signal duty ratio should meet the SR conduction ratio.

7.1.5.2 Rectifier mathematic modeling

Figure 7.12 shows the mathematic algorithm to obtain C_2 and V_{ref}. A set of C_2 is selected to calculate v_{C2} with the $v_{ds\_SR}(t)$ expression. Then a set of V_{ref} is selected to solve the equation $v_{C2}(t) = V_{ref}$. Therefore, the relationship among φ_s, D, C_2, and V_{ref} can be presented graphically by the mathematic modeling with the MATLAB® numerical calculation. Finally, the desired φ_s and D determined by the converter parameters can reversely induce the desired C_2 and V_{ref}.

In order to obtain $v_{C2}(t)$ filtered from $v_{ds\_SR}(t)$, $v_{ds\_SR}(t)$ should be derived. Figure 7.13 shows the HEMT rectifier equivalent circuit. The output is treated as the constant voltage source V_{out} and the input is treated as the sinusoidal voltage

Table 7.3 Component initial values

R_1	1 kΩ	R_2	1.5 kΩ
R_3	1 kΩ	C_1	56 pF

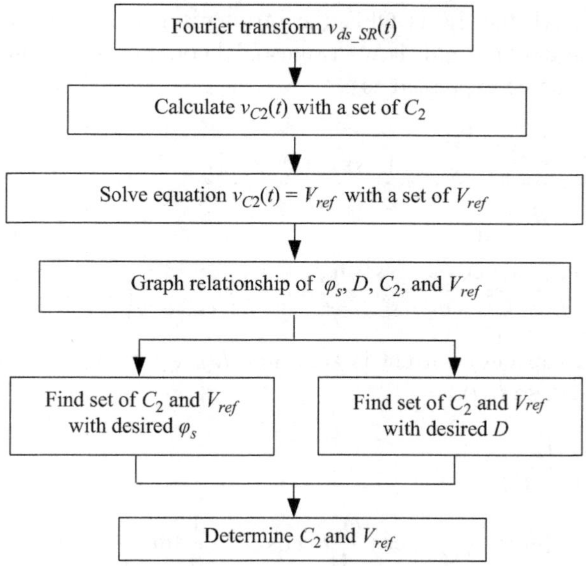

Figure 7.12 Algorithm to obtain C_2 and V_{ref}

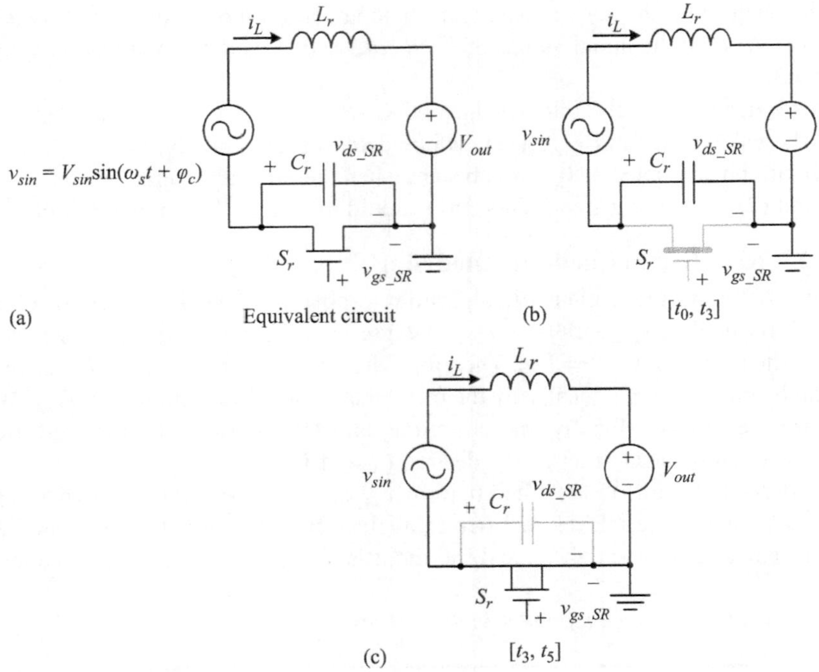

Figure 7.13 HEMT rectifier equivalent circuit

source $v_{sin} = V_{sin}\sin(\omega_s t + \varphi_c)$. φ_c is the phase angle between the initial phase of v_{ds_SR} and the zero crossing point of v_{sin} in Figure 7.14. V_{sin} is the v_{sin} amplitude.

During $[t_0, t_3]$, the SR is OFF, the series resonant circuit is driven by the sinusoidal voltage source and shows a sinusoidal current and voltages of L_r and C_r. The L_r current can be expressed as

$$I_L(s) = \frac{V_{sin}(s) - \dfrac{V_{out}}{s}}{sL + \dfrac{1}{sC}} = \frac{1}{L}\cdot\frac{sV_{sin}(s) - V_{out}}{s^2 + \omega_0^2}$$
$$= \frac{V_{sin}}{L}\cdot\frac{s\omega_s\cos\varphi_c - s^2\sin\varphi_c}{(s^2+\omega_s^2)(s^2+\omega_0^2)} - \frac{V_{out}}{L}\cdot\frac{1}{s^2+\omega_0^2} \tag{7.18}$$

At $\omega_s t = 0$, the capacitor current is zero and $dv_{ds\_SR}/dt = 0$. The inverse Laplace transform is used, so $i_L(t)$ is

$$i_L(t) = \frac{V_{sin}}{L}\cdot\frac{A^2}{(1-A^2)}$$
$$\cdot\frac{1}{\omega_s}\left(\cos(\omega_s t + \varphi_c) - \cos\frac{\omega_s}{A}t\cdot\cos\varphi_c + \frac{1}{A}\sin\frac{\omega_s}{A}t\cdot(\sin\varphi_c - M(1-A^2))\right) \tag{7.19}$$

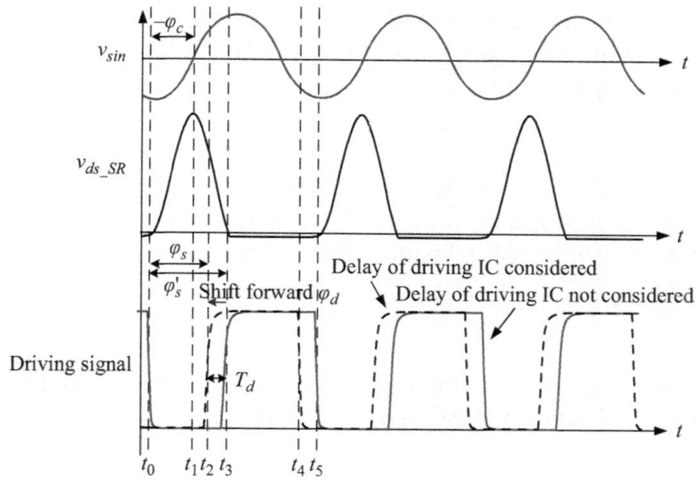

Figure 7.14 Phase relationship among v_{sin}, v_{ds_SR}, and driving signal

Therefore, the drain-to-source voltage can be obtained in (7.20):

$$v_{ds\_SR}(t) = \begin{cases} -V_o\left(\dfrac{1}{M(1-A^2)}\cdot\left(\sin(\omega_s t+\varphi_c)-A\sin\dfrac{\omega_s t}{A}\cos\varphi_c -A\cos\dfrac{\omega_s t}{A}\sin\varphi_c\right)+\cos\dfrac{\omega_s t}{A}-1\right) & 0<\omega_s t<2\pi(1-D) \\ 0 & 2\pi(1-D)<\omega_s t<\dfrac{2\pi}{\omega_s} \end{cases} \tag{7.20}$$

where $M = V_o/V_{sin}$, $A = \omega_s/\omega_0$, and $\omega_0 = 1/\sqrt{L_r C_r}$.

To calculate $v_{C2}(t)$ filtered from $v_{ds\_SR}(t)$ by the wave-shape circuit, $v_{ds\_SR}(t)$ is Fourier transformed and the first five items are extracted:

$$v_{ds\_SR}(t) = V_{ds\_SR\_0} + \sum_{k=1}^{5} V_{ds\_SR\_k}\cdot\cos(k\omega_s t+\theta_k) \tag{7.21}$$

where $V_{ds\_SR\_0}$ is the d.c. component of the $v_{ds\_SR}(t)$. $V_{ds\_SR\_k}$ and θ_k are the amplitude and phase angle of the kth harmonic voltage respectively. Then $v_{C2}(t)$ can be obtained with $v_{ds\_SR}(t)$ filtered by the wave-shape circuit:

$$v_{C2}(t) = V_{ds\_SR\_0}H(0) + \sum_{k=1}^{5} V_{ds\_SR\_k} \cdot |H(ki\omega_s)|\cdot\cos(k\omega_s t+\theta_k+\angle H(ki\omega_s)) \tag{7.22}$$

Solving (7.23) and the two solutions are the turn on and turn off instant of the obtained driving signal separately. They correspond to t_3 and t_5 in Figure 7.14 if the driving IC delay is not considered:

$$v_{C2}(t) = V_{ref} \tag{7.23}$$

The turn-on instant t_3 of the obtained driving signal without considering the driving IC delay corresponds to φ_s' in (7.24). It is the phase difference between v_{ds_SR} and the driving signal in Figure 7.14:

$$\varphi_s' = \frac{t_3 - t_0}{T_s} \times 360° \tag{7.24}$$

However, considering the driving IC delay T_d, the desired turn-on time should be shifted forward with the phase angle of φ_d in Figure 7.14:

$$\varphi_d = \frac{T_d}{T_s} \times 360° \tag{7.25}$$

Therefore, considering the driving IC delay, the desired phase angle φ_s is

$$\varphi_s = \varphi_s' - \varphi_d = \frac{t_2 - t_0}{T_s} \times 360° \tag{7.26}$$

and φ_s corresponds to the interval from t_0 to t_2 in Figure 7.14.

The eGaN SR duty ratio can be calculated:

$$D = \frac{t_5 - t_3}{T_s} \tag{7.27}$$

With a set of C_2 and V_{ref}, a corresponding set of φ_s and duty ratio D can be obtained with the mathematic model. With the desired φ_s and D, C_2 and V_{ref} can be derived.

7.1.5.3 Design example of C_2 and V_{ref}

The desired turn-on phase φ_s' and duty ratio are 181° and 0.49, respectively, depending on the converter parameters. With the driving IC propagation delay (include the comparator delay) of 23 ns, $\varphi_d = 58°$ from (7.25). Therefore, the desired φ_s is calculated to be 123° from (7.26).

Figure 7.15 shows the relationship among C_2, V_{ref}, and φ_s and Figure 7.16 shows the relationship among C_2, V_{ref}, and the duty ratio with the MATLAB® script. In Figure 7.15, the intersection line is the set of C_2 and V_{ref} corresponding to the desired $\varphi_s = 123°$. In Figure 7.16, the intersection line is the set of C_2 and V_{ref} corresponding to the desired $D = 0.49$. Figure 7.17 shows two sets of C_2 and V_{ref} corresponding to desire φ_s and D. The desired C_2 and V_{ref} are the intersection points. They are 36 pF and 3.05 V, respectively. Finally, the designed SR driving circuit parameters are given in Table 7.4.

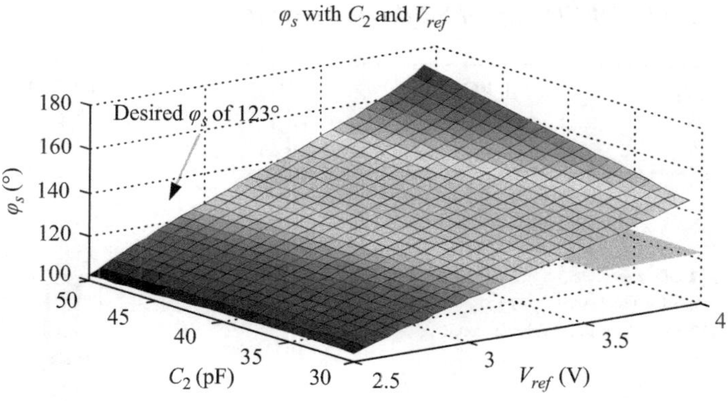

Figure 7.15 Relationship among C_2, V_{ref}, and φ_s

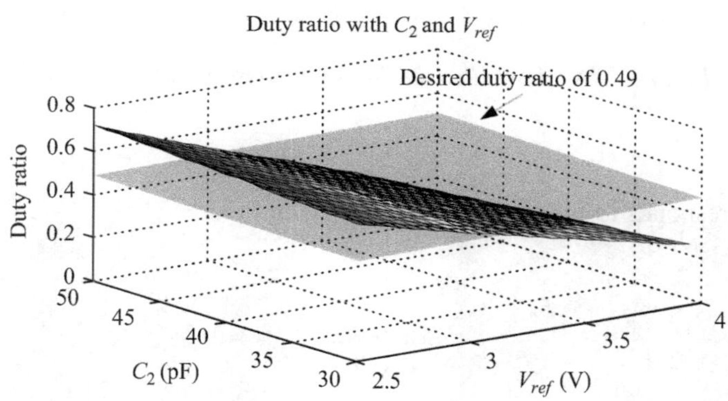

Figure 7.16 Relationship among C_2, V_{ref}, and D

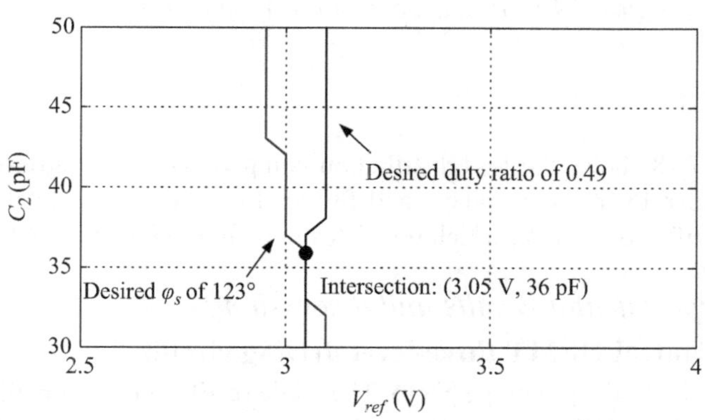

Figure 7.17 Desired φ_s and D

Table 7.4 Designed SR driving circuit parameters

R_1	1 kΩ	R_2	1.5 kΩ
R_3	1 kΩ	C_1	56 pF
C_2	36 pF	V_{ref}	3.05 V

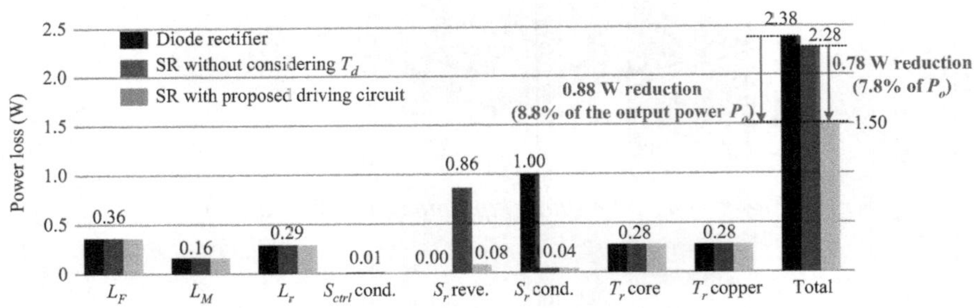

Figure 7.18 Loss distribution comparison under full load

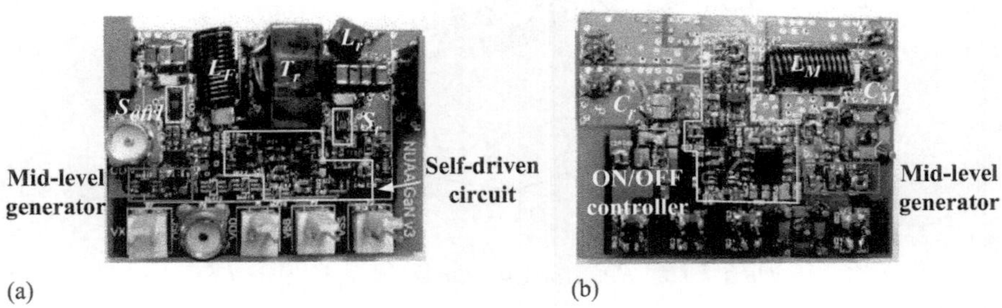

Figure 7.19 Prototype photo: (a) top and (b) bottom

Figure 7.18 shows the loss distribution comparison among different rectifier schemes under 18 V input voltage and full-load condition. The total power loss with SR rectifier is 1.50 W, which translates into an efficiency of 86.9%.

7.1.6 Experimental results and discussion

7.1.6.1 Control HEMT three-level driving circuit

Figure 7.19 shows the prototype photo. The diode rectification is initially used with PMEG4020 from NXP Corporation. The power-stage parameters are listed in Table 7.1, except that C_r is 4 nF with two diodes PMEG4020 and 3 nF with eGaN HEMT EPC2015. The driving circuit parameters are calculated in Section 7.2.4 and given in Table 7.2.

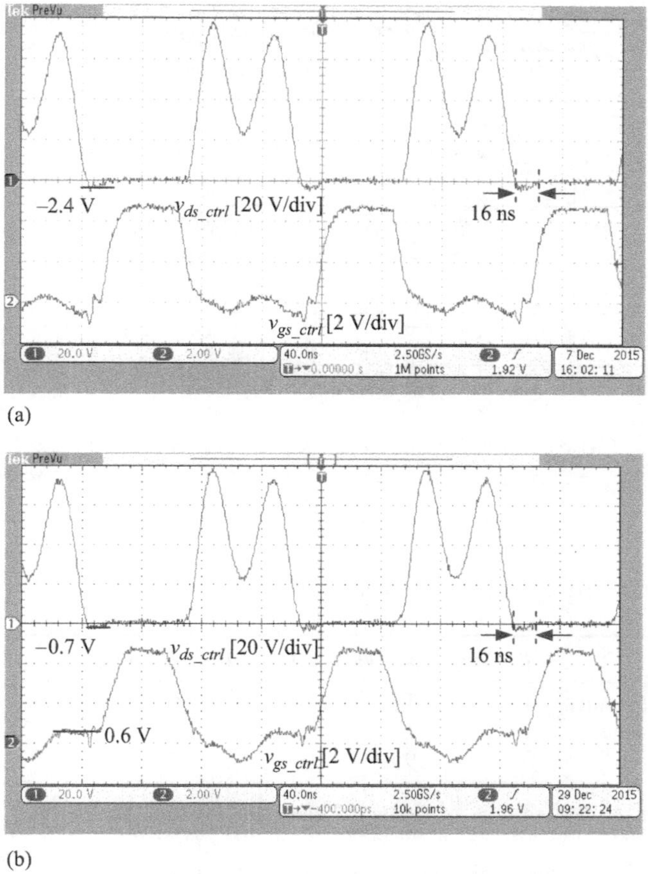

Figure 7.20 Control HEMT comparison: $V_{in} = 24$ V, $P_{out} = 10$ W, and $f_s = 7$ MHz; (a) without mid-level voltage and (b) with mid-level voltage

Figure 7.20(a) shows v_{gs_ctrl} and v_{ds_ctrl} without mid-level voltage under 24 V input voltage. The reverse conduction time is around 16 ns (11.2% of T_s) and reverse conduction voltage is about −2.4 V before ZVS turn-on. The reverse conduction loss is significant and should be minimized. As comparison, Figure 7.20(b) shows v_{gs_ctrl} and v_{ds_ctrl} with the mid-level voltage under 24 V input voltage. The reverse conduction time keeps around 16 ns. The positive bias around 0.6 V is generated on v_{gs_ctrl} before ZVS turn-on to reduce the reverse conduction voltage to −0.7 V. It means that the reverse conduction voltage is reduced of 71% by the positive bias.

Figure 7.21 shows closed-loop efficiency comparison between with and without mid-level voltage under 24 V input voltage. Under full-load condition, the measured efficiency without mid-level voltage is 72.7%, while with mid-level

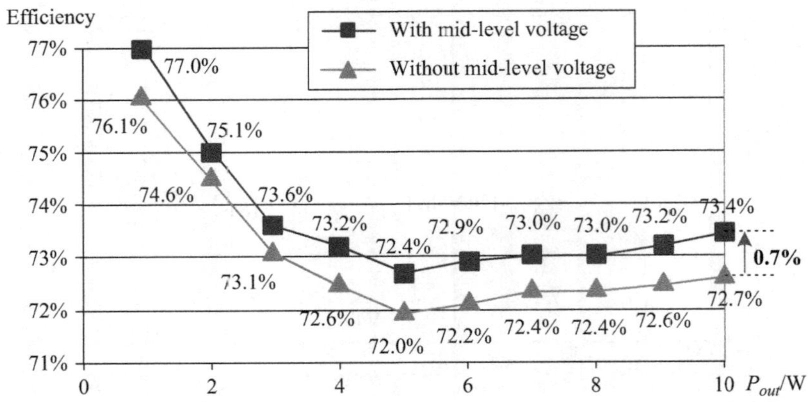

Figure 7.21 Closed-loop efficiency comparison: $V_{in} = 24$ V, $V_{out} = 5$ V, and $f_s = 7$ MHz

Table 7.5 SR driving circuit components value

	R_1 (kΩ)	R_2 (kΩ)	R_3 (kΩ)	C_1 (pF)	C_2 (pF)	V_{ref} (V)
Experimental results	1	1.5	1	56	39	3.1
Calculated results	1	1.5	1	56	36	3.05
Difference	0%	0%	0%	0%	7.7%	0.2%

voltage is 73.4%. The efficiency is improved of 0.7% by reducing the control HEMT reverse conduction voltage before ZVS turn-on.

7.1.6.2 SR HEMT driving circuit

Table 7.5 compares the experimental and calculated SR HEMT driving circuit value. The maximum difference appears in C_2 with 7.7%, which is acceptable because the capacitance tolerance is around 5% normally. The experimental parameters match well with calculated component value.

Figure 7.22 shows v_{gs_SR} and v_{ds_SR} using the proposed SR driving circuit. It can be observed that the SR driving signal matches well with the SR drain voltage. The driving IC delay is 23 ns (including the comparator). The desired phase shift $\varphi_s = 123°$ and duty ratio $D = 0.49$ are both obtained with the experimental component value in Table 7.5. The eGaN SR HEMT turns on when v_{ds_SR} resonates to zero at $t = t_2$, so that ZVS is achieved and reverse conduction mechanism is not triggered. As comparison, in Figure 7.23, the driving signal sets when v_{ds_SR} resonates to zero at $t = t_2$ without considering T_d. However, v_{gs_SR} turns on at $t = t_3$ due to T_d and the reverse conduction voltage is −1.9 V during from t_2 to t_3. The SR

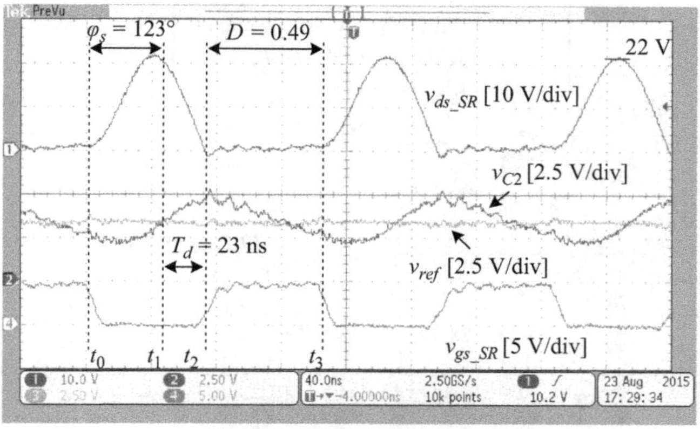

Figure 7.22 SR waveforms: $V_{in} = 18$ V, $P_{out} = 10$ W, and $f_s = 7$ MHz

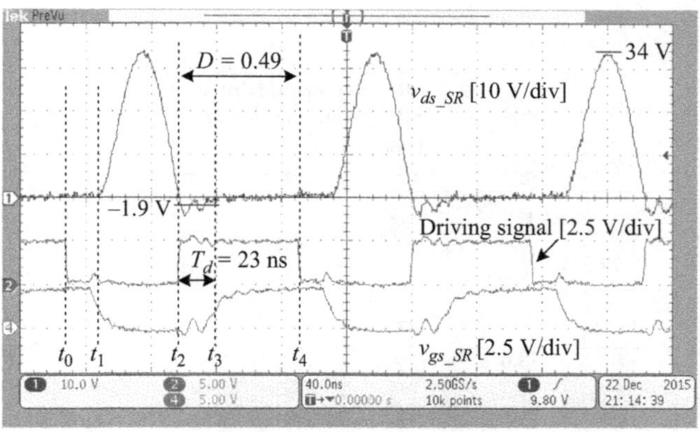

Figure 7.23 Without considering T_d: $V_{in} = 18$ V, $P_{out} = 10$ W, and $f_s = 7$ MHz

HEMT voltage stress is increased due to reduced resonant period from t_1 to t_2. v_{ds_SR} resonates to 34 V, which is much higher than 22 V in Figure 7.22.

Figure 7.24 shows closed-loop efficiency comparison among different rectifier schemes. Under 18 V input voltage and full-load conditions, the efficiency is increased from 79.5% using the diode rectifier to 79.9% without considering T_d, and is increased to 84.7% considering T_d (an improvement of 5.2%). On the other hand, the efficiency using the SR driving circuit without considering T_d is close to that using the bad performance diode rectifier because of the long SR reverse conduction time at multi-MHz.

Figure 7.25 shows the closed-loop efficiency comparison between different rectifier schemes. The converter with the SR is optimized under 18 V input voltage with a peak efficiency of 84.7% under full-load condition.

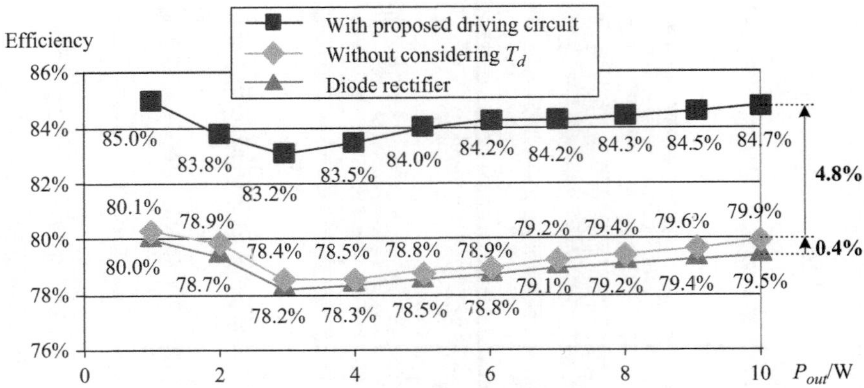

Figure 7.24 Closed-loop efficiency comparison: $V_{in} = 18$ V, $V_{out} = 5$ V, and $f_s = 7$ MHz

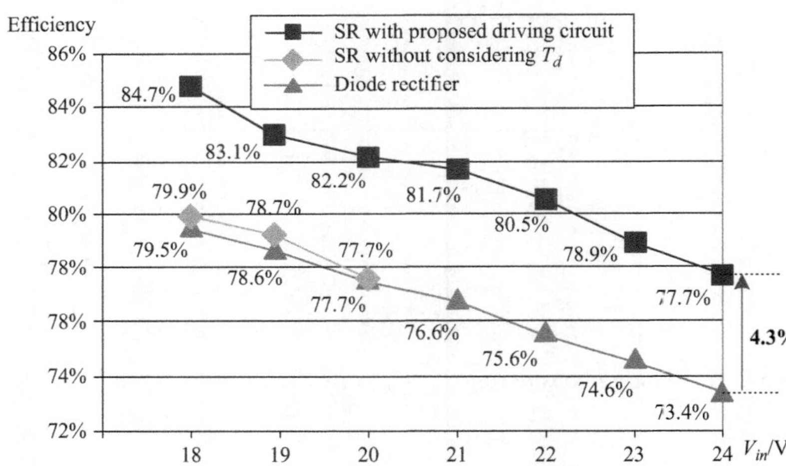

Figure 7.25 Closed-loop efficiency comparison: $V_{out} = 5$ V, $P_{out} = 10$ W, and $f_s = 7$ MHz

7.2 A digital adaptive driving scheme for eGaN HEMTs in VHF converters

7.2.1 Introduction

Very high frequency (VHF, 30–300 MHz) converters with reduced size, weight, and cost, as well as extremely fast dynamic performance have attracted many

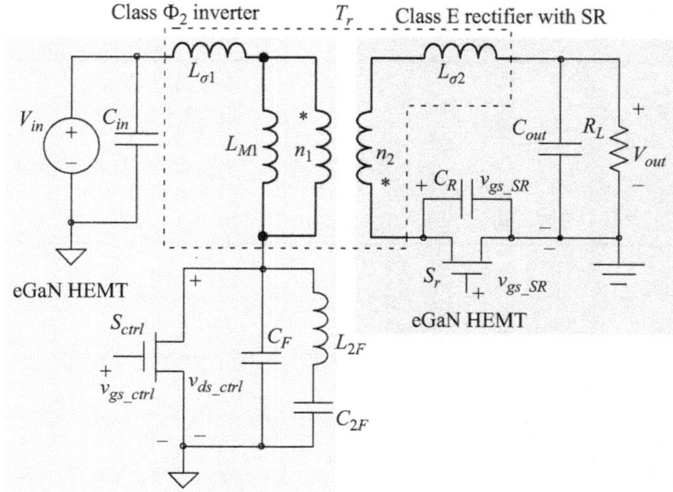

Figure 7.26 VHF class Φ_2 resonant flyback

research efforts. With excellent Figure of Merit, eGaN HEMTs are quite suitable in VHF converters. However, since there is no body-diode with eGaN HEMTs, the reverse current has to travel via the channel by the reverse conduction mechanism, resulting in much higher reverse conduction voltage (typically 1.9 V of EPC2036) compared to that of a Si MOSFET (0.7 V). Therefore, it is important to avoid this high reverse conduction loss.

VHF converters are normally designed and optimized with fixed driving signal under minimal input voltage so that high efficiency can be maintained over wide-load range owing to ON/OFF hysteretic control [17–19]. However, the control and SR HEMT gate-drive timing, including the phase relationship and duty ratio, varies widely over entire input voltage range. With the conventional driving scheme, the soft switching condition may lose and reverse conduction mechanism may be triggered. Due to high reverse conduction voltage, reverse conduction loss results in lower efficiency seriously. Therefore, the SR VHF converters can hardly maintain high efficiency over wide input voltage range.

To solve the reverse conduction and hard switching problem, a voltage following control is proposed in [12]. It is derived quantitatively that the switches gate timing relationship is fixed when the output voltage is proportional to the input voltage. Optimum switching condition can be maintained over wide input voltage range as DCXs. But this control scheme is limited to DCX applications while not suitable for the regulated output converters. The switch HEMTs are used in a regulated class Φ_2 resonant converter in [11]. The gate-drive timing is implemented with a RC delay circuit and tuned iteratively to match the simulation results. However, the desired gate-drive timing in class Φ_2 resonant converters has not been analyzed quantitatively yet. There are still no reported literature investigating the relationship quantitatively between the gate-drive timing and input voltage.

Table 7.6 Power-stage component values

$L_{\sigma 1}$	7.3 nH	C_F	150 pF (AVX)
L_{M1}	45.1 nH	C_{2F}	80 pF (AVX)
$L_{\sigma 2}$	8.9 nH	C_R	1 nF (AVX)
n_1/n_2	3/1	C_{in}	100 nF × 10 + 10 μF × 1 (Murata)
S_{ctrl}	EPC2036 (EPC, 100 V, 1 A)	C_{out}	100 nF × 8 + 10 μF × 1 (Murata)
S_r	EPC2014 (EPC, 40 V, 10 A)	L_{2F}	68 nH (Coilcraft)

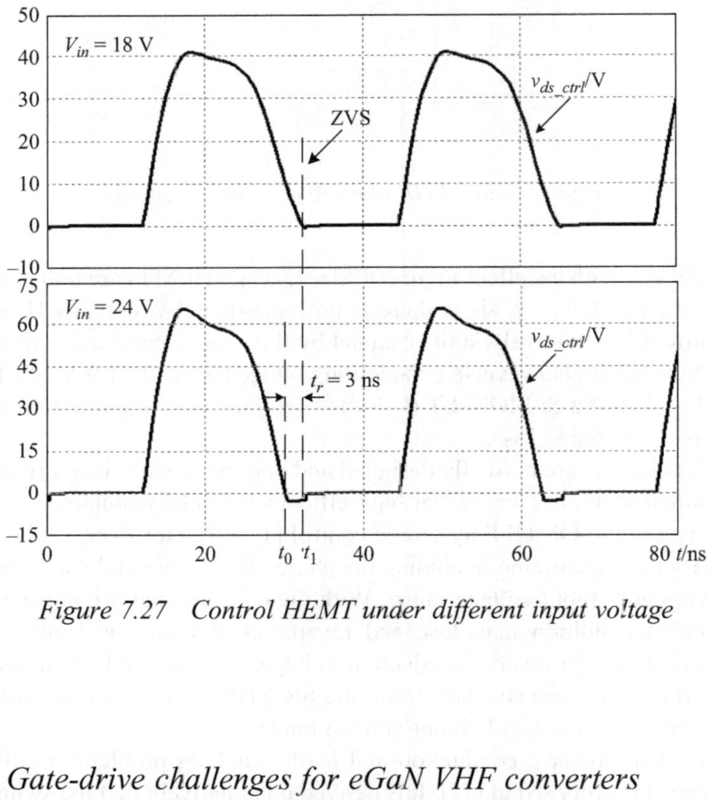

Figure 7.27 Control HEMT under different input voltage

7.2.2 Gate-drive challenges for eGaN VHF converters

7.2.2.1 VHF class Φ_2 resonant flyback converter

Figure 7.26 shows the VHF class Φ_2 resonant flyback with eGaN HEMTs. The converter consists of the class Φ_2 inverter, class E rectifier, and coreless transformer. The benefit is that the transformer absorbs the inductance $L_{\sigma 1}$, L_{M1}, and $L_{\sigma 2}$. Moreover the printed-circuit techniques ensure high uniformity [20].

The detailed power-stage design procedure can be found in [20]. With $V_{in} = 18$ V, $V_{out} = 5$ V, $I_{out} = 2$ A, and $f_s = 30$ MHz, the parameters are given in Table 7.6. The following analysis and experimental verification are regarding to the converter in Figure 7.26.

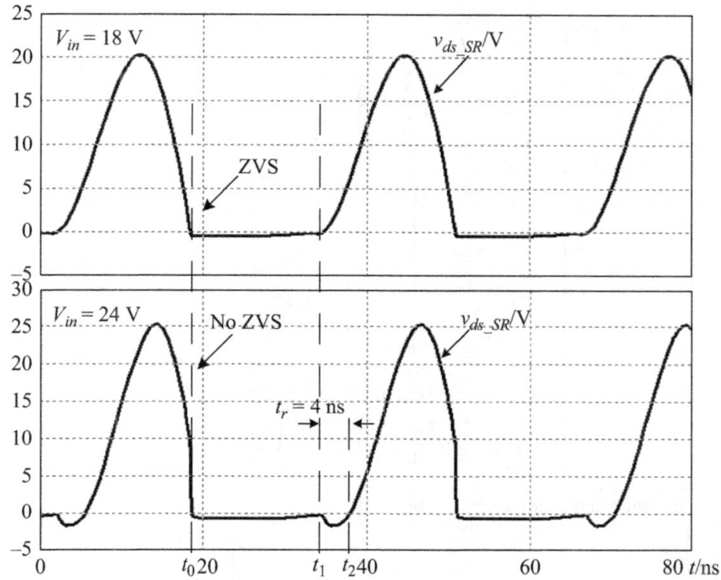

Figure 7.28 SR under different input voltage

7.2.2.2 Gate-drive timing challenge over input voltage range

Figure 7.27 shows the control HEMT drain waveforms under different input voltage. When $V_{in} = 18$ V, the control HEMT is optimized to turn on when v_{ds_ctrl} resonates to zero at $t = t_1$ and reverse conduction mechanism is not triggered. However when V_{in} increases to 24 V, v_{ds_ctrl} resonates to zero at $t = t_0$ and the control HEMT turns on under ZVS at $t = t_1$. Then, the reverse conduction mechanism is triggered and reverse conduction time t_r is 3 ns during $[t_0, t_1]$ (9% of $T_s = 33.3$ ns). The reverse conduction loss is 0.12 W (1.2% of the output power of 10 W) and should be minimized.

Figure 7.28 shows the SR drain waveforms under different input voltage. When $V_{in} = 18$ V, the SR is optimized to turn on when v_{ds_SR} resonates to zero at $t = t_0$ and reverse conduction mechanism is not triggered. However, when V_{in} increases to 24 V, the SR turns on at $t = t_0$ and hard switching happens. Then, the reverse conduction mechanism is triggered and reverse conduction time t_r is 4 ns during $[t_1, t_2]$ (12% of $T_s = 33.3$ ns). The calculated reverse conduction loss is 0.17 W (1.7% of the output power of 10 W). The high switching and reverse conduction loss hurts the efficiency seriously with the input voltage increased.

In summary, the challenge to drive the eGaN control and SR HEMTs is to: (1) analyze the relationship between the desired gate-drive timing and input voltage in high order and nonlinear resonant converters, and (2) generate the gate-drive signals with desired timing precisely according to input voltage to minimize the reverse conduction and switching loss.

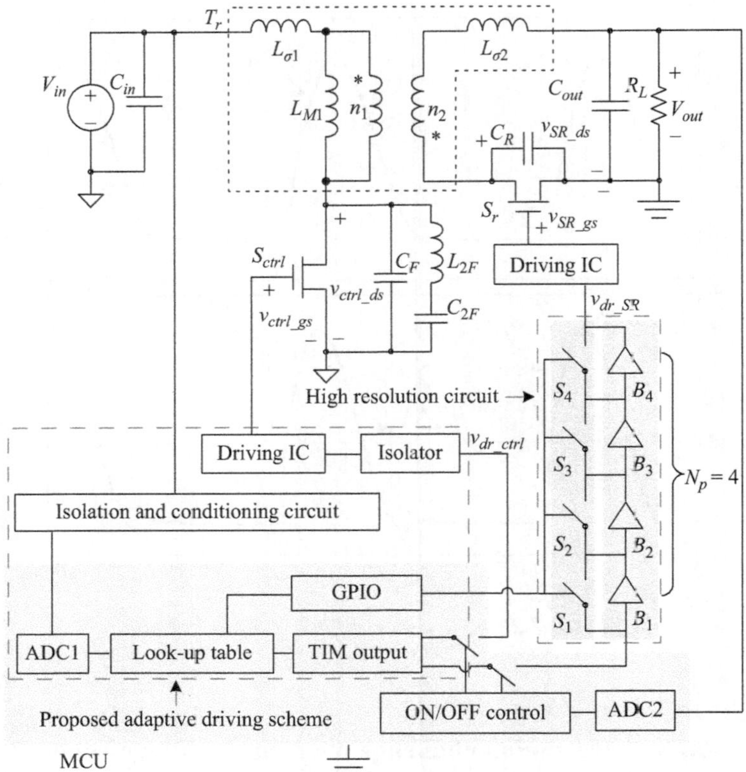

Figure 7.29 Proposed digital adaptive driving scheme

7.2.3 Proposed digital adaptive driving scheme

7.2.3.1 Implementation of digital adaptive driving circuit

Figure 7.29 shows the proposed digital adaptive driving scheme. The proposed driving scheme adapts gate-drive signals under different input voltage according to the gate-drive timing predicted by the converter state-space model. Therefore, both eGaN control and SR HEMTs can be driven with precise timing to minimize switching and reverse conduction loss over entire input voltage range. The relationship between the desired gate-drive timing and input voltage is predicted by the converter state-space model and stored in the Microcontroller Unit (MCU). The MCU samples the input voltage, looks up the relationship table, and generates the corresponding gate-drive signals by Timer (TIM) output module. It should be pointed that the driving IC propagation delay of the control and SR HEMTs counteract with each other and isolator propagation delay of the control HEMT can be compensated in the relationship table.

ARM STM32F746VG from ST Corporation with high CPU frequency is selected to obtain high-resolution gate-drive signals. It can operate up to 216 MHz and the time resolution T_m is 4.7 ns. However, the highest time resolution of 4.7 ns

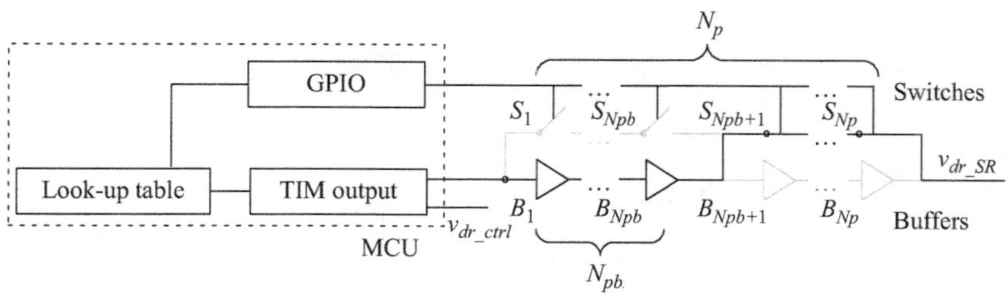

Figure 7.30 Delay selection in high time resolution circuit

is still 14.1% of one switching period $T_s = 33.3$ ns ($f_s = 30$ MHz), which means that the reverse conduction time below 4.7 ns cannot be reduced due to limited time resolution. This still causes relatively high reverse conduction loss. Therefore, increasing the gate-drive signal resolution is a big challenge to implement the driving scheme and further improve the efficiency.

7.2.3.2 Proposed high time resolution circuit

Figure 7.30 shows the proposed high time resolution circuit. It is a variation of the delay line technique in [21]. Following the MCU TIM output, the ultra-high speed buffers are cascaded to be selected to delay the gate-drive signal with smaller time step so that the time resolution can be improved. The buffer propagation delay T_d is much smaller than MCU time resolution T_m. Each buffer is in parallel with a near-zero propagation delay switch. The switches can be controlled by MCU General Purpose Input/Output (GPIO) to determine whether the buffer delay is required to tune SR gate-drive timing. The buffer number N_p is calculated as

$$N_p = \frac{T_m}{T_d} - 1 \tag{7.28}$$

Supposed the phase difference between the control and SR HEMTs is T_p, it can be expressed by:

$$T_p = N_{pm} \cdot T_m + N_{pb} \cdot T_d \tag{7.29}$$

where N_{pm} is the delay number needed in the MCU TIM module and N_{pb} is the delay number needed in the high time resolution circuit. The MCU would select the buffer number to N_{pb} in the high time resolution circuit from GPIO. Figure 7.30 shows the high time resolution circuit with the delay of $N_{pb} \cdot T_d$. The MCU switches S_1 to S_{Npb} off to make buffer B_1 to B_{Np} delay the input SR gate-drive signal, meanwhile turn on S_{Npb+1} to S_{Np} to bypass B_{Npb+1} to B_{Np}. The delay of $N_{pb} \cdot T_d$ can be obtained by the high time resolution circuit. It should be noted that the switches S_1 to S_{Np} can be controlled simultaneously by MCU GPIO. The minimum step change of gate timing is the switching transition time (typically less than 5 ns).

In the experiment verification, NC7SZ126 from Fairchild Corporation with around 1 ns propagation delay is selected as the buffers and SN74CBT3305C from TI Corporation is selected as the switches. The calculated N_p in (7.28) is 4 and the

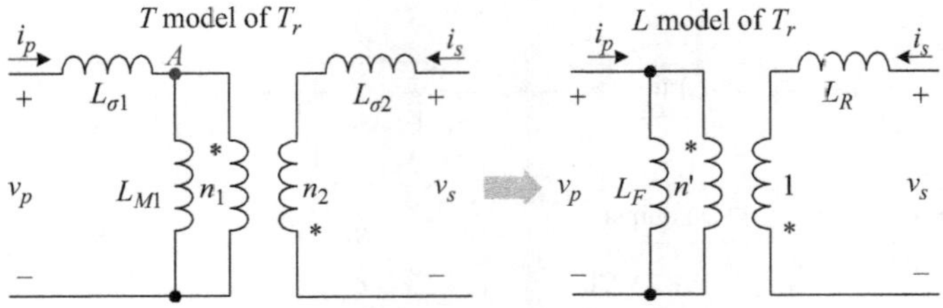

Figure 7.31 Transformation of T model to L model

timing signal resolution is increased to 1 ns (an improvement of 78.7%), which is about 3% of 33.3 ns (one switching period T_s).

7.2.4 Relationship analysis between gate-drive timing and input voltage

7.2.4.1 State-space modeling of VHF class Φ_2 resonant flyback converter

L *transformer model*

From Figure 7.31, a pure inductance cutset at node *A* is formed by $L_{\sigma1}$, $L_{\sigma2}$, and L_{M1} of *T* transformer model. To remove the pure inductance cutset, *T* model is transformed to *L* model. The total number of the inductance and capacitance is the circuit order. L_F, L_R, and n' in *L* model are calculated as

$$L_F = L_{\sigma1} + L_{M1} \tag{7.30}$$

$$L_R = L_{\sigma2} + \left(\frac{n_2}{n_1}\right)^2 (L_{M1} \| L_{\sigma1}) \tag{7.31}$$

$$n' = \left(\frac{n_1}{n_2}\right)\left(1 + \frac{L_{\sigma1}}{L_{M1}}\right) \tag{7.32}$$

State-space representation

Figure 7.32 shows the VHF flyback equivalent circuit with *L* model. r_p and r_s are the series equivalent transformer resistance. r_{ctrl} and r_{SR} represent the control and SR HEMTs. The load is modeled by a d.c. voltage source V_{out} due to ON/OFF control.

The equivalent circuit can be analyzed by the following general state-space representation:

$$\dot{X}(\omega t) = AX(\omega t) + BU(\omega t) \tag{7.33}$$

where *X*(ω*t*) is the state vector and contains six voltage and current states variables as

$$X(\omega t) = [v_{CF}(\omega t) \quad v_{CR}(\omega t) \quad v_{C2F}(\omega t) \quad i_{LF}(\omega t) \quad i_{LR}(\omega t) \quad i_{L2F}(\omega t)]^T \tag{7.34}$$

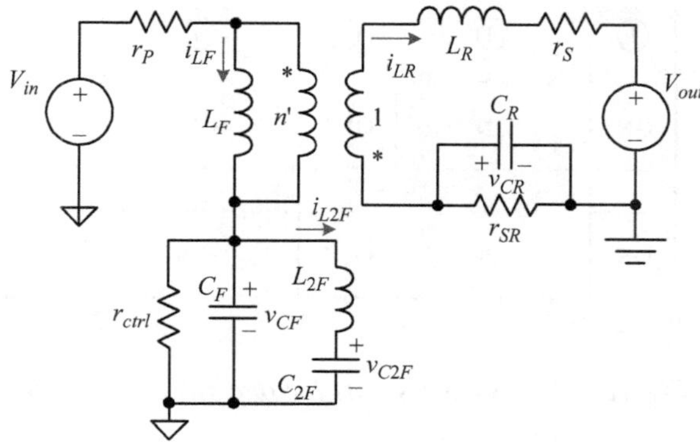

Figure 7.32 Equivalent circuit model

and $U(\omega t)$ is the input vector, which equals to a unit step function:

$$U(\omega t) = [\,V_{in} \quad V_{out}\,]^T \tag{7.35}$$

Using the node method, the matrices A and B are given

$$A = \begin{bmatrix} -\dfrac{1}{\omega C_F \cdot r_{ctrl}} & 0 & 0 & \dfrac{1}{\omega C_F} & -\dfrac{1}{n' \cdot \omega C_F} & -\dfrac{1}{\omega C_F} \\ 0 & -\dfrac{1}{\omega C_R \cdot r_{SR}} & 0 & 0 & -\dfrac{1}{\omega C_R} & 0 \\ 0 & 0 & 0 & 0 & 0 & \dfrac{1}{\omega C_{2F}} \\ -\dfrac{1}{\omega L_F} & 0 & 0 & -\dfrac{r_P}{\omega L_F} & \dfrac{r_P}{n' \cdot \omega L_F} & 0 \\ \dfrac{1}{n' \cdot \omega L_R} & \dfrac{1}{\omega L_R} & 0 & \dfrac{r_P}{n' \cdot \omega L_R} & -\dfrac{r_P}{n'^2 \cdot \omega L_R} - \dfrac{r_S}{\omega L_R} & 0 \\ \dfrac{1}{\omega L_{2F}} & 0 & -\dfrac{1}{\omega L_{2F}} & 0 & 0 & -\dfrac{r_{2F}}{\omega L_{2F}} \end{bmatrix} \tag{7.36}$$

$$B = \begin{bmatrix} 0 & 0 & 0 & \dfrac{1}{\omega L_F} & -\dfrac{1}{n' \cdot \omega L_R} & 0 \\ 0 & 0 & 0 & 0 & -\dfrac{1}{\omega L_R} & 0 \end{bmatrix}^T \tag{7.37}$$

Switching intervals and operating modes

Figure 7.33 shows four modes in one switching cycle depending on the switches status. The switches are either ON represented by ON resistances, or OFF represented by OFF resistance. The converter has four operation modes, which share the same equivalent circuit in Figure 7.32. The difference is that r_{ctrl} and r_{SR} depend on

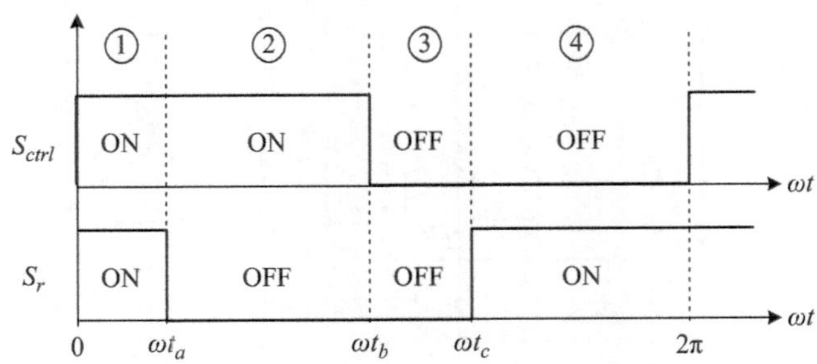

Figure 7.33 Switch status for one switching cycle

Table 7.7 Operation modes definition

Mode	Time	r_{ctrl}	r_{SR}
□	$0 < \omega t \leq \omega t_a$	r_{ON1}	r_{ON2}
□	$\omega t_a < \omega t \leq \omega t_b$	r_{ON1}	∞
□	$\omega t_b < \omega t \leq \omega t_c$	∞	∞
□	$\omega t_c < \omega t \leq 2\pi$	∞	r_{ON2}

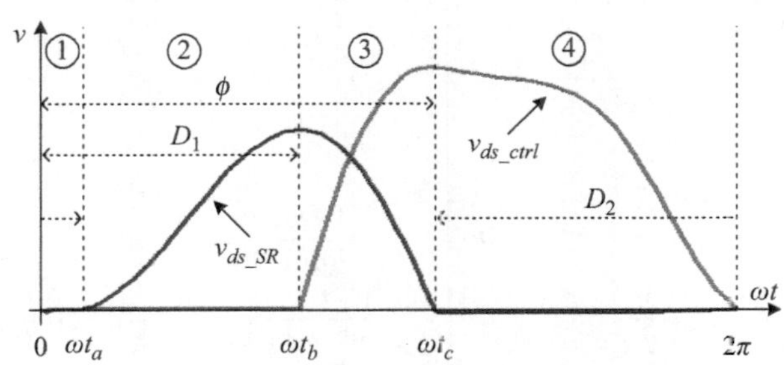

Figure 7.34 Control and SR HEMT drain node waveforms

switches status like ON or OFF. In Table 7.7, the resistances under four operation modes are defined. The switching period begins from Mode ①. In this mode, both HEMTs are on, and r_{ctrl} and r_{SR} are represented by ON resistances r_{ON1} and r_{ON2} respectively. At $\omega t = \omega t_a$, the converter switched into Mode ②. The SR turns off and the equivalent resistance is infinite. At $\omega t = \omega t_b$, the converter switched into Mode ③ and the control HEMT turns off. At $\omega t = \omega t_c$, the converter switched to Mode ④. The SR turns on and control HEMT remains off. Four operation modes lead to four different A matrices.

Figure 7.34 shows the control and SR HEMTs drain node waveforms. The control and SR HEMTs duty ratio D_1 and D_2 are

$$D_1 = \frac{\omega t_b}{2\pi} \tag{7.38}$$

$$D_2 = \frac{2\pi + \omega t_a - \omega t_c}{2\pi} \tag{7.39}$$

and the phase difference $\mathbf{\Phi}$ between D_1 and D_2 is defined as

$$\phi = \omega t_c \tag{7.40}$$

Determining states initial condition

The general solution to (7.33) is

$$X(\omega t) = X_n(\omega t) + X_f(\omega t) \tag{7.41}$$

where function $X_n(\omega t)$ is the natural response matrix, or zero-input response matrix. It is equal to

$$X_n(\omega t) = e^{A\omega t} \cdot X(0) \tag{7.42}$$

where e is the matrix exponential function and $X(0)$ is the initial condition matrix.

Function $X_f(\omega t)$ is the forced response matrix, or zero-state response matrix. It is equal to

$$X_f(\omega t) = \omega \int_0^t e^{A\omega(t-\tau)} \cdot B \cdot U(\omega\tau)d\tau = A^{-1} \cdot (e^{A\omega t} - I) \cdot B \cdot U \tag{7.43}$$

where I is the 6×6 identity matrix.

Substituting (7.42) and (7.43) to (7.41), the general solution of (7.33) is

$$X(\omega t) = e^{A\omega t} \cdot X(0) + A^{-1} \cdot (e^{A\omega t} - I) \cdot B \cdot U \tag{7.44}$$

The initial condition matrix $X(0)$ can be determined from the continuity condition of the voltage and current as the converter switches from the previous mode to next as follows:

$$X_{②}(0) = X_{①}(\omega t_a) \tag{7.45}$$

$$X_{③}(0) = X_{②}(\omega t_b - \omega t_a) \tag{7.46}$$

$$X_{④}(0) = X_{③}(\omega t_c - \omega t_b) \tag{7.47}$$

$$X_{①}(0) = X_{④}(2\pi - \omega t_c) \tag{7.48}$$

Combining (7.45)–(7.48) with (7.44), (7.49)–(7.52) can be obtained:

$$X_{②}(0) = e^{A_{①}\omega t_a} \cdot X_{①}(0) + A_{①}{}^{-1} \cdot (e^{A_{①}\omega t_a} - I) \cdot B \cdot U \tag{7.49}$$

$$X_{③}(0) = e^{A_{②}\omega(t_b - t_a)} \cdot X_{②}(0) + A_{②}{}^{-1} \cdot (e^{A_{②}\omega(t_b - t_a)} - I) \cdot B \cdot U \tag{7.50}$$

$$X_{④}(0) = e^{A_{③}\omega(t_c-t_b)} \cdot X_{③}(0) + A_{③}{}^{-1} \cdot (e^{A_{③}\omega(t_c-t_b)} - I) \cdot B \cdot U \tag{7.51}$$

$$X_{①}(0) = e^{A_{④}(2\pi-\omega t_c)} \cdot X_{④}(0) + A_{④}{}^{-1} \cdot (e^{A_{④}(2\pi-\omega t_c)} - I) \cdot B \cdot U \tag{7.52}$$

To obtain initial condition, all terms with the initial condition in (7.49)–(7.52) are rearranged to left side of the equation in (7.53):

$$\begin{bmatrix} -e^{A_{①}\cdot\omega t_a} \cdot X_{①}(0) + X_{②}(0) \\ -e^{A_{②}\cdot\omega(t_b-t_a)} \cdot X_{②}(0) + X_{③}(0) \\ -e^{A_{③}\cdot\omega(t_c-t_b)} \cdot X_{③}(0) + X_{④}(0) \\ -e^{A_{④}\cdot(②\pi-\omega t_c)} \cdot X_{④}(0) + X_{①}(0) \end{bmatrix} = \begin{bmatrix} A_{①}^{-1} \cdot (e^{A_{①}\cdot\omega t_a} - I) \\ A_{②}^{-1} \cdot (e^{A_{②}\cdot\omega(t_b-t_a)} - I) \\ A_{③}^{-1} \cdot (e^{A_{③}\cdot\omega(t_c-t_b)} - I) \\ A_{④}^{-1} \cdot (e^{A_{④}\cdot(2\pi-\omega t_c)} - I) \end{bmatrix} \cdot B \cdot U \tag{7.53}$$

Extracting initial condition terms coefficient, the initial condition in four modes are

$$\begin{bmatrix} X_{①}(0) \\ X_{②}(0) \\ X_{③}(0) \\ X_{④}(0) \end{bmatrix} = \begin{bmatrix} -e^{A_{①}\cdot\omega t_a} & I & 0 & 0 \\ 0 & -e^{A_{②}\cdot\omega(t_b-t_a)} & I & 0 \\ 0 & 0 & -e^{A_{③}\cdot\omega(t_c-t_b)} & I \\ I & 0 & 0 & -e^{A_{④}\cdot(2\pi-\omega t_c)} \end{bmatrix}^{-1} \cdot \begin{bmatrix} A_{①}^{-1} \cdot (e^{A_{①}\cdot\omega t_a} - I) \\ A_{②}^{-1} \cdot (e^{A_{②}\cdot\omega(t_b-t_a)} - I) \\ A_{③}^{-1} \cdot (e^{A_{③}\cdot\omega(t_c-t_b)} - I) \\ A_{④}^{-1} \cdot (e^{A_{④}\cdot(2\pi-\omega t_c)} - I) \end{bmatrix} \cdot B \cdot U \tag{7.54}$$

Solving for optimum switching condition

For a designed VHF converter, the inductance and capacitance value is determined. The aim is to calculate the gate-drive timing, including the duty ratio and phase relationship between the control and SR HEMTs, so that optimum switching condition under different input voltages can be achieved. In Figure 7.34, the optimum control HEMT ZVS condition occurs in Mode ① at $\omega t = 0$. The optimum SR ZVS condition occurs in the Mode ④ at $\omega t = \omega t_c$ and ZCS condition occurs in the Mode ② at $\omega t = \omega t_a$. These constrains can be summarized as

$$v_{ds\_ctrl}(0) = 0 \rightarrow v_{CF}(0) = 0 \rightarrow X_{①1}(0) = 0 \tag{7.55}$$

$$v_{ds\_SR}(\omega t_c) = 0 \rightarrow v_{CR}(\omega t_c) = 0 \rightarrow X_{④2}(0) = 0 \tag{7.56}$$

$$i_{LR}(\omega t_a) = 0 \rightarrow X_{②5}(0) = 0 \tag{7.57}$$

where $X_{①1}$ represents the first variable of $X(\omega t)$ with the converter in the Mode ①, $X_{④2}$ represents the second variable of $X(\omega t)$ in the Mode ④ and $X_{②5}$ represents the fifth variable of $X(\omega t)$ in the Mode ②.

Using (7.54) with three constrains of (7.55)–(7.57), ωt_a, ωt_b, and ωt_c in the initial condition expressions can be solved numerically by a MATLAB® script.

Table 7.8 Gate-drive timing with input voltage

V_{in} (V)	D_1 (%)	D_2 (%)	Φ (°)
18	37.4	51.8	203
19	38.8	52.7	208
20	40.0	53.6	212
21	41.1	54.3	215
22	42.1	54.9	218
23	43.0	55.2	221
24	43.9	55.5	223

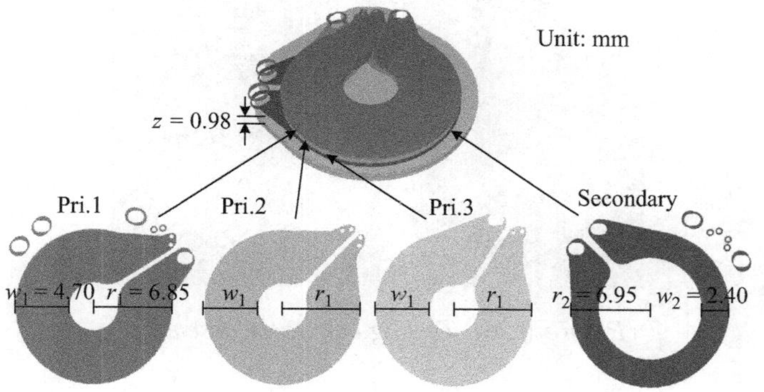

Figure 7.35 Designed transformer

7.2.4.2 Relationship between gate-drive timing and input voltage

Equations (7.54)–(7.57) were programmed in MATLAB® to solve value of ωt_a, ωt_b, and ωt_c under different input voltages. Table 7.8 shows the relationship between the gate-drive timing and input voltage. The gate-drive timing includes the phase difference and duty ratios the control and SR HEMTs in Figure 7.34. The control HEMT ZVS optimum condition and SR ZVS and zero-current switching (ZCS) optimum condition can be realized when the gate-drive signals are adapted to the input voltage according to the calculated relationship.

7.2.5 Air-core transformer design

The general idea to design the air-core transformer is to enumerate a set of geometries and select the ones with desired magnetic parameters by MATLAB® script [20]. Figure 7.35 gives the designed PCB transformer. The primary winding comprises three circular turns connected in series with one turn in each layer. The Primary coil width w_1 and radius r_1 are 6.85 and 4.70 mm, respectively. With a single turn at the PCB bottom layer as the secondary winding, a 3:1 transformer is built. The Secondary coil width w_1 and radius r_1 are 6.95 and 2.40 mm,

Figure 7.36 Testing converter assembly

respectively. The copper thickness is 70 μm (2 oz) considering the skin effect. The total PCB thickness z is 0.98 mm.

7.2.6 Experimental results and discussion

A 30-MHz class Φ_2 resonant flyback prototype was built. Figure 7.29 shows the complete schematic. The parameters are the same as the design example in Section 2 and in Table 7.6. The output voltage is regulated with digital ON/OFF controller. Figure 7.36 shows the testing converter assembly.

7.2.6.1 Gate-drive signals with proposed high time resolution circuit

Figure 7.37 shows the continuous gate-drive voltage modulation with the proposed high resolution circuit. It is observed that the highest time resolution is improved to around 1 ns (an improvement of 78.7%) with ultra-high speed buffer NC7SZ126.

7.2.6.2 VHF flyback with proposed driving scheme

Figure 7.38 shows the eGaN control and SR HEMT waveforms under optimum condition when $V_{in} = 18$ V. In Figure 7.38(a), the MCU generates the control HEMT gate-drive signal with $D = 37.0\%$, which corresponds to the timing under $V_{in} = 18$ V (37.4%) in Table 7.8. A duty ratio difference of 0.4% exists due to the digital control quantization. The control HEMT turns on when v_{ds_ctrl} resonates to zero at $t = t_0$ and ZVS is achieved and reverse conduction mechanism is not

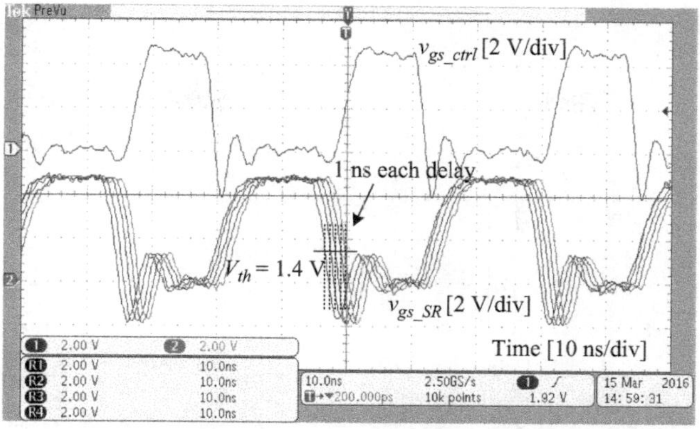

Figure 7.37 Gate-drive signals with proposed high resolution circuit

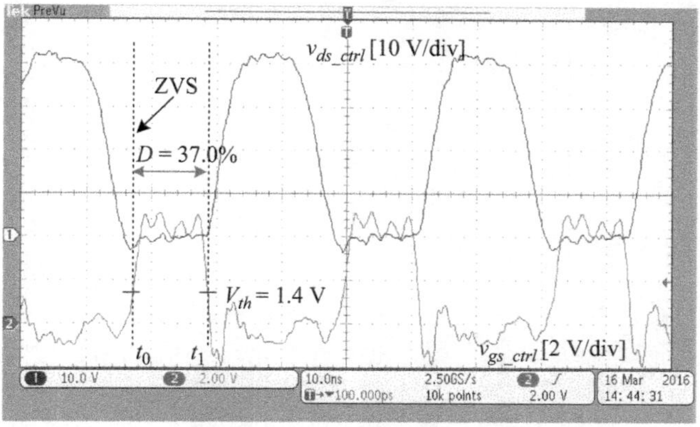

(a) v_{gs_ctrl} and v_{ds_ctrl}

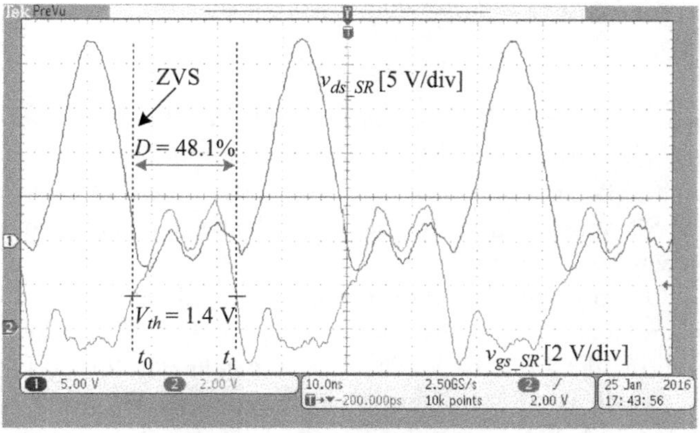

(b) v_{gs_SR} and v_{ds_SR}

Figure 7.38 Control and SR HEMTs waveforms: V_{in} = 18 V and f_s = 30 MHz

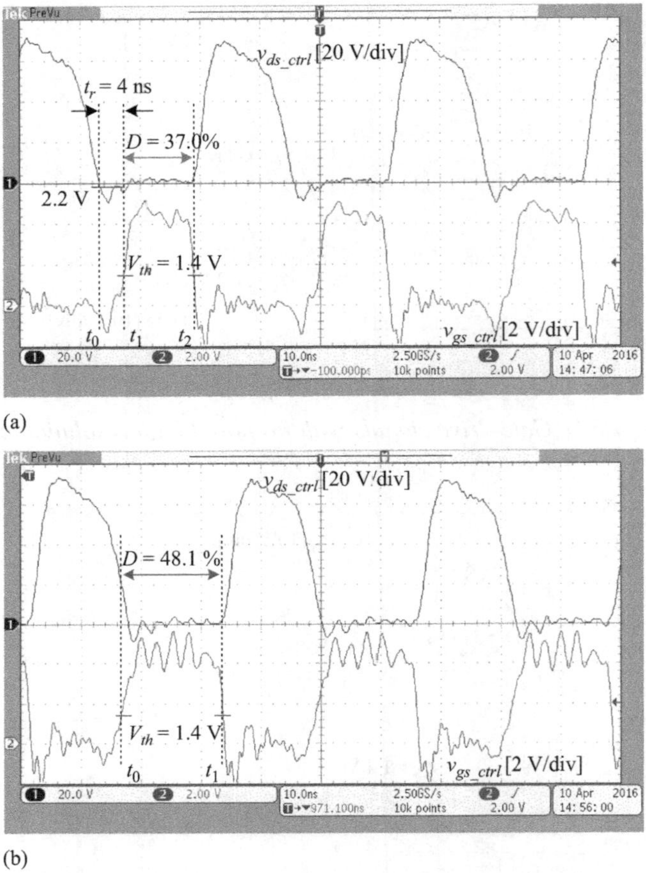

Figure 7.39 Control HEMT comparison: V_{in} = 24 V and f_s = 30 MHz; (a) conventional and (b) proposed

triggered. In Figure 7.38(b), the SR gate drive $D = 48.1\%$. It corresponds to $D = 51.8\%$ (a duty ratio difference of 3.7% due to quantization) and $\Phi = 203°$ under $V_{in} = 18$ V. The SR turns on when v_{ds_SR} resonates to zero at $t = t_0$, and ZVS is achieved. The SR turns off at $t = t_1$ when v_{ds_SR} begins to resonate and the reverse conduction mechanism is not triggered.

Figure 7.39 shows the waveform comparison of the control HEMT at 24 V input voltage. In Figure 7.39(a), the gate-drive signal $D = 37.0\%$, which is the same as the duty ratio under $V_{in} = 18$ V. The reverse conduction time t_r is around 4 ns from t_0 to t_1, which is 12.3% of one switching period. The reverse conduction voltage is about 2.2 V before ZVS turn-on. As comparison, in Figure 7.39(b), the MCU generates the control HEMT gate-drive signal with $D = 48.1\%$ according to the gate-drive timing in Table 7.8 under $V_{in} = 24$ V ($D = 43.9\%$, a duty ratio difference of 4.2% due to quantization). The reverse conduction time is eliminated to

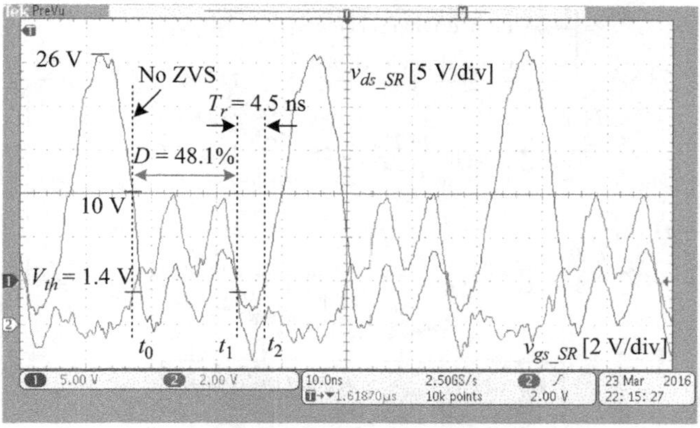

(a)

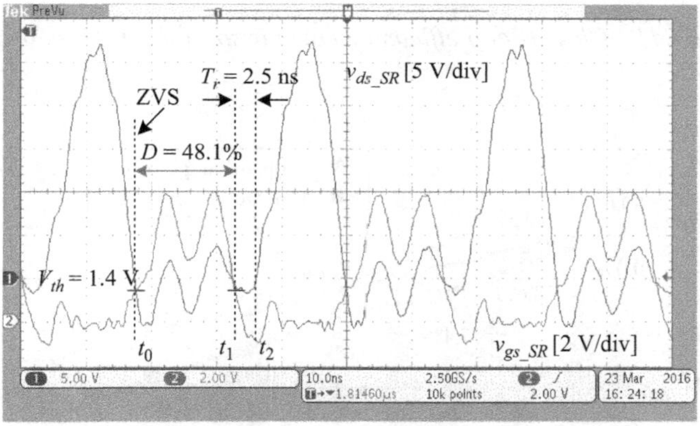

(b)

Figure 7.40 SR comparison: $V_{in} = 24$ V and $f_s = 30$ MHz; (a) conventional and (b) proposed

minimize the reverse conduction loss. It should be pointed that the duty ratio resolution can be improved by the method in [21] or using high resolution (around 1 ns) FPGAs such as Stratix V series from Altera Corporation [22]. Then, ZVS of the control HEMT can be realized better so that the efficiency can be further improved.

Figure 7.40 shows SR waveform comparison at 24 V input voltage. In Figure 7.40(a), the same SR gate-drive signal is given as the condition under 18 V input voltage with $D = 48.1\%$ and $\Phi = 203°$. The SR turns on at $t = t_0$ when v_{ds_SR} decreased to 10 V and ZVS is not achieved. The gate-drive signal turns off at $t = t_1$ and reverse conduction time t_r is around 4.5 ns from t_1 to t_2 (13.9% of one switching period). As comparison, in Figure 7.40(b) the SR gate drive $D = 48.1\%$. It corresponds to the timing with $D = 55.5\%$ (a difference of 7.4% due to quantization) and $\Phi = 223°$ (2 ns delay compared to $\Phi = 203°$). ZVS turn-on is achieved

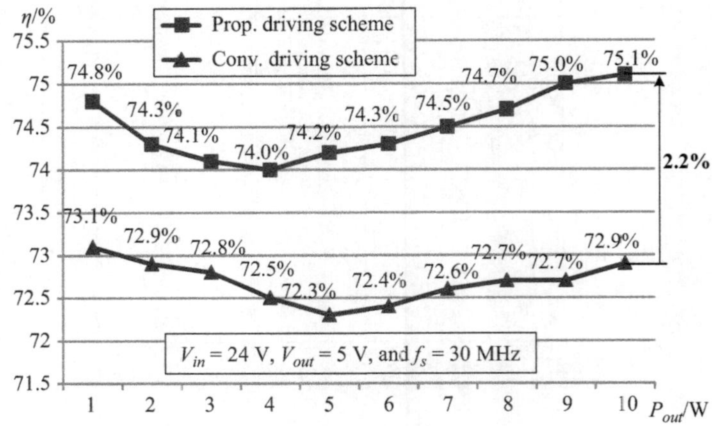

Figure 7.41 Closed-loop efficiency comparison (driving loss included)

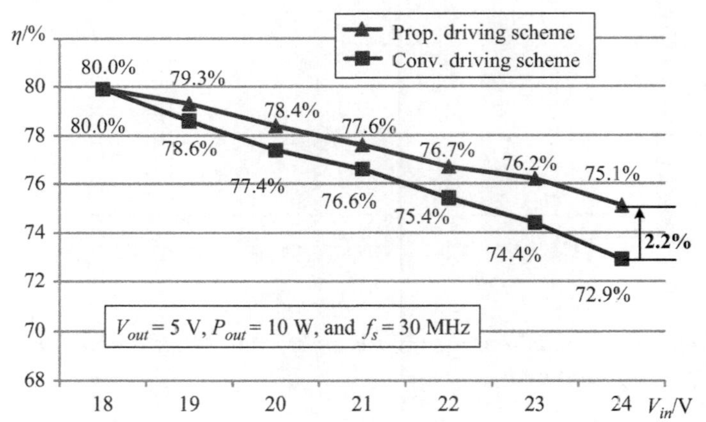

Figure 7.42 Closed-loop efficiency comparison (driving loss included)

at $t = t_0$ when v_{ds_ctrl} decreases to zero and the reverse conduction time is reduced to 2.5 ns from t_1 to t_2 (a reduction of 44.4%). The duty ratio resolution can be increased to further reduce the reverse conduction time and improve the efficiency.

Figure 7.41 shows the closed-loop efficiency comparison at 24 V input voltage. Under 24 V input voltage and full-load condition, the measured efficiency with the conventional scheme is 72.9%, while with the proposed scheme, the efficiency is 75.1% (an improvement of 2.2%).

Figure 7.42 shows closed-loop efficiency comparison under different input voltage. The converter is optimized at 18 V input voltage. As the input voltage increases, the reverse conduction time of the control and SR HEMTs increases and

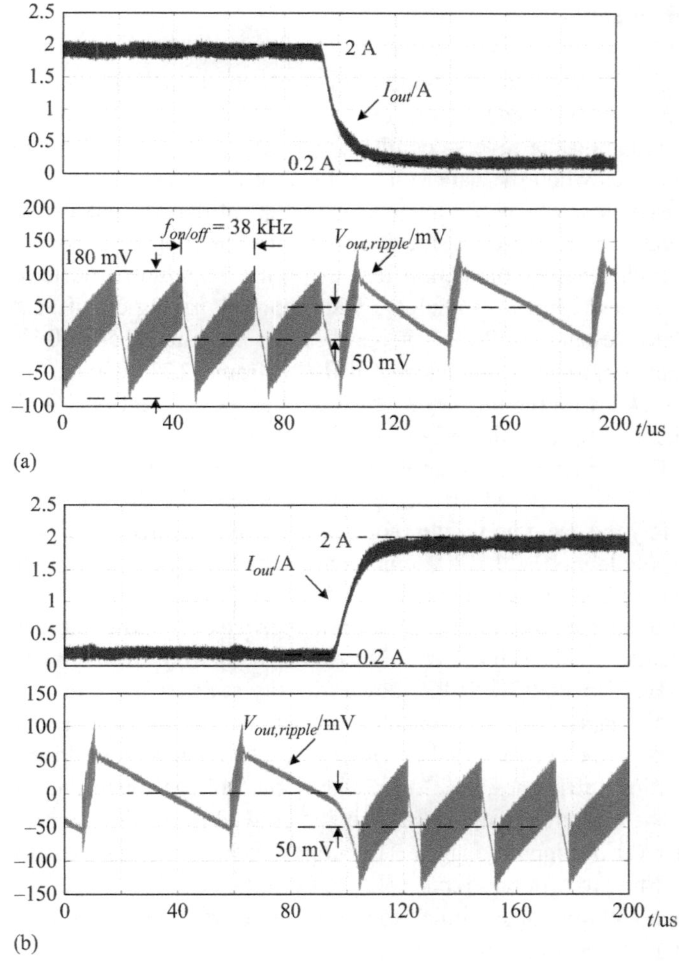

Figure 7.43 Load transient response waveforms; (a) Load step-down from 2 to 0.2 A and (b) Load step-up from 0.2 to 2 A

efficiency reduces gradually with the conventional scheme. With the proposed scheme, the efficiency is improved from 72.9% to 75.1% (an improvement of 2.2%) under full-load condition at 24 V input voltage.

Figure 7.43 shows the load transient performance using ON/OFF control. Near zero settling time and no voltage undershoot exist for both the step-up and step-down transient. This is because during the dynamics, the converter responses at 30 MHz. In Figure 7.43(b), with the load steps up from 0.2 to 2 A, a voltage drop of 50 mV (1% of V_{out} = 5 V) happens, which is acceptable in most applications. The output ripple at 5 V output is about 180 mV (3.6% of V_{out}) and the ON/OFF frequency is about 38 kHz with 2 A output.

7.3 Summary

A three-level driving circuit is proposed for eGaN control HEMTs with multi-MHz. The mid-level voltage reduces the reverse conduction loss before ZVS turn-on interval by reducing the reverse conduction voltage since that the source-to-drain voltage decreases when the gate voltage increases. The driving circuit layout and driving resistance selection are investigated to quantitatively analyze the negative gate bias requirement to prevent the false turn-on. The proposed driving circuit can reduce the high reverse conduction loss using the constructed positive gate bias, while achieve desired *dv/dt* immunity. The proposed driving circuit is applied to a 7 MHz isolated resonant SEPIC converter. With 24 V input and 5 V/10 W output, the driving circuit improves the efficiency of 0.7% (from 72.7% without the mid-level voltage to 73.4% with the mid-level voltage) over the conventional driving circuit.

In order to improve the efficiency further, an SR HEMT driving circuit is proposed. The driving IC is applied to meet the strict gate driving voltage requirements. The proposed driving circuit considers the driving IC propagation delay to satisfy the accurate timing requirements based on accurate HEMT rectifier mathematic modeling. With 18 V input and 5 V/10 W output, the proposed eGaN SR driving circuit improves the efficiency from 79.5% using the diode rectification to 84.7% (an improvement of 5.2%) and improves the efficiency from 79.9% without considering the driving IC delay to 84.7% (an improvement of 4.8%). The proposed eGaN isolated SR SEPIC power density realizes 20 W/in^3.

A digital adaptive driving scheme is proposed for eGaN VHF converters. In VHF converters, the conventional driving scheme with fixed gate-drive signals suffers from serious efficiency drop over wide input voltage range due to gate-drive mismatch. The converter state-space model is derived to predict the precise driving timing under various input voltages. The proposed driving scheme adapts gate-drive signals according to different input voltage. Both the eGaN control and SR HEMTs can be driven with optimized timing to minimize reverse conduction loss over entire input voltage range. Moreover a high time resolution circuit to improve the time resolution by selecting cascaded high speed buffers is proposed. The gate-drive signal resolution is improved of 78.7%. The proposed driving scheme was applied to a 30-MHz class Φ_2 resonant flyback converter. With 18 V input, 5 V/2 A output, the efficiency realizes 80.0%. With 24 V input, 5 V/2 A output, the efficiency is improved from 72.9% with conventional to 75.1% with proposed scheme (an improvement of 2.2%). The driving scheme can be extended to other resonant converters suffering the gate-drive signals variation over wide range, such as LLC, etc.

References

[1] X. Ren, D. Reusch, S. Ji, Z. Zhang, M. Mu, and F. C. Lee, "Three-level driving method for GaN power transistor in synchronous buck converter," in *Proceedings on IEEE Energy Conversion Congress and Exposition*, pp. 2949–2953, 2012.

[2] T. Wu, "Cdv/dt induced turn-on in synchronous buck regulators," in *Proceedings on Applied Manual-International Rectifier*, pp. 1–6, 2012.

[3] S. Jahdi, O. Alatise, J. A. Ortiz Gonzalez, R. Bonyadi, L. Ran, and P. Mawby, "Temperature and switching rate dependence of crosstalk in Si-IGBT and SiC power modules," *IEEE Transactions on Industrial Electronics*, vol. 63, no. 2, pp. 849–863, 2015.

[4] Z. Zhang, K. D. T. Ngo, and J. L. Nilles, "A 30-W flyback converter operating at 5 MHz," in *Proceedings on IEEE Applied Power Electronics Conference*, pp. 1415–1421, 2014.

[5] J. A. Santiago-Gonzalez, K. M. Elbaggari, K. K. Afridi, and D. J. Perreault, "Design of class E resonant rectifiers and diode evaluation for VHF power conversion," *IEEE Transactions on Power Electronics*, vol. 30, no. 9, pp. 4960–4972, 2015.

[6] W. Feng, F. Lee, P. Mattavelli, and D. Huang, "A universal adaptive driving scheme for synchronous rectification in LLC resonant converters," *IEEE Transactions on Power Electronics*, vol. 27, no. 8, pp. 3775–3781, 2012.

[7] D. Wang and Y. F. Liu, "A zero-crossing noise filter for driving synchronous rectifiers of *LLC* resonant converter," *IEEE Transactions on Power Electronics*, vol. 29, no. 4, pp. 1953–1965, 2014.

[8] X. Ren, Y. Zhou, W. Dong, X. Zou, and Z. Zhang, "A 10-MHz isolated class-Φ_2 synchronous resonant DC–DC converter," *IEEE Transactions on Power Electronics*, vol. 31, no. 12, pp. 8317–8328, 2016.

[9] J. Liao, J. Wang, J. Zhang, and Z. Qian, "A novel current driving scheme for LLC resonant converter with synchronized voltage-doubler rectifier," in *Proceedings on IEEE Applied Power Electronics Conference*, pp. 566–570, 2011.

[10] Z. Zhang, J. Fu, Y. F. Liu, and P. C. Sen, "Discontinuous current source drivers for high frequency power MOSFETs," *IEEE Transactions on Power Electronics*, vol. 25, no. 7, pp. 1863–1876, 2010.

[11] L. Gu, W. Liang, L. C. Raymond, and J. Rivas-Davila, "27.12 MHz GaN bi-directional resonant power converter," in *Proceedings of IEEE Workshop on Control and Modeling for Power Electronics*, pp. 1–7, 2015.

[12] Z. Zhang, X. W. Zou, Z. Dong, Y. Zhou, and X. Ren, "A 10-MHz eGaN isolated class-Φ_2 DCX," *IEEE Transactions on Power Electronics*, vol. 32, no. 3, pp. 2029–2040, 2017.

[13] Z. Zhang, J. Y. Lin, Y. Zhou, and X. Ren, "Analysis and decoupling design of a 30 MHz resonant SEPIC converter," *IEEE Transactions on Power Electronics*, vol. 31, no. 6, pp. 4536–4548, 2016.

[14] J. M. Rivas, Y. Han, O. Leitermann, A. D. Sagneri, and D. J. Perreault, "A high-frequency resonant inverter topology with low-voltage stress," *IEEE Transactions on Power Electronics*, vol. 23, no. 4, pp. 1759–1771, 2008.

[15] S. Aldhaher, P. C. K. Luk, K. El Khamlichi Drissi, and J. F. Whidborne, "High-input-voltage high-frequency class E rectifiers for resonant inductive links," *IEEE Transactions on Power Electronics*, vol. 30, no. 3, pp. 1328–1335, 2015.

[16] R. Pilawa-Podgurski, A. D. Sagneri, J. M. Rivas, D. I. Aderson, and D. J. Perreault, "Very high frequency resonant boost converter," *IEEE Transactions on Power Electronics*, vol. 24, no. 6, pp. 1654–1665, 2009.

[17] J. Lu, D. J. Perreault, D. M. Otten, and K. K. Afridi, "Impedance control network resonant DC–DC converter for wide-range high-efficiency operation," *IEEE Transactions on Power Electronics*, vol. 31, no. 7, pp. 5040–5056, 2016.

[18] H. Nguyen, R. Zane, and D. Maksimovic, "ON/OFF control of a modular DC–DC converter based on active-clamp LLC modules," *IEEE Transactions on Power Electronics*, vol. 30, no. 7, pp. 3748–3760, 2015.

[19] J. Zhao and Y. Han, "A novel switched-capacitor based partial power architecture for a 20 MHz resonant SEPIC," in *Proceedings of IEEE Electronics and Applied Conference and Exposition*, pp. 1442–1449, 2015.

[20] A. D. Sagneri, "Design of miniaturized radio-frequency dc–dc power converters," Ph.D. dissertation, Department of Electrical Engineering and Computer Science, Massachusetts Institute of Technology, Cambridge, MA, USA, 2012.

[21] R. F. Foley, R. C. Kavanagh, W. P. Marnane, and M. G. Egan, "An area-efficient digital pulsewidth modulation architecture suitable for FPGA implementation," in *Proceedings on IEEE Applied Power Electronics Conference*, pp. 1412–1418, 2005.

[22] Altera, Straix V, (2012). [Online] Available: https://www.altera.com.cn/content/dam/altera-www/global/zh_CN/pdfs/literature/hb/stratix-v/uf_01002.pdf

Index

accurate switching loss model, with current source drivers 69
  analytical modeling and simulation results 73–5
  experimental verification and discussion 79–85
  proposed MOSFET loss model with current source resonant driver 69
    analytical modeling of main switching transition 71–3
    circuit and basic assumption 70–1
  proposed optimal design with 75–9
adaptive current source driver for buck converters 159
  continuous current source driver
    optimal design and adaptive control with 160–1
    synchronous buck converter with 159
  discontinuous current source drivers 161
    optimal design and adaptive control with 161–3
    synchronous buck converter with 161
  experimental results and discussion 164–6
  implementation 163–4
adaptive current source driver for power factor correction application 166
  adaptive gate-drive current of power MOSFET 167–8
  application extension 168
  experimental results and discussion 172–3
  loss analysis of power stage 168–70
  loss comparison, example of 170–2
  proposed CSD for boost power factor correction converter 166
air-core transformer design 263–4
analog to digital converter (ADC) 182
auxiliary winding 191–3, 195–6, 198–9

body diode conduction loss 9–10, 54, 67, 93, 226
boost converter 56, 117–18, 120–1
boost power factor correction converter
  close-loop control of, with digital adaptive current source driver 180
  loss analysis of proposed adaptive CSD for 168
    experimental results and discussion 172–3
    loss comparison, example of 170–2
    power stage, loss analysis of 168–70
  proposed CSD for 166
  under critical conduction mode (CRM) 176–7

Center for Power Electronics Systems (CPES) 4
clamped inductive load 70
complementary metal oxide semiconductor (CMOS) technology 87, 190

complex programmable logic device (CPLD) 87, 96, 127, 139, 155
control HEMT driving circuit, design example of 240–1
control HEMT three-level driving circuit 248–50
control MOSFET 6–7, 9, 63, 65, 86–7, 89, 91–2, 96, 98, 126, 136–9, 141
conversion efficiency 4
coupled inductor resonant gate driver 15–16
current diversion problem 145
current source driver (CSD) 13, 45, 103
  accurate switching loss model with 69
    analytical modeling and simulation results 73–5
    experimental verification and discussion 79–85
    proposed MOSFET loss model with current CSD 69–73
    proposed optimal design with the accurate switching loss model 75–9
  concept of 17–19
  current doubler with 54
  current fed push-pull converter with 55
  discontinuous: *see* discontinuous current source driver
  high-side and low-side 86
    experimental results and discussion 96–101
    current-source gate driver with integrated magnetics 94–6
    problem of 86–7
    proposed decoupled 87–94
  high-side and low-side continuous CSD 59
    advantages 63–5
    experimental results 67–8
    loss analysis 65–7
    operating principle of proposed CSD 59–63
  major objectives of 45–6
  practical and accurate switching loss model 19
    parasitic inductance and load current, impact of 20–4
    switching loss model 24–5
    turn-off switching loss model 32–7
    turn-on switching loss model 25–32
    voltage source drive model, experimental validation of 38–41
    voltage source drive model verification 37–8
  resonant gate drivers 13–17
  two-channel low-side continuous CSD 45
    advantages of proposed CSD 52–4
    applications 54–6
    experimental results 56–9
    loss comparison 49–52
    operating principle 45–9
  voltage fed push-pull converter with 55

d.c. transformer (DCX) 232, 253
dead time control in the proposed RGD 200
digital adaptive driving circuit, implementation of 256–7
digital adaptive power factor correction 173
  design procedure and implementation 176
    close-loop control of interleaved boost power factor correction converter 180–2
    design procedure of adaptive-drive current 178–80

interleaved boost power factor correction converter under CRM 176–7
loss comparison between constant-drive current and adaptive-drive current 180
experimental results and discussion 182
of digital adaptive discontinuous CSD 183–5
loss reduction of digital adaptive discontinuous CSD 185–6
full-bridge current source driver, topology of 175–6
proposed pre-charge time based 176
digital signal processor (DSP) 181, 183
discontinuous current source driver 103
CSD design procedure 115–16
current diversion 145–6
experimental results and discussions 154–7
proposed switching loss model 147–54
design example 116
driver loss analysis 107–15
experimental results 117–23
high-side
design example 125
experimental results 125–31
proposed CSD for synchronous buck converter and operation 123–5
optimal design and adaptive control with 161–3
proposed low-side 103–7
synchronous buck converter with 161
with reduced inductance
experimental results 139–45
proposed CSD and principle of operation 132–7
proposed high-side CSD and hybrid gate-drive scheme 137–8
driver waveforms 122–4

eGaN field-effect transistors 2
energy storage requirement, reducing 1, 189
enhancement mode gallium nitride (eGaN) high electron mobility transistors (HEMTs) 2, 10, 231
control HEMT driving circuit, design example of 240–1
digital adaptive driving scheme for 252
air-core transformer design 263–4
experimental results and discussion 264–9
gate-drive challenges 254–5
proposed digital adaptive driving scheme 256–8
relationship analysis between gate-drive timing and input voltage 258–63
driving requirements for, in resonant SEPIC converters 232
control HEMT driving consideration 234–5
multi-MHz isolated resonant SEPIC converter 232–3
reverse conduction mechanism 233–4
SR HEMT driving consideration 235
experimental results and discussion 248
control HEMT three-level driving circuit 248–50
SR HEMT driving circuit 250–1
negative gate bias during turn-off transition 239–40
operation principle 237–8

proposed three-level gate drivers for eGaN HEMTs 235–7
rectifier mathematic modeling and design 241–8
three-level gate drivers 231–51

field-effect transistors (FETs) 53, 159
figure of merit (FOM) 2, 231, 253
frequency-dependent loss 3–5, 79
full-bridge (FB) converters, candidates of the RGDs for 208
full-bridge (FB) CSD structure 166

gate charge and gate-drive loss, relationship between 223
gate-drive challenges for eGaN VHF converters 254
over input voltage range 255
VHF class F2 resonant flyback converter 254
gate drive loss 10–11, 52–3, 56–7, 141, 190, 222
with the proposed RGD 211–13
gate-drive loss comparison 65–6
between the proposed RGD and conventional VSD 213–14
gate-drive timing and input voltage, relationship analysis between 253, 255–6, 258–63

high electron mobility transistors (HEMTs) 2, 231
see also enhancement mode gallium nitride (eGaN) high electron mobility transistors (HEMTs)
high-frequency-dependent loss 4–10
high-side (HS)/low-side resonant gate driver 15
high-side and low-side continuous current source drivers 59
advantages 63–5
experimental results 67–8
loss analysis 65
gate-drive loss comparison 65–6
switching 66–7
operating principle 59–63
high-side and low-side current source drivers 86
experimental results and discussion 96–101
current-source gate driver with integrated magnetics 94–6
problem of 86–7
proposed decoupled 87
advantages of proposed current-source gate driver 89–91
application extension 94
drive circuit design analysis 91–2
optimal design 92–3
principle of operation 87–9
high-side discontinuous CSD
design example 125
experimental results 125–31
proposed CSD for synchronous buck converter and operation 123–5
high time resolution circuit, proposed 257–8
delay selection in 257
gate-drive signals with 264

inductor current 15, 46, 60, 63, 106–8, 128, 132, 139, 148, 209
root mean square (RMS) value of 49
isolated synchronous class-Φ2 resonant converter 190

Kirchhoff's voltage law (KVL) equation 108

level-shift circuit 193, 199–200
level-shifted RGD
design of 198–200
auxiliary winding 198–9
resonant tank 199–200
problems of 193–7
proposed ON-OFF controlled 197
self-driven 192–3

LM5114 235
low-side discontinuous CSD 103–7
L transformer model 258

Mathcad software 74, 171, 216
Microcontroller Unit (MCU) 256, 264, 266
  MCU General Purpose Input/Output (GPIO) 257
Miller capacitor 32, 89
Miller plateau 6, 53, 57, 151–2
Moore's law 3
multi-MHz isolated resonant SEPIC converter 232–3
multi-MHz isolated synchronous class-Φ2 converter 190

ON-OFF controlled level-shifted RGD 197–8

parasitic inductance and load current, impact of 20–4
piecewise linear inductor current 111–14
power factor correction application, adaptive current source driver for 166
  adaptive gate-drive current of power MOSFET 167–8
  application extension 168
  experimental results and discussion 172–3
  loss analysis of power stage 168–70
  loss comparison, example of 170–2
  proposed CSD for boost power factor correction converter 166
power MOSFET 3, 11, 13–17, 42, 45, 49–52, 54, 59, 69, 76, 91, 103, 106–7, 110, 115, 118, 123, 135–7, 157, 163, 209, 211, 213, 215, 218, 221–3
  adaptive gate-drive current of 167–8
PowerPAK SO-8 9
power system on a chip (PowerSoC) 1
printed circuit board (PCB) 1, 20, 53, 118, 164, 239
pulse width modulation (PWM) 56, 106, 176

Quad Flat No-lead (QFN) package 87

rectifier mathematic modeling 241–8
  component initial value in wave-shape circuit, determining 241–3
  design example 246–8
resonant gate drivers (RGD) 13–17, 189
  for FB converters 208–10
  for multi-megahertz isolated resonant converters 189
    challenges for the SR drive 190–2
    experimental verification and discussion 202–6
    proposed self-driven RGD for the SR 192–202
  for ZVS full-bridge converters 206
    comparison between RGD and previous gate-drive circuits 220–3
    experimental results and discussion 224–6
    gate-drive circuits 207–8
    loss analysis and optimal design 211–20
resonant inductor current 57, 68, 80, 82, 86, 89, 98, 176, 209, 225
resonant pulse gate driver 14
resonant pulse technique 13
resonant switching transition 225
resonant tank 190, 192–3, 195, 199
resonant transition gate driver 14
reverse conduction mechanism 2, 231, 233–4, 255
reverse recovery loss 10, 80, 101, 141, 177

rising time, defined 218
root mean square (RMS) value 49, 108

second-order RC filter transfer function 242
self-driven level-shifted RGD 192–3, 195
self-oscillating resonant gate driver 13
single-ended primary-inductor converters (SEPICs) 189, 232
small outline integrated circuit (SOIC) package 87
SR HEMT driving circuit 250–1
SR MOSFET CSD design parameters 127
Surface mount (SMT) power inductors 96
switching frequency 1–4, 6, 9, 51–2, 56
switching loss 5–9, 19, 77, 145, 166, 216
  analysis 66–7
  reduction 53
switching loss model 24–5
  turn-off 32–7
  turn-on 25–32
synchronous buck converter 19, 53, 118, 139, 159, 161, 168, 173, 186–7, 215–16
  with continuous current source driver 159
  with discontinuous current source driver 161
synchronous buck HS MOSFET waveforms 25–6, 28, 30
synchronous buck voltage regulator with parasitic inductances 21
synchronous MOSFET 65, 86–7, 92–3, 96, 98, 157
synchronous rectification 189
  field effect transistors (FETs) 189–91
synchronous rectifier (SR) 9, 17, 45
turn-off switching loss model 32–7
turn-off transition 6–8, 32, 152–4
  negative gate bias during 239–40
turn-on switching loss model 25–32
turn-on transition 1, 5, 8, 148–52
two-channel low-side continuous current source drivers 45
  advantages 52
    body diode conduction time reduction 53
    gate-drive loss saving 52–3
    high noise immunity and alleviation of *dv/dt* effect 54
    less impact of parasitic inductance 54
    switching loss reduction 53
  applications 54–6
  experimental results 56–9
  loss comparison 49–52
  operating principle 45–9

very high frequency (VHF) converters 252
  digital adaptive driving scheme for 252
    air-core transformer design 263–4
    experimental results and discussion 264–9
    gate-drive challenges 254–5
    proposed digital adaptive driving scheme 256–8
    relationship analysis between gate-drive timing and input voltage 258–63
Virginia Power Electronics Center (VPEC) 4
VLSI (very large scale integration) technologies 3
voltage drop, defined 212
voltage regulator (VR) 3, 20–1, 181
voltage regulator modules (VRMs) 3
voltage source drive model
  experimental validation of 38–41
  verification 37–8

voltage source drivers (VSDs) 10, 40, 74, 103, 139, 143, 161, 163, 206, 220–2
voltage source gate driver 4–10

zero-current switching (ZCS) 1, 19, 21, 50, 103
zero-voltage switching (ZVS)
  comparison between RGD and previous gate-drive circuits 220–3
  experimental results and discussion 224–6
  gate-drive circuits 207–8
  loss analysis and optimal design 211–20
  RGD with low gate-drive loss for ZVS full-bridge converters 206
  turn-on interval 231
ZVZCS converters 223